# RADIOACTIVITY IN THE TERRESTRIAL ENVIRONMENT

RADIOACTIVITY IN THE ENVIRONMENT

A companion series to the Journal of Environmental Radioactivity

Series Editor
M.S. Baxter
Ampfield House
Clachan Seil
Argyll, Scotland, UK

---

Volume 1: Plutonium in the Environment
(A. Kudo, Editor)

Volume 2: Interactions of Microorganisms with Radionuclides
(F.R. Livens and M. Keith-Roach, Editors)

Volume 3: Radioactive Fallout after Nuclear Explosions and Accidents
(Yu.A. Izrael, Author)

Volume 4: Modelling Radioactivity in the Environment
(E.M. Scott, Editor)

Volume 5: Sedimentary Processes: Quantification Using Radionuclides
(J. Carroll and I. Lerche, Authors)

Volume 6: Marine Radioactivity
(H.D. Livingston, Editor)

Volume 7: The Natural Radiation Environment VII
(J.P. McLaughlin, S.E. Simopoulos and F. Steinhäusler, Editors)

Volume 8: Radionuclides in the Environment
(P.P. Povinec and J.A. Sanchez-Cabeza, Editors)

Volume 9: Deep Geological Disposal of Radioactive Waste
(R. Alexander and L. McKinley, Editors)

Volume 10: Radioactivity in the Terrestrial Environment
(G. Shaw, Editor)

# RADIOACTIVITY IN THE TERRESTRIAL ENVIRONMENT

Editor

**George Shaw**

*Division of Agricultural and Environmental Sciences*
*University of Nottingham*
*Nottingham, U.K.*

ELSEVIER
AMSTERDAM – BOSTON – HEIDELBERG – LONDON – NEW YORK – PARIS
SAN DIEGO – SAN FRANCISCO – SINGAPORE – SYDNEY – TOKYO

Elsevier
The Boulevard, Langford Lane, Kidlington, Oxford OX5 1GB, UK
Radarweg 29, PO Box 211, 1000 AE Amsterdam, The Netherlands

First edition 2007



**British Library Cataloguing in Publication Data**
A catalogue record for this book is available from the British Library

**Library of Congress Cataloging-in-Publication Data**
A catalog record for this book is available from the Library of Congress

ISBN-13: 978-0-08-043872-6
ISBN-10: 0-08-043872-5
ISSN: 1569-4860



Transferred to digital print 2007
Printed and bound by CPI Antony Rowe, Eastbourne

# Contents

# Introduction

George Shaw

*Division of Agricultural & Environmental Sciences, University of Nottingham, University Park, Nottingham, NG7 2RD, UK*

## 1. Radionuclides in the environment

### 1.1. *The pre-nuclear world*

Until relatively recently the environment was taken for granted as the satisfier of all mankind's needs, providing food, mineral and biological resources and a seemingly limitless facility for the disposal of waste materials. Today, it is increasingly evident that the capacity of natural systems to absorb the wastes and contaminants of modern human existence is limited. These limits are both quantitative and qualitative. Some of our wastes are naturally occurring and even essential substances, such as $CO_2$, but are being produced in ever increasing quantities which simply overload the capacity of biogeochemical systems to maintain the steady state chemical compositions of the atmosphere, hydrosphere and lithosphere. Other substances are produced in small quantities but are potentially so detrimental that their presence in the environment, even at low levels, is sufficient to cause concern. Prominent among this latter category are radionuclides, some of which are natural components of the environment but many of which are present largely as a result of man's activities.

Widespread, low-level contamination of the environment from anthropogenic sources can be traced back at least to Roman times, approximately 2000 years before present, when lead and silver mining and smelting were carried out at several major sites across Europe (Rosman et al., 1997). The environmental record for lead deposition is particularly revealing. In northern Scandinavia, for instance, Brannvall et al. (1999) have detected the presence of lead from the Roman period but have also shown the beginnings of a steady increase in lead concentrations in lake sediments which can be traced back to 900 AD. However, while the Roman and medieval metal industries made a permanent mark on the environment it was the beginnings of the Industrial Revolution in Europe which marked a step change in the rate and type of contaminants being introduced into the environment (Markham, 1994). Environmental evidence gathered over recent decades indicates the impact which emissions from large scale coal combustion had on the chemistry and ecology of the environment, not just in the immediate vicinity of centres of industrialisation but also farther afield. In locations remote from

RADIOACTIVITY IN THE ENVIRONMENT
VOLUME 10 ISSN 1569-4860/DOI: 10.1016/S1569-4860(06)10001-7

any major industrial activities, such as Galloway (south western Scotland), the beginnings of a gradual reduction in environmental quality, as measured by the presence of acidifying pollutants such as sulphur and heavy metals, can be traced to the middle of the $19^{th}$ century (Battarbee et al., 1985). Further north, the sediment record from Loch Lomond has revealed other key events such as the introduction of leaded petroleum in 1929 (Farmer et al., 1996) which marked a new source of contamination directly linked with technological progress at that time.

The phenomenon of radioactivity was discovered in 1896 by Henri Becquerel, one year after X-rays were discovered by Konrad Roentgen. Experiments to determine the effects of X-rays on biological materials were conducted as early as 1896 (Tarkhanov, 1896). An early 'nuclear industry' thrived in several countries across the world in the 1920s, based not on uranium but on the extraction of naturally occurring radium from ores (Landa, 1982). Environmental measurements were carried out by Vernadsky (1929) who examined the concentration of radium by living organisms, work which was to contribute to the development of the modern discipline of biogeochemistry. Prior to the 1940s, however, the world inventory of radioactive materials was limited to those found in nature, with the exception of a few mega becquerels ($10^6$ Bq) produced in cyclotrons in the late 1930s. One of the substances created at this time was the first 'man-made' element, technetium, the isotopes of which are exclusively radioactive and which had previously been 'missing' from the Periodic Table.

### **1.2.** *The Atomic Age*

If the environmental effects of early industrialisation were at first gradual and imperceptible, the enormous technological developments during the second World War were rapid and, quite literally, earth shattering. These developments culminated in the detonation of the first atomic bomb at the Trinity test site in New Mexico on 16th July, 1945. This device had an explosive yield of 19 kiloton and resulted in a radioactive plume within the atmosphere spreading hundreds of kilometres to the north east. The Trinity test marked the arrival of the 'Atomic Age' and a period of atmospheric weapons testing in which exotic and potentially hazardous radioactive substances would be introduced into the environment until 1963, when the Partial Test Ban Treaty came into force. Chief among the substances of concern were $^{137}Cs$ and $^{90}Sr$, two uranium fission products with physical (radioactive) half lives of 30 and 28 years, respectively, sufficient to ensure their persistence in the environment for many decades and several human generations. Plutonium isotopes were also released, notably $^{239}Pu$, a product of neutron capture by $^{238}U$ during nuclear fission, with a physical half life of 24,110 years. The geographical locations at which nuclear weapons were tested in the 1950s and early 1960s were deliberately remote (see Figure 1 of Chapter 8), but the radioactive fallout from these tests was easily measurable thousands of miles away from the points of detonation. Despite the limited number of test sites, it rapidly became evident that radioactive fallout from weapons tests was worldwide (Figure 1 of Chapter 5). Evidence for the rapidity with which global contamination occurred as a result of atmospheric nuclear weapons tests is provided by a recent analysis of grass samples collected in southern England in the 1950s and archived until the present day. Ratios of plutonium isotopes ($^{240}Pu$:$^{239}Pu$) in these samples reveal that they originated from tests in the Nevada desert in 1952 and 1953 and that they crossed the continental USA and the Atlantic Ocean within as little as five days (Warneke et al., 2002).

The Trinity Test could not have taken place without a substantial infrastructure of uranium processing and fuel manufacture which was constructed as part of the Manhattan Project (Asimov, 1972). Each of the steps in the newly devised nuclear fuel cycle resulted in the production of wastes which, given the urgency of the project, inevitably found their way into the environment as a 'free' waste disposal route. The contamination of the Columbia River in the north west USA by plutonium-containing effluents from the Hanford reactors is well documented (Whicker and Schultz, 1982) and the Hanford Reservation has since been dubbed 'the dirtiest place on Earth' (Fishlock, 1994). The technology developed at Hanford for the extraction of plutonium from irradiated uranium reactor fuel subsequently found its way to other locations, most notably to the Sellafield reprocessing complex in the north west of England. Nuclear operations began at Sellafield (formerly called Windscale) in 1951 with the commissioning of two air cooled reactors, followed in 1952 by the first reprocessing operations (Gray et al., 1995). A major fire in one of these reactors ('Pile I') in October 1957 resulted in widespread contamination of the local environment, primarily with $^{131}$I, 740 tera Bq ($7.4 \times 10^{14}$ Bq) of which were released and rapidly found their way into the atmosphere and food chain via the pasture-cow-milk pathway. The two original Windscale reactors were closed down immediately after the fire although reprocessing operations have continued at Sellafield to the present day, with accompanying discharges to the sea and to the atmosphere. A complete chronology of liquid and atmospheric discharges from the Sellafield site from 1952 to 1992 has been provided by Gray et al. (1995). Liquid discharges to the Irish Sea of $^{137}$Cs, $^{90}$Sr, $^{106}$Ru, $^{241}$Am and plutonium isotopes peaked in the 1970s and have since declined dramatically with the introduction of improved separation and recovery processes. Marine discharges of previously stored $^{99}$Tc, however, increased in 1994 and reached a peak in 1995–96 (Copplestone et al., 2004) with the potential for increased transfer of this radionuclide from the marine to the terrestrial environment (Webster et al., 2003).

In the Soviet Union, a major nuclear accident also occurred in 1957 at Kyshtym near Chelyabinsk. This accident was not officially recognised by the Soviet government until 21 years later, with an announcement by TASS[1] in December 1988 that a major chemical explosion had occurred in the southern Urals in September 1957. Details of this accident began to emerge only in the 1970s when Medvedev (1977, 1979) revealed a substantial body of Soviet ecological research papers relating to the contamination of lakes, soils, animals and plants by the radionuclides $^{90}$Sr, $^{106}$Ru, $^{137}$Cs, $^{144}$Ce and $^{95}$Zr. A fault in the cooling system of a chemical storage tank is now known to have been responsible for an explosion involving ammonium nitrate which resulted in the release of 0.74 exa Bq ($7.4 \times 10^{17}$ Bq) of long- and short-lived radionuclides in the form of a plume 65 miles long and 5 to 6 miles wide (Trabalka and Auerbach, 1991). This episode led to a vigorous and sustained scientific effort within the Soviet Union devoted to understanding the environmental behaviour and impacts of radionuclides released as the result of the exploitation of nuclear energy (Sokolov et al., 1993).

For most people today, the event which links the former Soviet Union with radioactive contamination of the environment is the Chernobyl accident of 1986. This involved the most catastrophic loss to the environment of reactor core materials ever experienced. Three to four % of the reactor fuel (representing 6.7 tonnes of core materials) and associated radionuclides were lost to the atmosphere in a steam explosion which lifted the 1000 tonne cover from the

[1] Telegraph Agency of the Soviet Union.

reactor and exposed the reactor core (UNSCEAR, 2000). Materials expelled included irradiated fuel fragments (so-called 'hot particles'; Kasparov et al., 1995), noble gases and volatile radionuclides of iodine, tellurium and caesium amounting to a total release of approximately 14 exa Bq ($1.4 \times 10^{19}$ Bq) of radioactivity over a period of 10 days. The relatively large and heavy hot particles deposited close to the reactor, within a radius of several tens of km. The volatile radionuclides, however, especially $^{131}$I and $^{137}$Cs, were carried in the atmosphere for thousands of kilometres, mostly to the north and west, where they deposited to the ground surface over an area greater than 200,000 km$^2$ in countries across Europe and Scandinavia. A characteristic feature of the accident was that 'hot spots' of deposited radioactivity were created as the radioactive cloud encountered rainstorms during its passage over the land surface (ApSimon et al., 1989). The Chernobyl accident provided a dramatic demonstration that radionuclides released to the atmosphere are subject to rapid and widespread migration. Furthermore, some initial predictions in western Europe that the effects of $^{137}$Cs on agriculture would be relatively short-lived proved optimistic and, in some areas even in western Europe, residual contamination is still sufficiently high to require ongoing restrictions (Bell and Shaw, 2005). Recent reviews by Smith and Beresford (2005) and the IAEA (2006) provide detailed analyses of the ongoing effects of Chernobyl in both the former Soviet Union and other affected countries 20 years after the accident.

### **1.3.** *Radionuclides and radioecology*

Sixty years after the detonation of the first atomic bomb mankind occupies a terrestrial environment which is ubiquitously contaminated, to a greater or lesser degree, by the fission and activation products released as a result of the exploitation of nuclear energy. Radionuclides from nuclear weapons tests, reprocessing of nuclear fuels and the Chernobyl accident can be detected from the arctic (AMAP, 1997) to the south pole (Pourchet et al., 2003) and at all latitudes in between. However, while we can detect these radionuclides in environmental samples with ever improving analytical methods, most locations are contaminated at extremely low levels. In most areas of the Earth the radiation signature of soils, waters and the air is dominated by the naturally occurring radionuclides described in Chapter 10. Nevertheless, some of the events described in section 1.2 have given rise to localised levels of artificial radioactive contamination which are undesirable but not necessarily harmful, when judged using our best current understanding of the fate and impact of radionuclides in environmental systems. Although such judgements are often contentious, they are based on primary international standards of radiation protection as recommended by the International Commission on Radiological Protection. These recommendations cover human individuals and populations (ICRP, 1991) and, more recently, non-human species (ICRP, 2003).

Most of this sounds bleak and the perception of the nuclear industry by the general public continues to be negative. However, an extremely large body of scientific evidence has been amassed during the past 60 years in an attempt to understand and predict the behaviour and impact of radionuclides introduced into environmental systems. The field of science underlying these efforts is generally known as 'Radioecology', although this name does not adequately reflect the extremely broad range of scientific disciplines involved in this endeavour. The research activities of radioecologists have generally accelerated in response to the need to understand a new source term of environmental radioactivity; hence numbers of publications burgeoned during the 1950s and 1960s during the atomic weapons tests and, again, in 1986 and the two decades following the Chernobyl accident.

Benjamin Franklin (1706–1790) famously stated that 'nothing in this world can be said to be certain but death and taxes'. Had he known anything about Henri Becquerel's discovery of 1896 he could have justifiably added radioactivity to his short list of life's inevitables. We now know that radioactivity is omnipresent. This, and the almost absolute reliability of radioactive decay, have provided us with powerful tools to understand the biogeochemical flows of materials within the global environment as well as the chronology of past events through radiometric dating. Radioactivity is useful because it is ubiquitous. However, it is also potentially harmful and most readily associated by the general public with the nuclear weapons programmes which gave rise to the Atomic Age. Radioecologists have, since the mid-1940s, attempted to evaluate the significance and usefulness of environmental radioactivity in an objective way, often arriving at surprising conclusions.

## 2. Sources and attributes of radionuclides

### 2.1. *Sources*

Man-made or 'artificial' radionuclides are produced either as a result of the fission of uranium ($^{235}U$) or plutonium ($^{239}Pu$) or as the result of the activation of initially stable (non-radioactive) nuclides following neutron bombardment within reactors or during nuclear detonations. Collectively these are referred to as 'fission and activation products'. About 350 of these have been detected in the environment, the principal sources of which can be identified as:

- global fallout from nuclear weapons testing (accounting for approximately 1000 times the radioactive emissions to the environment from non-military nuclear sources)
- low-level atmospheric and liquid discharges from nuclear installations
- wastes from research establishments and hospitals
- solid radioactive waste repositories
- accidental releases to the environment.

Prior to the release of large quantities of radioactive fission and activation products into the environment during the atmospheric weapons tests of 1945–1963, there was limited scientific interest in environmental radioactivity as the world inventory of artificial radionuclides was very small. As these releases grew the science of radioecology evolved, primarily in response to society's requirement to assess the impact of the nuclear industry on the environment and human health (Shaw, 2005). In practice, an erratic history of unique discharge events has resulted in a range of specific source terms at different times and geographical locations around the world. Sheppard (2003) surveyed topics in radioecology which had been significant in the period 1995–2002, using publications within the Journal of Environmental Radioactivity (JER) as a yardstick. JER was first issued in 1984 and has recently been fully electronically archived to allow key word searches via the Science Direct[2] resource. Using this facility, the

[2] www.ScienceDirect.com

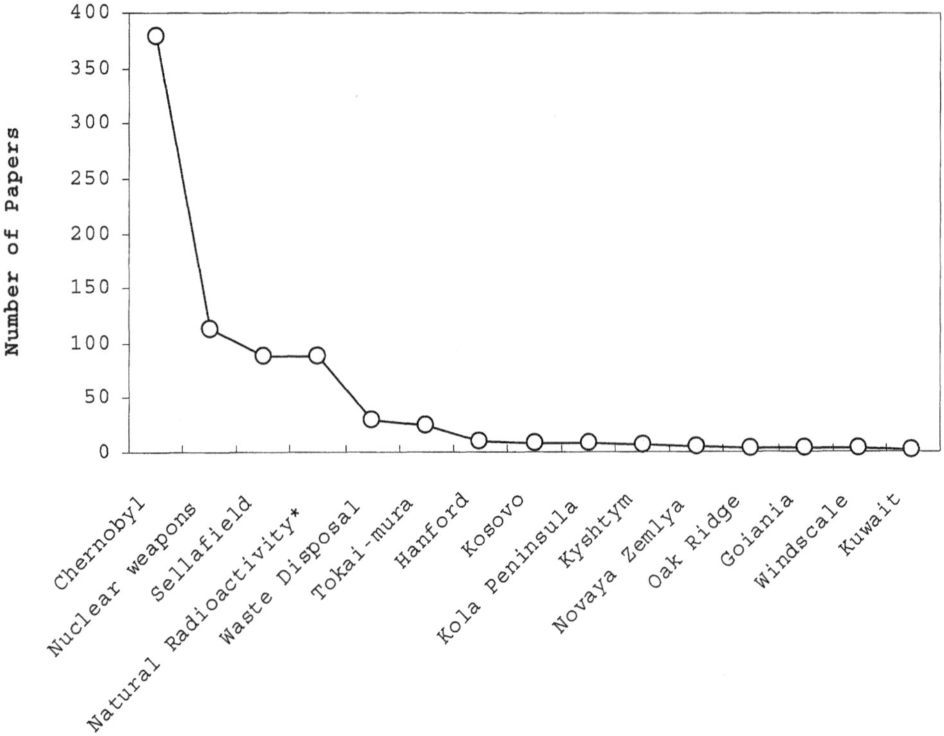


Fig. 1. Frequency with which specific release events, sources or locations are mentioned in the title, keywords or abstract in papers published in the Journal of Environmental Radioactivity from 1984–2006. (*Naturally occurring radioactivity, NORM).

numbers of publications in JER addressing key events, sources and locations can be quantified. Figure 1 shows that Chernobyl accounted for just over 50% of papers in JER throughout the journal's history, while nuclear weapons, Sellafield and natural radioactivity accounted for 12–15% each. Hence, it is the behaviour of radionuclides artificially introduced into the environment by accidents, nuclear weapons and discharges associated with nuclear fuel reprocessing which forms the basis of most radioecological research. However, naturally occurring radionuclides, especially $^{222}Rn$ (radon gas), continue to be of major interest due to the fact that they are often of greater radiological significance than artificial radionuclides.

Figure 2 illustrates the principal sources of artificial radionuclides released to the environment and the complex pathways by which these sources are redistributed and ultimately impact on organisms. Radionuclides released to the atmosphere and marine waters can be transported over large distances and eventually find their way to the terrestrial environment, often with surprising rapidity, as described in section 1.2. Once terrestrial ecosystems become contaminated through such routes, the residence times and environmental impacts of individual radionuclides within specific ecological compartments is a function of the physical half life of the radionuclide, the chemistry of the element to which it belongs and the nature of the compartment itself. The majority of this book (Chapters 2 to 10) is devoted to detailed

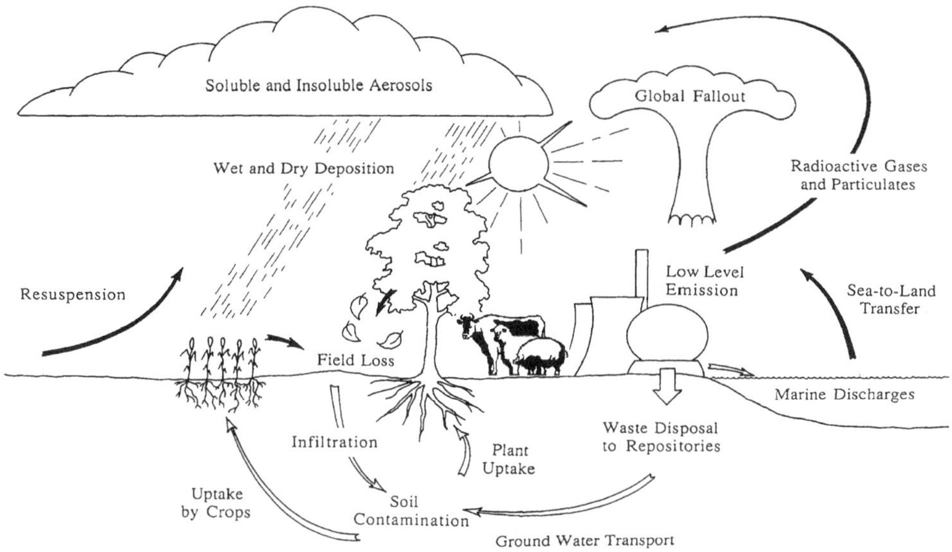


Fig. 2. Pathways leading to redistribution throughout the environment of artificial radionuclides from nuclear fuel cycle facilities, atomic weapons and radioactive waste disposal sites (redrawn and adapted from Kennedy et al., 1990. Reproduced by permission of the Natural Conservancy Council, UK).

Table 1
Periodic table showing thirty elements (in shaded boxes) of interest to radioecologists. Based on frequency of publications within the Journal of Environmental Radioactivity, fifteen of these elements are of primary interest and are detailed in Table 2.

| Group → / Period ↓ | 1 | 2 | | 3 | 4 | 5 | 6 | 7 | 8 | 9 | 10 | 11 | 12 | 13 | 14 | 15 | 16 | 17 | 18 |
|---|---|---|---|---|---|---|---|---|---|---|---|---|---|---|---|---|---|---|---|
| 1 | 1 H | | | | | | | | | | | | | | | | | | 2 He |
| 2 | 3 Li | 4 Be | | | | | | | | | | | | 5 B | 6 C | 7 N | 8 O | 9 F | 10 Ne |
| 3 | 11 Na | 12 Mg | | | | | | | | | | | | 13 Al | 14 Si | 15 P | 16 S | 17 Cl | 18 Ar |
| 4 | 19 K | 20 Ca | | 21 Sc | 22 Ti | 23 V | 24 Cr | 25 Mn | 26 Fe | 27 Co | 28 Ni | 29 Cu | 30 Zn | 31 Ga | 32 Ge | 33 As | 34 Se | 35 Br | 36 Kr |
| 5 | 37 Rb | 38 Sr | | 39 Y | 40 Zr | 41 Nb | 42 Mo | 43 Tc | 44 Ru | 45 Rh | 46 Pd | 47 Ag | 48 Cd | 49 In | 50 Sn | 51 Sb | 52 Te | 53 I | 54 Xe |
| 6 | 55 Cs | 56 Ba | * | 71 Lu | 72 Hf | 73 Ta | 74 W | 75 Re | 76 Os | 77 Ir | 78 Pt | 79 Au | 80 Hg | 81 Tl | 82 Pb | 83 Bi | 84 Po | 85 At | 86 Rn |
| 7 | 87 Fr | 88 Ra | ** | 103 Lr | 104 Rf | 105 Db | 106 Sg | 107 Bh | 108 Hs | 109 Mt | 110 Ds | 111 Rg | | | | | | | |

| | | | | | | | | | | | | | | | |
|---|---|---|---|---|---|---|---|---|---|---|---|---|---|---|---|
| *Lanthanides | * | 57 La | 58 Ce | 59 Pr | 60 Nd | 61 Pm | 62 Sm | 63 Eu | 64 Gd | 65 Tb | 66 Dy | 67 Ho | 68 Er | 69 Tm | 70 Yb |
| **Actinides | ** | 89 Ac | 90 Th | 91 Pa | 92 U | 93 Np | 94 Pu | 95 Am | 96 Cm | 97 Bk | 98 Cf | 99 Es | 100 Fm | 101 Md | 102 No |

descriptions of the behaviour of specific radionuclides in specific ecosystems but it is useful, first, to survey the range of elements and their radionuclides which have been considered, over recent decades, to warrant scrutiny by radioecologists.

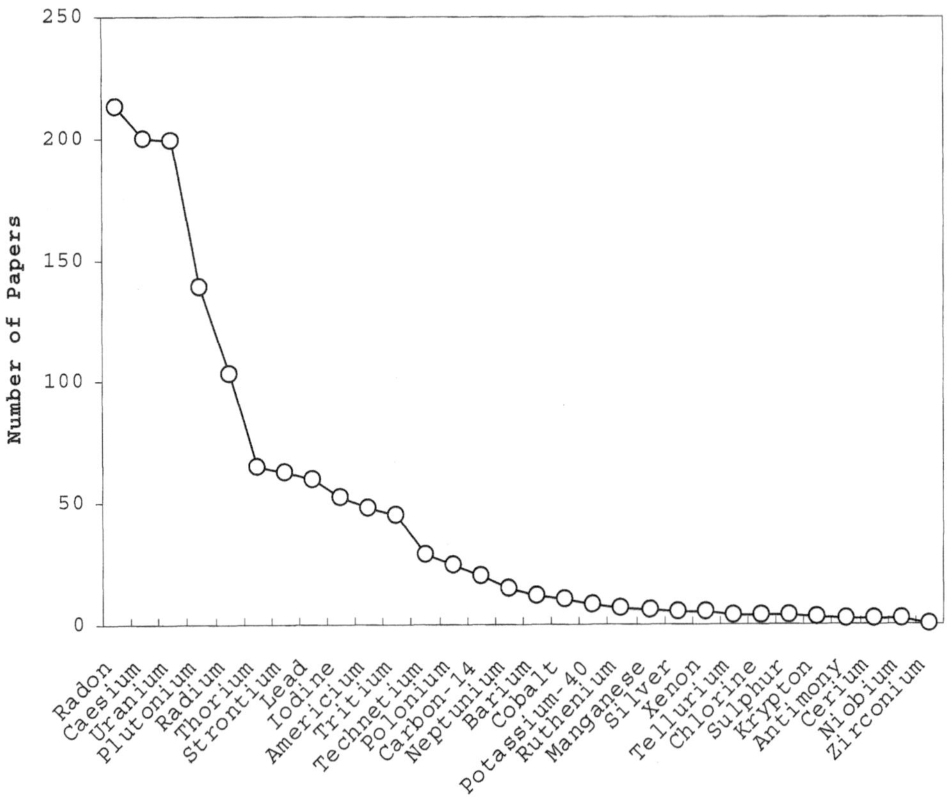


Fig. 3. Frequency with which specific elements and associated radionuclides are mentioned in the title, keywords or abstract in papers published in the Journal of Environmental Radioactivity from 1984–2006.

Table 1 shows the elements within the Periodic Table which possess radionuclides of major environmental significance. The radioecologist is concerned with a range of elements from the lightest (hydrogen) to the heaviest naturally occurring element, uranium, and several man-made transuranic elements heavier than uranium (Np, Pu and Am). At least thirty elements of radioecological significance can be identified (Table 1), but it can be seen from Figure 3 that only four of these elements (Rn, Cs, U and Pu) account for more than 50% of publications in JER. Half of the elements shown in Figure 3 account for more than 95% of publications in JER and can reasonably be identified as the most significant to the radioecologist. The most important radionuclides of these fifteen elements, their sources and physical half lives are identified in Table 2.

## 2.2. *Special features of radionuclides as contaminants*

One of the main features of a radionuclide which dictates its importance as an environmental contaminant is its physical half life. It is reasonable to assume that the longer the physical half life, the longer the residence time within ecological systems and, consequently, the greater

Table 2
Elements of principal interest to radioecologists, with environmentally significant radionuclides and physical half lives.

| Element name | Key radionuclide(s) | Physical half life (years)* | Source (see key below) |
|---|---|---|---|
| Hydrogen | $^{3}H$ | $1.233 \times 10^{1}$ | Nat$^{Cos}$, NFC, NW, RP |
| Carbon | $^{14}C$ | $5.730 \times 10^{3}$ | Nat$^{Cos}$, NFC, NW, RP |
| Strontium | $^{90}Sr$ | $2.879 \times 10^{1}$ | NFC, NW |
| Technetium | $^{99m}Tc$ | $6.861 \times 10^{-4}$ | RP |
| | $^{99}Tc$ | $2.111 \times 10^{5}$ | NFC, NW |
| Antimony | $^{125}Sb$ | $2.758 \times 10^{0}$ | NFC, NW |
| Iodine | $^{129}I$ | $1.570 \times 10^{7}$ | NFC, NW, RP |
| | $^{131}I$ | $2.197 \times 10^{-2}$ | NFC, NW |
| Caesium | $^{134}Cs$ | $2.065 \times 10^{0}$ | NFC, NW |
| | $^{137}Cs$ | $3.007 \times 10^{1}$ | NFC, NW |
| Lead | $^{210}Pb$ | $2.230 \times 10^{1}$ | Nat$^{Dec}$ |
| Polonium | $^{210}Po$ | $3.791 \times 10^{-1}$ | Nat$^{Dec}$ |
| Radon | $^{219}Rn$ ('actinon') | $1.256 \times 10^{-7}$ | Nat$^{Dec}$ |
| | $^{220}Rn$ ('thoron') | $1.763 \times 10^{-6}$ | Nat$^{Dec}$ |
| | $^{222}Rn$ ('radon') | $1.048 \times 10^{-2}$ | Nat$^{Dec}$ |
| Radium | $^{226}Ra$ | $1.600 \times 10^{3}$ | Nat$^{Dec}$ |
| Thorium | $^{230}Th$ | $7.538 \times 10^{4}$ | Nat$^{Dec}$ |
| | $^{232}Th$ | $1.405 \times 10^{10}$ | Nat$^{Pri}$ |
| Uranium | $^{234}U$ | $2.455 \times 10^{5}$ | Nat$^{Dec}$, NFC |
| | $^{235}U$ | $7.038 \times 10^{8}$ | Nat$^{Pri}$, NFC |
| | $^{238}U$ | $4.468 \times 10^{9}$ | Nat$^{Pri}$, NFC |
| Neptunium | $^{237}Np$ | $2.144 \times 10^{6}$ | NFC, NW |
| Plutonium | $^{238}Pu$ | $8.770 \times 10^{1}$ | NFC, NW |
| | $^{239}Pu$ | $2.411 \times 10^{4}$ | NFC, NW |
| | $^{240}Pu$ | $6.563 \times 10^{3}$ | NFC, NW |
| Americium | $^{241}Am$ | $4.322 \times 10^{2}$ | NFC, NW |

**Key**: Nat—natural, primordial (Pri) or produced by either cosmic radiation in the atmosphere (Cos) or radioactive decay of uranium ($^{238}U$) or thorium ($^{232}Th$) series (Dec); NFC—nuclear fuel cycle (including fuel extraction/manufacture, reactor operations/accidents, reprocessing and waste disposal); NW—nuclear weapons; RP—radiopharmaceuticals (including discharges from hospitals, research establishments and manufacturing facilities).
*Physical half life data obtained from LBNL Isotopes Project (2004).

the impact on that system. However, the environmental impact of a radionuclide is usually measured in terms of its radiological impact which is controlled, to a large extent, by the environmental pathways and processes which result in exposure of organisms (usually man) to the radionuclide in question. In this respect, some of the most short-lived radionuclides in Table 2 have disproportionately high environmental impacts. $^{222}Rn$ (half life = 3.82 days) poses a significant radiological hazard in confined spaces, including mines, caves and buildings, due to its continuous emanation into the atmosphere resulting from the decay of $^{226}Ra$ in rocks, soils and building materials (Chapter 10) as well as the highly hazardous alpha radiation emitted by its very short lived progeny, $^{218}Po$ (half life = 3.1 minutes) and $^{214}Po$ (half life = 164 $\mu$s). $^{131}I$ (half life = 8.02 days) is one of the most significant radionuclides in the aftermath of a nuclear detonation or accidental release from a reactor. This is due to its rapid and efficient biological incorporation from the environment, especially into the milk of lactating animals and thyroid glands of children (see Chapter 4). Analysis of human epidemiology 20 years after the Chernobyl accident has shown that the only cancers directly attributable to the release of radionuclides in 1986 are thyroid cancers (Cardis et al., 2006) as a result of exposure to $^{131}$I, even though this radionuclide was no longer measurable in the environment some 3 months after the accident.

Longer-lived radionuclides pose environmental hazards on different timescales. Hence, $^{137}$Cs and $^{90}$Sr (half lives = 30 years and 28 years, respectively) deposited on land surfaces around the world at the peak of atmospheric weapons testing in 1963 are still present in the environment and are likely to remain detectable for another 150 years. $^{239}$Pu (half life = 24,110 years) deposited at the same time will contaminate the environment in perpetuity, when judged on the time scale of human civilisation. Radionuclides with physical half lives of the order of thousands or millions of years are significant when considering disposal of wastes from the nuclear fuel cycle since even slow migration processes within the geosphere will allow the eventual emergence of some of these substances at locations within the biosphere, where exposures to humans and other organisms may occur (Crossland et al., 2005). When considering the potential impacts of long-lived, man-made radionuclides, however, it must be remembered that primordial radionuclides such as $^{238}$U and $^{40}$K (half lives = 4.47 and 1.28 billion years, respectively) have, by definition, been components of Earth's environment since its formation and all living organisms have evolved in the continuous presence of radiation from these sources.

As well as controlling the residence time within different ecological compartments, the half life is also important in determining the physical quantity of a radionuclide which is present. For any radionuclide, the radioactivity per unit mass can be calculated using the following relationship:

$$\frac{A}{m} = \frac{\ln(2) N_{\mathrm{A}}}{T_{1/2} A'}$$

in which $A$ is the radioactivity in Bq (disintegrations $s^{-1}$), m is the mass of the radionuclide (g), $N_{\mathrm{A}}$ is Avogadro's number ($mol^{-1}$), $T_{1/2}$ is the radioactive half life of the radionuclide (s) and $A'$ is the relative atomic mass ($g\,mol^{-1}$) of the radionuclide. This relationship dictates that the radioactivity per unit mass of any radionuclide is inversely proportional to its physical half life, as illustrated in Figure 4. Thus, $^{131}$I has a specific activity of $4.6 \times 10^{15}$ $Bq\,g^{-1}$ while $^{238}$U has a specific activity of $1.2 \times 10^{4}$ $Bq\,g^{-1}$. Clark and Smith (1988) reported an activity concentration for $^{131}$I of 4000 $Bq\,l^{-1}$ in prolonged rain in Cumbria, UK, in May 1986. From the specific activity stated above, this corresponds to a physical (mass) concentration of approximately $0.9 \times 10^{-12}$ $g\,l^{-1}$. Thus, short-lived radionuclides at even relatively high environmental activity concentrations ($Bq\,kg^{-1}$, $Bq\,l^{-1}$) are present at infinitesimally low mass concentrations ($g\,kg^{-1}$, $g\,l^{-1}$). Conversely, at a typical soil activity concentration of 24 $Bq\,g^{-1}$ (Bowen, 1979) the corresponding mass concentration of $^{238}$U is 2 $mg\,kg^{-1}$ which is of the same order of magnitude as many other 'trace' elements in soils and significantly higher than some toxic elements such as mercury and cadmium.

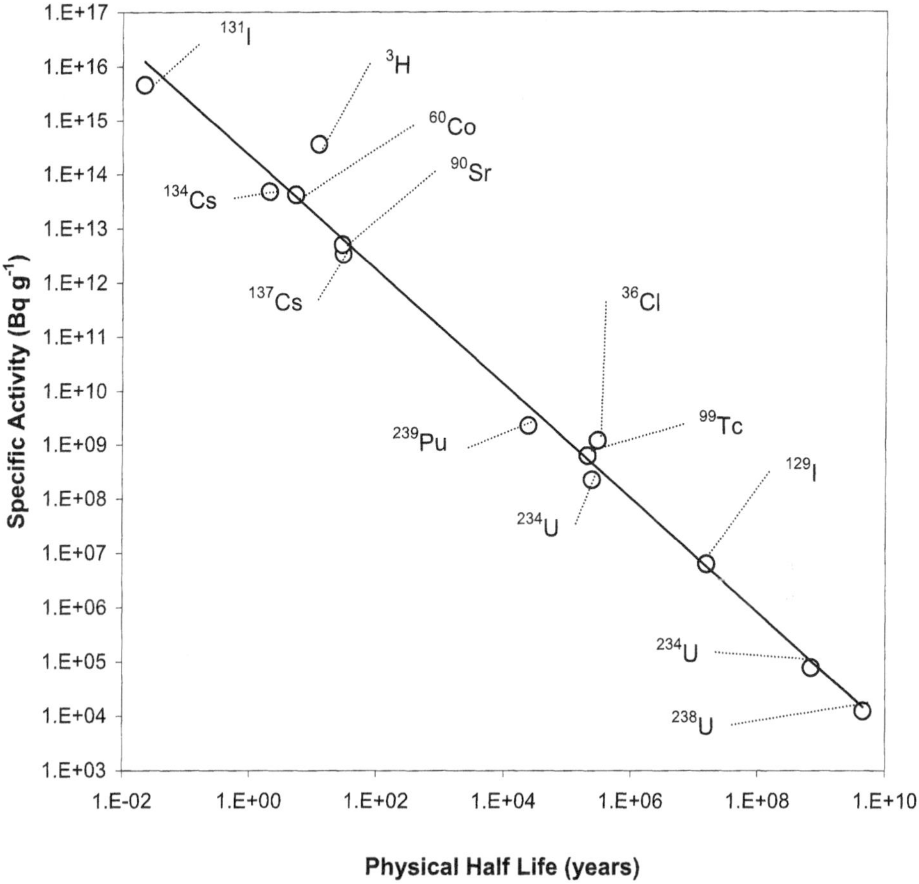


Fig. 4. Relationship between physical half lives (years) and specific activities (Bq $g^{-1}$) of selected radionuclides.

A major consequence of the specific activity versus half life relationship is that relatively short lived radionuclides will invariably be present in the environment at sub- or ultra-trace concentrations. Their biogeochemical behaviour is, therefore, often controlled not by their own physico-chemical attributes but by stable isotopes (e.g. $^{127}I$ in the case of $^{131}I$) or homologous elements or ions (e.g. K and $NH_4^+$ in the case of $^{137}Cs$, Ca in the case of $^{90}Sr$) which invariably exist at much higher mass concentrations in environmental media. The large discrepancy between mass concentrations of radionuclides in the environment and concentrations of stable isotopes or homologues can be exploited to help mitigate the degree of transfer of radionuclides into sensitive ecological compartments. Thus, fertilisers (particularly K- and Ca-containing fertilisers) can be applied to 'dilute' $^{137}Cs$ and $^{90}Sr$ and reduce their entry into food crops (see section 6 of Chapter 2) and non-radioactive iodine treatments may be given as a prophylactic to reduce uptake of $^{131}I$ by animals or even humans (see Chapter 4).

Another consequence of the extremely low mass concentrations of radionuclides is that quantifying their transfers from one ecological compartment to another is, in theory, made easier. Partitioning of substances between solid and liquid phases in soils and sediments is a critical process controlling their potential biological availability and migration (section 2 of Chapter 2). Similarly, incorporation of substances into organisms is controlled by passage across biological membranes. Both of these processes are strongly influenced by the concentration of the substance being considered. In the case of solid-liquid partitioning in soils, the availability of binding sites on the soil's solid phase is finite, which leads to saturation of these sites as the concentration of a substance within the solution phase is increased. This leads to typically curvilinear sorption isotherms (Sposito, 1989), an effect which has been implicated in the idea of the 'chemical time bomb' which recognises that the capacity of soils and sediments to sequester contaminants is finite (Stigliani et al., 1991). Similarly, ion uptake rates across membranes such as those in plant root cells are subject to saturation at high ion concentrations, leading to non-linear uptake kinetics with respect to concentration (Shaw and Bell, 1989). It was recognised in the 1960s that the extremely low mass concentrations of radionuclides in the environment were unlikely to result in saturation of sorption or membrane transport mechanisms so the concept of the 'transfer coefficient' was proposed to quantify the degree of movement of a radionuclide from one ecological compartment to another (Ward et al., 1965). The transfer coefficient is simply a ratio of activity concentrations in 'donor' and 'receiver' compartments, e.g. soil and plant (section 4.1 of Chapter 2), or feedstuffs and animal tissues (section 4.3 of Chapter 4), which assumes that radionuclide activity concentrations encountered in such compartments will never be sufficiently high to cause a deviation from a more-or-less constant ratio between compartments. While the transfer coefficient is non-mechanistic and considered by many radioecologists to be of limited predictive value, large databases exist which provide a ready reference source when making quantitative assessments of radionuclide behaviour in environmental systems (e.g. IAEA, 1994). Indeed, transfer coefficients lie at the heart of most radioecological models and provide the focus for many practical radioecological studies.

Some radioactive isotopes are so long-lived that they can almost be considered to occupy an area somewhere between the stable and radioactive domains in terms of their environmental behaviour and biological impacts. $^{232}Th$, $^{235}U$ and $^{238}U$, all of primordial origin and with extremely long physical half lives (Table 2, Figure 1 of Chapter 10), occupy an island of relative stability with respect to the heavier actinides (Np, Pu and Am). For these and other very long-lived radionuclides such as $^{129}I$ (half life = 15.7 million years), the question of chemical versus radiotoxicity arises since the ratio of mass per unit radioactivity is high (Sheppard, 2003). As an example, it is recognised that the environmental and human health impacts of contamination derived from depleted uranium munitions, which has recently received considerable attention, is a combination of both radiological and chemical toxicity (Bleise et al., 2003).

## 3. The aims and methods of radioecology

As mentioned above, the discipline of radioecology grew as a response to society's need for the best possible scientific understanding of the environmental impacts of radioactive materials released from the nuclear industry after 1945. It is a complex hybrid of the primary

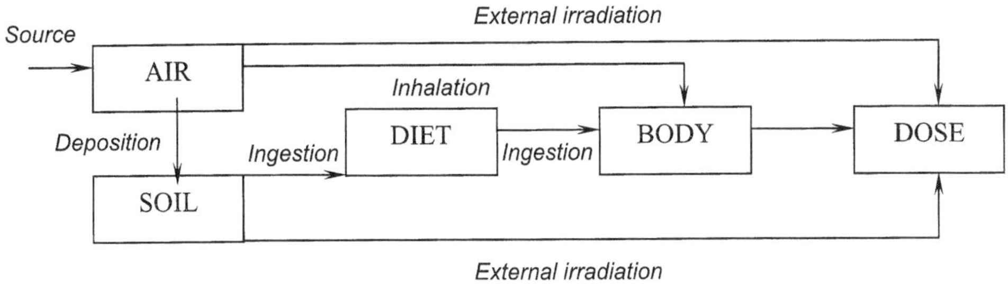


Fig. 5. Transfer and exposure pathways commonly evaluated in radioecological models (after UNSCEAR, 2000; permission requested).

disciplines of physics, chemistry, biology and geology and, as a result, borrows many methods from each of these disciplines. Radioecology has the probable distinction of being the first science devoted to understanding the environmental impact of man's industrial activities. As such it was probably the pioneer in the much broader area of pollution science which, from the 1960s until the present day, has sought to understand, predict and mitigate the effects of a multitude of chemical contaminants on the health of our planet.

The primary aim of radioecology is to assist in the protection of human health by developing an understanding of the impacts of radionuclides in the environment on human individuals and populations. Since this impact is measured in terms of radiological dose, radioecology is essentially the science of dose-assessment within the context of environmental exposure pathways. Environmental dose assessment methodologies are well established (Thorne, 2001) and typically follow the approach illustrated in Figure 5. Radioecology is ultimately a predictive science and research effort is strongly directed towards the construction and validation of models of radionuclide transfers through environmental pathways (Scott, 2002). Models are used to estimate activity concentrations of radionuclides in the human diet and in environmental compartments, such as soil and atmosphere, in which exposure to radionuclides by inhalation and direct irradiation may occur. While models are often used to carry out such assessments, they are underpinned by empirical data in the form of key parameters such as transfer coefficients (section 2.2) as well as measurements of environmental activity concentrations which are essential for the validation of model predictions. As in other branches of the environmental sciences, radioecology uses an iterative process of model formulation, data collection and model verification to develop an objective understanding of the environmental behaviour and impacts of radionuclides.

Model development has become increasingly sophisticated since the Chernobyl accident, especially with the advent of advanced information technologies such as Geographic Information Systems which have revolutionised spatial analysis and prediction (Dubois and De Cort, 2001). Nevertheless, the importance of an empirical understanding of radionuclide behaviour in specific ecosystems can never be overstated and practical measurements of radionuclides in environmental systems are still an essential part of radioecology. A feature of many radionuclides is the relative ease with which they can be measured within environmental matrices, as illustrated by the vast number of measurements of $^{137}Cs$ amassed since the Chernobyl accident. This applies particularly to gamma-emitting radionuclides which can be measured in complex environmental media with minimal or no sample preparation. In situ measurements of gamma-emitting radionuclides are also possible and form the basis of rapid ground-based or even aircraft-based surveys of radionuclide distributions (Dickson, 2004). Analysis of alpha and beta emitters is slower and technically more demanding, which has resulted in relatively small data sets (compared with those for $^{137}Cs$) being obtained for radionuclides such as plutonium isotopes and $^{90}Sr$. As a result there has been a recent trend towards using mass spectrometry techniques such as Inductively Coupled Plasma Mass Spectrometry (ICPMS) for certain alpha and beta emitters. For radionuclides with extremely low abundance (e.g. cosmogenic $^{36}Cl$) or for which conventional radiometric analysis is unsuitable (e.g. in discriminating isotopes such as $^{239}Pu$ and $^{240}Pu$) Accelerator Mass Spectrometry (AMS) is now the method of choice.

The presence in environmental systems of natural cosmogenic tracers such as $^{14}C$ and $^{36}Cl$, and primordial tracers such as $^{40}K$ and $^{210}Pb$, has been exploited by several branches of the environmental sciences which are concerned with quantifying natural processes such as soil erosion, sedimentation or biogeochemistry, rather than radiological dose assessment. So, the radiological sciences have provided a powerful array of methods by which measurements can be made of fundamental aspects of the behaviour of environmental systems. Similarly, radioecology contributed directly to 'systems ecology' at the pioneering phase in its development when Odum (1953) used artificial radionuclides as newly-available tracers to study the flow of energy and materials in natural ecosystems. The link between radioecology and our understanding of the functioning of ecological systems appears to have been overlooked in recent times.

## 4. Key topics in terrestrial radioecology

This book does not set out to describe and review specific events such as the Chernobyl accident and its aftermath. This has been done elsewhere by Warner and Harrison (1993), Smith and Beresford (2005) and, most recently, by the United Nations Chernobyl Forum (IAEA, 2006). Rather, the aim of the following chapters is to deal with generic issues in radioecology by reviewing key topics, some of which are long-standing problems but some of which did not assume any significant level of importance until the years following the Chernobyl accident.

Chapter 2 deals with one of the most important and most studied topics in terrestrial radioecology, namely the behaviour of radionuclides in soil/crop systems following contamination. The soil's role as the principal long-term reservoir of radionuclides has been recognised since the 1950s when radioactive inventories of soils, particularly in the northern hemisphere, were growing rapidly. The Chernobyl accident added fresh impetus to our need to understand the transfer of radionuclides from soils to plants and, more recently, the idea of using plants to clean-up contaminated soils has developed considerable interest. Chapter 3 provides an up to date review of this latter field which is relevant to the phytoremediation of soils contaminated with radionuclides and metals—it also suggests possible new directions for radioecology in the 21$^{st}$ century. Chapter 4 deals with the transfer of radionuclides in animal production systems which, like soil/crop systems, have been the subject of much research since the beginnings of radioecology because of their importance in the human food-chain.

The next five chapters describe the behaviour of radionuclides in specific types of natural and semi-natural ecosystems. The importance of these types of ecosystems in enhancing the radiological exposures of certain groups of human consumers was recognised in the 1960s when the specific diets of northern reindeer herders were studied intensively in countries such as Sweden and Canada (Hanson, 1967; section 3.5 of Chapter 7). Since 1986, however, it has also become apparent that many other human groups, including some in developed cultures, can be exposed to radionuclides via dietary and non-dietary pathways in ecosystems including grassland, heath, mire and mountains, forests, arctic and arid ecosystems and, finally, in tropical and sub-tropical environments. Data collections on the behaviour of radionuclides, especially radiocaesium, have grown significantly for some of these types of ecosystem in recent years. For others, however, data are still relatively scarce. Tropical and sub-tropical agriculture, in particular, supports a very large number of people in the countries of the Pacific rim and South East Asia yet the behaviour of key radionuclides in production systems such as rice paddies is only beginning to be elucidated (Chapter 9).

While the main emphasis of this book is on anthropogenically produced radionuclides in the terrestrial environment, the principal source of radiological dose for most people in the world remains the numerous radioactive isotopes associated with the decay chains of uranium and thorium present naturally in rocks and soils. Chapter 10 describes these naturally occurring radioactive materials (NORM) from both natural and industrial sources. Finally, Chapter 11 covers a topic which has emerged in recent years as a complete shift in emphasis in radiological protection. Safeguarding non-human species from ionising radiation within the terrestrial environment was a very controversial idea before 2000. Six years later it is one of the major challenges in radioecology which is attracting both the interest and the funding of regulatory agencies in many countries. This topic provides an example of how a fundamental understanding of the behaviour of radionuclides in ecosystems, based on a sound quantitative understanding of environmental exposure pathways, some of which might be considered esoteric, can be used to solve hitherto undefined problems.

## References

AMAP (1997). Arctic Pollution Issues: A State of the Arctic Environment Report. Arctic Monitoring and Assessment Programme, Oslo, Norway. 188 pp.

ApSimon, H.M., Wilson, J.J.N., Simms, K.L. (1989). Analysis of the dispersal and deposition of radionuclides from Chernobyl across Europe. *Proceedings of the Royal Society of London A* **425**, 365–405.

Asimov, I. (1972). Worlds Within Worlds: The Story of Nuclear Energy. Volumes I, II and III. Office of Information Services, US Atomic Energy Commission, Washington DC.

Battarbee, R.W., Flower, R.J., Stevenson, A.C., Rippey, B. (1985). Lake acidification in Galloway: a palaeoecological test of competing hypotheses. *Nature* **314**, 350–352.

Bell, J.N.B., Shaw, G. (2005). Ecological lessons from the Chernobyl accident. *Environment International* **31**, 771–777.

Bleise, A., Danesi, P.R., Burkart, W. (2003). Properties, use and health effects of depleted uranium (DU): a general overview. *Journal of Environmental Radioactivity* **64**, 93–112.

Bowen, H.J.M. (1979). *Environmental Chemistry of The Elements*. Academic Press, London.

Brannvall, M.L., Bindler, R., Renberg, I., Emteryd, O., Bartnicki, J., Billstrom, K. (1999). The medieval metal industry was the cradle of modern large-scale atmospheric lead pollution in northern Europe. *Environmental Science and Technology* **33**, 4391–4395.

Cardis, E., Howe, G., Ron, E., et al. (2006). Cancer consequences of the Chernobyl accident: 20 years on. *Journal of Radiological Protection* **26**, 127–140.

Clark, M.J., Smith, F.B. (1988). Wet and dry deposition of Chernobyl releases. *Nature* **332**, 245–249.

Copplestone, D., Jackson, D., Hartnoll, R.G., Johnson, M.S., McDonald, P., Wood, N. (2004). Seasonal variations in activity concentrations of $^{99}$Tc and $^{137}$Cs in the edible meat fraction of crabs and lobsters from the central Irish Sea. *Journal of Environmental Radioactivity* **73**, 29–48.

Crossland, I.G., Pinedo, P., Kessler, J.H., Torres-Vidal, C., Walters, B. (2005). "Reference Biospheres" for solid radioactive waste disposal: the BIOMASS Methodology. *Journal of Environmental Radioactivity* **84**, 135–149.

Dickson, B.L. (2004). Recent advances in aerial gamma-ray surveying. *Journal of Environmental Radioactivity* **76**, 225–236.

Dubois, G., De Cort, M. (2001). Mapping $^{137}$Cs deposition: data validation methods and data interpretation. *Journal of Environmental Radioactivity* **53**, 271–289.

Farmer, J.G., Eades, J., Mackenzie, A.B., Kirika, A., Bailey-Watts, A.E. (1996). Stable lead isotope record of lead pollution in Loch Lomond sediments since 1630 A.D. *Environmental Science and Technology* **30**, 3080–3083.

Fishlock, D. (1994). The dirtiest place on Earth. *New Scientist* **1913**, 34–37.

Gray, J., Jones, S.R., Smith, A.D. (1995). Discharges to the environment from the Sellafield Site, 1951–1992. *Journal of Radiological Protection* **15**, 99–131.

Hanson, W.C. (1967). Radioecological concentration processes characterizing arctic ecosystems. In: Aberg, B., Hungate, F.P. (Eds.), *Radioecological Concentration Processes*. Pergamon Press, Oxford, pp. 183–191.

IAEA (1994). Handbook of parameter values for the prediction of radionuclide transfer in temperate environments. TECDOC 364, International Atomic Energy Agency, Vienna.

IAEA (2006). Environmental Consequences of the Chernobyl Accident and Their Remediation: Twenty Years of Experience. Report of the Chernobyl Forum Expert Group 'Environment'. Radiological Assessment Reports Series, International Atomic Energy Agency, Vienna.

ICRP (1991). 1990 Recommendations of the International Commission on Radiological Protection, ICRP Publication 60, Annals of the ICRP, Vol. 21, pp. 1–201.

ICRP (2003). A framework for assessing the impact of ionising radiation on non-human species. ICRP Publication 91, Annals of the ICRP, Vol. 33, No. 3.

Kasparov, V.A., Ivanov, Y.A., Zvarisch, S.I., Protsak, V.P., Khomutinin, Y.V., Kurepin, A.D., Pazukhin, E.M. (1995). Formation of hot particles during the Chernobyl nuclear power plant accident. *Nuclear Technology* **114**, 246–253.

Kennedy, V.H., Horrill, A.D., Livens, F.R. (1990). Radioactivity and wildlife. Nature Conservancy Council (NCC) Focus on Nature Conservation series, No. 24.

Landa, E.R. (1982). The first nuclear industry. *Scientific American* **247**, 180–193.

LBNL Isotopes Project (2004). Lund University, Sweden, WWW Table of Radioactive Isotopes. http://ie.lbl.gov/toi/.

Markham, A. (1994). *A Brief History of Pollution*. Earthscan publications, London.

Medvedev, Z.A. (1977). Facts behind the Soviet nuclear disaster. *New Scientist* **74**, 761–764.

Medvedev, Z.A. (1979). *Nuclear Disaster in The Urals*. Norton, New York.

Odum, E.P. (1953). *Fundamentals of Ecology*, 1st edn. W.B. Saunders Co., Philadelphia.

Pourchet, M., Magand, O., Frezzotti, M., Ekaykin, A., Winther, J.-G. (2003). Radionuclides deposition over Antarctica. *Journal of Environmental Radioactivity* **68**, 137–158.

Rosman, K.J.R., Chisholm, W., Hong, S., Candelone, J.P., Boutron, C.F. (1997). Lead from Carthaginian and Roman Spanish mines isotopically identified in Greenland ice dated from 600 B.C. to 300 A.D. *Environmental Science and Technology* **31**, 3413–3416.

Scott, E.M. (Ed.) (2002). *Modelling Radioactivity in the Environment*. Elsevier Science Publishers, Amsterdam.

Shaw, G. (2005). Applying radioecology in a world of multiple contaminants. *Journal of Environmental Radioactivity* **81**, 117–130.

Shaw, G., Bell, J.N.B. (1989). The kinetics of caesium absorption by roots of winter wheat and the possible consequences for the derivation of soil-to-plant transfer factors for radiocaesium. *Journal of Environmental Radioactivity* **10**, 213–231.

Sheppard, S.C. (2003). An index of radioecology: what has been important? *Journal of Environmental Radioactivity* **68**, 1–10.

Smith, J.T., Beresford, N.A. (2005). *Chernobyl: Catastrophe and Consequences*. Springer/Praxis publishers, Chichester, UK.

Sokolov, V.E., Krivolutski, D.A., Voronov, A.G., Shilova, S.A. (Eds.) (1993). *Ecological After-Effects of The Radioactive Contamination in The Southern Urals*. Nauka, Moscow.

Sposito, G. (1989). *The Chemistry of Soils*. Oxford University Press.

Stigliani, W.M., Doelman, P., Salomons, W., Schulin, R., Smidt, G.R.B., Van der Zee, S.E.A.T.M. (1991). Chemical time-bombs. *Environment* **33**, 26–30.

Tarkhanov, I.R. (1896). Experiments on the action of Roentgen's X-rays on the animal organism. *Izv. St. Petersburg Biol. Lab.* **1** (3), 47 (in Russian).

Thorne, M.C. (2001). Assessing the radiological impact of releases of radionuclides to the environment. In: Van der Stricht, E., Kirchmann, R. (Eds.), *Radioecology: Radioactivity and Ecosystems*, pp. 391–446.

Trabalka, J.R., Auerbach, S.I. (1991). One western perspective of 1957 Soviet nuclear accident. In: *Comparative assessment of the environmental impact of radionuclides realeased during three major accidents: Kyshtym, Windscale and Chernobyl*. CEC, Luxembourg, pp. 41–69. Report EUR 13574.

UNSCEAR (2000). Sources and effects of ionizing radiation. United Nations Scientific Committee on the Effects of Atomic Radiation, Report to the General Assembly, with scientific annexes.

Vernadsky, V.I. (1929). Concentration of radium by living organisms. *Dokl. Acad. Nauk SSSR, Series A* **2**, 33 (in Russian).

Ward, G.M., Johnson, J.E., Stewart, H.F. (1965). Deposition of fallout 137Cs on forage and its transfer to cows' milk. Proceedings of the Second AEC Symposium on Fallout, Oak Ridge, Tennessee.

Warneke, T., Croudace, I.W., Warwick, P.E., Taylor, R.N. (2002). A new ground-level fallout record of uranium and plutonium isotopes for northern temperate latitudes. *Earth Planet. Sci. Lett.* **203**, 1047–1057.

Warner, F., Harrison, R.M. (1993). *Radioecology After Chernobyl: Biogeochemical Pathways of Artificial Radionuclides, SCOPE 50*. John Wiley and Sons, Chichester, UK.

Webster, S., Salt, C.A., Howard, B.J. (2003). Sea-to-land transfer of technetium-99 through the use of contaminated seaweed as an agricultural soil conditioner. *Journal of Environmental Radioactivity* **70**, 127–137.

Whicker, F.W., Schultz, V. (1982). *Radioecology: Nuclear Energy and The Environment*, vol. I, CRC Press, Boca Raton Florida.

# Behaviour of radionuclides in soil/crop systems following contamination

Friederike Strebl[a], Sabine Ehlken[b,c], Martin H. Gerzabek[d], Gerald Kirchner[b,e]

[a]*Austrian Research Centers GmbH, A-2444 Seibersdorf, Austria*
[b]*University of Bremen, Dept. of Physics/FB 1, D-28334 Bremen, Germany*
[c]*Klinikum Bremen-Mitte, St.-Jürgen-Str. 1, Abt. Ökologie, D-238205 Bremen, Germany*
[d]*University of Agricultural Sciences, Institute for Soil Research, Peter-Jordan Strasse 82, A-1190 Vienna, Austria*
[e]*Federal Office for Radiation Protection, Köpenicker Allee 120–130, 10318 Berlin, Germany*

## 1. Introduction

About 1300 different radionuclides exist, partly originating from natural sources, partly anthropogenically produced. Natural radionuclides can be generated by activation of stable isotopes via cosmic radiation, for example $^{3}$H (12.3 years physical half-life—$t_{1/2\text{phys}}$), $^{7}$Be (53.3 days), $^{14}$C (5730 y), $^{35}$S (87.5 d), or they originated during the creation of the universe. The latter are known as primordial radionuclides and include $^{40}$K ($t_{1/2\text{phys}}$: $1.3 \times 10^{9}$ y) and the isotopes of uranium and thorium which give rise to various daughter nuclides, including $^{226}$Ra (1600 y), $^{222}$Rn (3.8 d), $^{210}$Pb (22.3 y) (see Chapter 10).

Artificial radionuclides have entered the human environment as a consequence of atmospheric nuclear weapon tests. Global fallout of fission products in the 1950s and 1960s involved deposition of $^{137}$Cs ($t_{1/2\text{phys}}$: 30.17 y), $^{90}$Sr (28.5 y), $^{89}$Sr (50.5 d), $^{3}$H, $^{54}$Mn (312 d), $^{65}$Zn (244 d), $^{95}$Zr (64 d), $^{103/106}$Ru (39.4 d/368 d), $^{129}$I ($1.6 \times 10^{7}$ y) and $^{144}$Ce (284.8 d). At present the most important sources are routine releases from nuclear power (NPP) and reprocessing plants which can include $^{3}$H, $^{14}$C, $^{60}$Co ($t_{1/2\text{phys}}$: 5.27 years), $^{54}$Mn, $^{89}$Sr, $^{90}$Sr, $^{95}$Zr, $^{85}$Kr (10.8 y), $^{238}$Pu (87.7 y), $^{239/240}$Pu (24,000 y/6,600 y), $^{129}$I, $^{95}$Nb (35 d), $^{103/106}$Ru, $^{110m}$Ag (249.9 d), $^{125}$Sb (2.77 y), $^{134/137}$Cs (2.06 y/30.17 y), $^{140}$Ba (12.8 d), $^{144}$Ce. Very high emissions of these isotopes can occur in the very rare circumstances of NPP accidents such as occurred at Chernobyl (Ukraine) in 1986. This accident increased $^{134/137}$Cs inventories of many European countries considerably. Some typical values of $^{137}$Cs and $^{90}$Sr soil inventories (kBq $m^{-2}$) in Austria and other selected locations are reported in Table 1. Other, mainly short-lived isotopes are used in medicine as tracers for diagnostics or as agents for radiotherapy and include $^{99m}$Tc ($t_{1/2\text{phys}}$: 6 h), $^{131}$I, $^{51}$Cr (27.7 d), $^{54}$Mn and $^{60}$Co. Additional risks of radionuclide contamination of the environment may arise from final radioactive waste depositories.

RADIOACTIVITY IN THE ENVIRONMENT
VOLUME 10 ISSN 1569-4860/DOI: 10.1016/S1569-4860(06)10002-9

Table 1
Typical ranges of radionuclide content in Austrian soils and comparison to other regions (all artificial radionuclides corrected for decay to 1st May 1986; ChNPP: Chernobyl Nuclear Power Plant)

| | Source | Average topsoil (0–20 cm) content ($Bq\,kg^{-1}$) | Soil inventory ($kBq\,m^{-2}$) |
|---|---|---|---|
| $^{14}C$ | Natural | 330 | |
| $^{40}K$ | Natural | 1–1000 | |
| $^{210}Pb$ | Natural | 5–50 | |
| $^{226}Ra$ | Natural | 5–200 | |
| $^{238}U$ | Natural | 5–150 | |
| $^{234}U$ | Natural | 9–120 | |
| $^{90}Sr$ | Bomb tests | | Austria: range: 1–4 |
| | Nuclear accident | | ChNPP 30-km-zone: 18–1800 |
| | | | Kyshtym region: 2000–4000 |
| $^{103}Ru$ | Nuclear accident | | Austria: 3.7–74 |
| $^{137}Cs$ | Bomb tests | | Austria: 23.4 |
| | Nuclear accident | | Denmark: 3.7 |
| | | | ChNPP 30-km-zone: 1500–5000 |
| $^{239/240}Pu$ | Bomb tests | | Austria: $1–15 \times 10^{-2}$ |
| | Nuclear accident | | ChNPP 30-km-zone: > 3.7 |

Source: Bunzl (1997), Aarkrog et al. (1992), Ageets (1999), and United Nations Scientific Committee on the Effects of Atomic Radiation (UNSCEAR) (1988).

Besides the total amount deposited, the radiological importance of artificial radionuclides is determined by their radiotoxicity and radiation type ($\alpha$, $\beta$, $\gamma$), their bioavailability and behaviour within the foodchain. Radionuclides from natural decay chains, despite short physical half-lives, can be of high radiological importance because they are continuously produced and thus remain in the environment at a constant level. Some radioactive isotopes, like those of iodine, cannot be discriminated from the stable forms and readily enter living systems. Others show close chemical similarity to nutrients (Cs–K; Sr, Ra–Ca), which can also lead to a considerable uptake into the food-chain.

## 2. Physical and chemical behaviour of relevant radionuclides in arable soils

In the case of a radionuclide deposition event during the cropping season (late spring, summer, early autumn), or in case of a chronic atmospheric release, direct contamination of plant surfaces poses the main radiological problem. However, over the long term soil is the major source for radionuclides entering the food chain or for leaching of radionuclides and contamination of groundwater. Aside from the mass/activity relationships described in Chapter 1 there are no general differences between radioactive and stable isotopes of a given chemical element with respect to their behaviour in the environment. Soil parameters which exert a major impact on radionuclide mobility are, according to Bunzl (1997), (i) the composition of the soil solution (pH, concentration of inorganic ions, redox potential, concentration of organic substances), (ii) physical and chemical soil properties (species/characteristics and contents of clay minerals, oxides and organic matter, surface and charges of particles), (iii) micro-organisms and fungi (mycorrhiza) and (iv) temperature. Most radionuclides in soil are present in cationic forms. In general, therefore, low pH-values, low clay contents and low cation exchange capacities lead to an increase in radionuclide mobility within the soil profile and favour plant uptake of radionuclides. Changes in soil organic matter content can yield different effects, depending on the ability of the respective radionuclide to form organic complexes or not (Frissel et al., 1990).

The vertical migration velocity of radionuclides is determined by the site-specific water balance. Vertical water movements in soils result from the balance between precipitation, which provides an input of water at the soil surface, and evapotranspiration and runoff which can remove this meteoric water either before or during its passage through the soil's pore space. Net infiltration of water into the soil (i.e. precipitation minus evapotranspiration and runoff) is able to transport radionuclides down the soil profile in both soluble form and bound to soil particles (colloids). The latter process is especially important for radionuclides which tend to bind strongly to soil solids. Sorption of radionuclides on the soil solid phase is a key process in retarding the migration of radionuclides in the solution phase of the soil. For modelling purposes sorption is quantified using the solid–liquid distribution coefficient ($K_d$), which is simply the ratio of the concentrations at equilibrium of a specified radionuclide (RN) in the solid and liquid phases of the soil, respectively.

$$K_d = \frac{\text{RN concentration in solid phase (Bq g}^{-1}\text{)}}{\text{RN concentration in liquid phase (Bq ml}^{-1}\text{)}} \quad (\text{ml g}^{-1}) \tag{1}$$

For the migration of different radionuclides to a soil depth of one metre, Bunzl (1987) reported typical time spans of 150–4500 years for Cs, 40–200 years for Sr and 0.5–20 years for Tc. The migration rates of individual radionuclides in soils are inversely related to their degree of sorption or fixation to the soil solid phase, as measured by the $K_d$. Experimental results have shown that the sorption of radionuclides in agricultural soils increases in the following order: $^{95}Tc < {}^{131}I < {}^{85}Sr < {}^{103}Ru < {}^{109}Cd \sim {}^{65}Zn < {}^{57}Co < {}^{141}Ce < {}^{137}Cs$ (Bunzl and Schimmack, 1988). Table 2 summarises the effects of key soil parameters on the mobility of some elements with important radionuclides and the specific behaviour of these elements is described below.

*Caesium*-137 is one of the most common radionuclides contaminating soils world-wide. It has been known for several decades that radiocaesium can be strongly fixed in soils containing clay minerals and, in such soils, it shows a very low degree of vertical mobility and availability for plant uptake. Since the late 1980s the mechanism by which Cs interacts with clay minerals, specifically illite, has been elucidated. Clay mineral particles consist of repeating stacks of alumina and silica sheets which are interspersed with ions or water molecules, depending on the specific mineral. Micaceous clays such as illite are characterised by the presence of potassium ions ($K^+$) in their interlayer spaces as well as so-called frayed-edge-sites (FES) at the edges of their mineral particles (Figure 1). These FES can bind a variety of ions including $K^+$, $NH_4^+$ and $Cs^+$ which all have similar ionic radii and which can lose their hydration shells relatively easily. However, the selectivity of FES for $Cs^+$ is approximately 100× that for $NH_4^+$ and 1000× that for $K^+$. As a result, sorption of caesium on the FES is much greater than sorption of ammonium or potassium. Cremers et al. (1988) devised a laboratory method which

Table 2
Influence of soil characteristics on the behaviour of different radionuclides in soils (compiled from Bunzl, 1987; Coughtrey et al., 1983; IAEA, 1994; Mortvedt, 1994). ( Key: ↑ increased mobility; ↓ decreased mobility; ⇔ no clear effect)

| | Cs | Sr, Ra | I | U, Pu | Ru |
|---|---|---|---|---|---|
| Similar ions | K, Rb, $NH_4$ | Ca, Ba | | Np | Tc, Mo, Nb, Zr, Y, Co, Fe, Mn, Cr |
| Chemical form in soil | $Cs^+$ | $Sr^{2+}$ | $I_2$, $CH_3I$, $I^-$, $IO_3^-$ | $PuO_2^{2+}$, $Pu(NO_3)^{3+}$ | $RuNO^{3+}$, $Ru^{3+}$, $RuO_4^-$ |
| | Changes in RN mobility and plant availability if... | | | | |
| pH-value decreases | ↑ | ↑ | | ↑ | ↑ |
| Clay content decreases | ↑ | ↑ | | ↑ | ↑ |
| Sand content decreases | ↓ | ↓ | | ↓ | ↓ |
| Humus content low | ⇔ | ↓ | ↑ | ↓ | ⇔ |
| CEC decreases | ↑ | ↑ | | | ↑ |
| $K_d$-value (l $kg^{-1}$) | $10^2$–$10^3$ | $10^1$–$10^2$ | $10^{-1}$–$10^2$ | $10^2$–$10^3$ | $10^1$–$10^3$ |

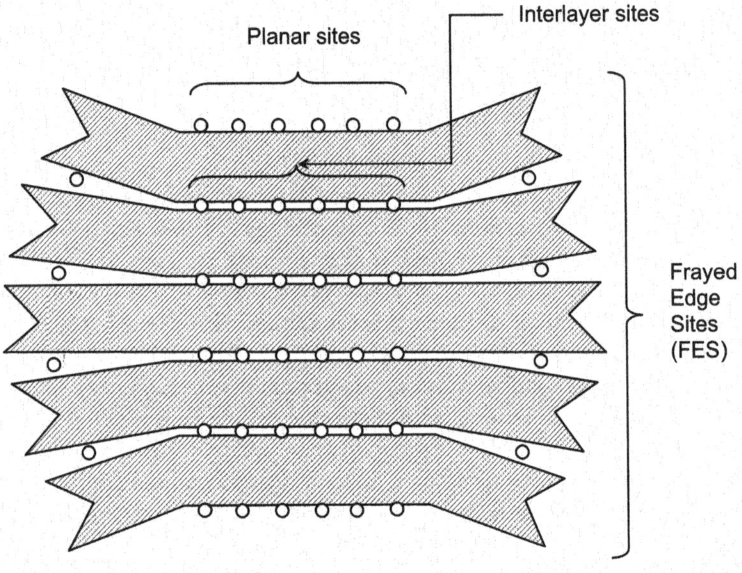


Fig. 1. Generalised structure of illite, a common micaceous clay mineral, which plays an important role in fixing potassium and caesium in soils. Binding of $K^+$ and/or $Cs^+$ ions (O) can occur at planar, frayed edge and interlayer sites.

quantifies the abundance and the Cs-specificity of the FES which they termed the Specific Radiocaesium Interception Potential (SRIP). The SRIP is the product of the Cs to K selectivity coefficient in the FES and the abundance of the FES (measured in meq $kg^{-1}$; Sweeck et al., 1990) and, essentially, provides a measure of the cation exchange capacity of the soil which is able to exchange $Cs^+$ ions with high specificity. Despite the tendency for Cs sorption to clays to be highly specific it is still reduced by competition with the $NH_4^+$ and $K^+$ ions in the soil solution. Hilton and Comans (2001) showed how the solid–liquid $K_d$ for Cs could be predicted based on a knowledge of the abundance of FES ([FES]), the $NH_4^+$ and $K^+$ ion concentrations and the $Cs^+/K^+$ ($K_c(Cs^+, K^+) = 1000$) and $NH_4^+/K^+$ ($K_c(NH_4^+/K^+) = 5$) selectivity coefficients, as follows.

$$K_d^{Cs} = \frac{K_c(Cs^+, K^+) \times [FES]}{[K^+] + K_c(NH_4^+, K^+) \times [NH_4^+]} \quad (ml\,g^{-1}) \tag{2}$$

At a general level, Livens and Loveland (1988) summarised the soil properties which contributed to the immobilisation of radiocaesium in upland soils of the UK after the Chernobyl accident. These are the clay mineral content and type, organic matter (humus) content, pH, ammonium and potassium content. The general effects of these factors on caesium mobility in soils are summarised in Table 2.

*Strontium* is chemically similar to calcium and often shows much higher mobility than caesium. Gerzabek et al. (1992a, 1992b) reported a fraction between 18–62% of the total Sr soil inventory being in exchangeable form in Austrian soils. The portion of mobile radiostrontium (and other elements) can be influenced by the presence of hot particles (small fragments of irradiated reactor fuel or other particles with abnormally high radionuclide contents) from nuclear power plant accidents or nuclear weapons testing (Gastberger et al., 2000a). In such particles radionuclides are strongly fixed over long periods. Sorption of Sr is increased at higher pH-values in parallel with the formation of Sr-humic complexes (Juo and Barber, 1970). Calcium competes with strontium for exchange sites. According to the low fixation strength in soils (low $K_d$ value, see Table 2), migration rates for Sr of up to several cm per year are observed in agricultural soils (Haunold et al., 1987). Recently it was reported that, in a similar fashion to Cs, Sr is efficiently retained in alpine soils and recycling within the vegetation cover is likely (Gastberger et al., 2000b).

*Radium*, a divalent cation, also behaves in a similar way to calcium in soil and plants. In comparison with Cs, Ra mobility in soils is high although it decreases with increasing pH and with increasing contents of clay and hydroxides of Mn and Fe. Plant availability and plant uptake of radium is distinctly influenced by the essential plant nutrient Ca (Gerzabek et al., 1998). For more information on Ra see Chapter 10.

Some *iodine* isotopes are very short-lived ($^{131}I$, $t_{1/2phys} = 8.02$ days) and, as a result, are irrelevant to the soil–plant pathway. Only $^{129}I$ has some relevance with respect to soil–plant transfer due to its extremely long physical half life. In soil it occurs in close interaction with soil organic matter (Gerzabek et al., 1999) and microbial biomass can also fix iodine very effectively (Dertinger et al., 1986). As an anion, iodine can be sorbed to Fe and Al hydroxides and clay minerals (Whitehead, 1984). A positive correlation between clay content and iodine concentration was found for Austrian soils (Gerzabek et al., 1999).

In soil *ruthenium* can be observed as a trivalent cation ($Ru^{3+}$) and the ruthenate anion ($RuO^{4-}$). The mobility of Ru is slightly higher than that of Cs and typical $K_d$ values for Ru are between those of Cs and Sr (Bunzl, 1987).

*Plutonium* has an extremely high radiotoxicity: $^{239}Pu$ ingested in the human diet imparts a dose which is approximately 5× greater than that imparted by $^{137}Cs$ (IAEA, 1996). In soil it can be found in five different oxidation states (III, IV, V, VI, VII), although at common soil pH values Pu(IV) is likely to be the most stable (Dahlman et al., 1976). The other most common oxidation state of plutonium under environmental conditions is Pu(VI) but this is likely to be reduced to Pu(IV) by complexation with soil humic substances. Significant amounts of Pu in soils seem to be complexed with humic substances (Bunzl, 1997). $K_d$ values are similar to those of Cs, as are migration rates in soil, as observed by Shaw and Wang (1996) in the close vicinity of the Chernobyl NPP. Bunzl et al. (1995) also observed similar migration rates of $^{239+240}$Pu and $^{137}$Cs in German forest soils contaminated by global nuclear weapons fallout.

For *cobalt* at pH $< 5$ the free cation $Co^{2+}$ is abundant in the soil solution. At higher pH values, soluble organic complexes are formed and, at pH $> 6$, more than 90% of Co in the soil solution is bound in organic complexes (Scheffer and Schachtschabel, 1989). Cobalt adsorption occurs on Fe and Mn oxides and occlusion in sesquioxide coatings on soil particles is possible. Lysimeter experiments have indicated that cobalt mobility in various Austrian agricultural soils is significantly higher than radiocaesium mobility (Gerzabek et al., 1996).

Agricultural management practices typical for crop production in temperate latitudes lead to lower radionuclide mobility in soil as compared with semi-natural environments, primarily due to the following.

- Topsoils are regularly ploughed, which severely disturbs the vertical distribution of superficially deposited radionuclides. Meisel et al. (1991) reported a minimum number of three ploughing steps to obtain a uniform radionuclide distribution within the plough layer (0–20 cm). This mixing leads to a dilution of radionuclide concentration and enhances contact between deposited particles and the soil matrix, resulting in increased sorption.
- Applications of fertilisers containing nitrogen, phosphorus and potassium (N, P and K) and lime ($CaCO_3$) maintain favourable crop growth conditions in soils, avoiding nutrient deficiencies and low pH values. Low soil nutrient status and acid pH depress plant yields and tend to favour a higher degree of radionuclide uptake by vegetation, as observed in many semi-natural environments.
- Poorly drained and organic soils, which often favour increased radionuclide soil–plant transfer, are commonly cultivated as grassland because they are not suitable for the use of heavy equipment and only sub-average yields can be expected.

## 3. Physiological aspects of crop root-uptake and translocation of radionuclides in arable systems

In soil–plant systems, radionuclides are present in sub-trace quantities, even at contamination levels which are typical of major releases from nuclear installations. For example, activity concentrations of $^{90}$Sr and $^{137}$Cs in soil solutions of 1 $kBq\,l^{-1}$, as measured in the territories heavily contaminated by the Chernobyl fallout, correspond to concentrations of approximately $2 \times 10^{-12}$ $mol\,l^{-1}$ which are lower by about nine orders of magnitude than concentrations of major cations in soil solution (Robson and Pitman, 1983). As a result, when considering the root uptake of many radionuclides their absolute concentrations in the soil–plant system often are of less importance than the corresponding concentrations of other micro- and macronutrients which are generally present at very much higher levels.

Concentrations in solution of most elements are determined by cation exchange reactions with the soil matrix. In general, our understanding of the chemistry of trace radionuclides in soils is adequate to explain bulk concentrations in the soil. However, evidence is growing that, within the rhizosphere, radionuclide concentrations available for plant uptake may differ considerably from those in the bulk soil (Merckx et al., 1983; Marschner et al., 1986). For example, it has been shown that roots may actively increase the absorption of radiocaesium on clay minerals (Guivarch et al., 1999), but they may also remobilize radiocaesium fixed in interlayer sites of clay minerals (Thiry, 1997; Delvaux et al., 2000).

Compared with the bulk soil, the rhizosphere contains elevated numbers of bacteria and soil fungi which also accumulate radionuclides. It has been shown that in organic and in forest soils a significant fraction of the radiocaesium present in the soil–plant system may be stored and potentially immobilised by soil fungi (Dighton et al., 1991; Brückmann and Wolters, 1994; Nikolova et al., 2000). Despite these findings, direct evidence of the impact of roots and of rhizosphere organisms on the availability of radionuclides for plant uptake seems to be limited.

Transport of solutes, including radionuclides, to plant roots is brought about by convective mass flow and by diffusion (Tinker and Nye, 2000), but variations in soil moisture influence both the concentration and the mobility of solutes. A lower soil water content may decrease (Pendleton and Uhler, 1960; D'Souza and Mistry, 1980) or increase (Shalhevet, 1973; Tikhomirov, 1988) root uptake of radionuclides depending on whether soil drying mainly restricts the mobility of a radionuclide in solution or increases its concentration (Shalhevet, 1973). Under natural environmental and climatic conditions, periods of low soil water contents are associated with elevated plant root uptake rates of radiocaesium and radiostrontium by meadow vegetation growing on various soil types (Ehlken and Kirchner, 1996).

Radionuclides as well as other minerals are taken up by plant roots in ionic form. After passing the plasma membrane of the epidermal or cortical cells the ions move into the cell cytoplasm (symplasm) in which they are free to move through the plasmodesmata which connect adjoining cells. By crossing the plasma membrane for a second time, ions move from the symplasm into the xylem vessels. In recent years, it has been shown that ion transport through the plasma membranes of root cortical and xylem parenchyma cells is facilitated by ion pumps (Michelet and Boutry, 1995), carriers (Tanner and Caspari, 1996) and ion channels (White, 1997).

Solution culture experiments have demonstrated that competitive and inhibitory interactions are of prime importance in root uptake and translocation of alkaline metal (Epstein and Hagen, 1952; Shaw and Bell, 1991; Smolders et al., 1997) and alkaline earth (Epstein and Leggett, 1954) elements, as well as of many other metals (Kawasaki and Moritsugu, 1987). For the alkaline metals, selectivity sequences observed with intact roots are in good agreement with selectivities measured at isolated plasma membranes both at low ion concentrations when transport by carriers dominates and at high concentrations when movement across the membrane is through so-called potassium channels mainly (Maathuis and Sanders, 1995, 1996). A second type of channel (so-called calcium channels) present in plasma membranes of plant root cells have been shown to mediate the translocation of alkaline earth elements, but also of other divalent cations ($Mn^{2+}$, $Co^{2+}$, $Ni^{2+}$, $Cu^{2+}$, $Zn^{2+}$ and $Cd^{2+}$), although with lower selectivities (White, 1998). Moreover, Sr:Ca permeability ratios observed for root cell calcium channels (White, 1998) are in good agreement with empirical data on the discrimination of these elements during root uptake ('observed ratios') which have been traditionally used to characterise the plant uptake of radiostrontium (Russell and Newbould, 1966).

The term translocation describes the redistribution of radionuclides within a plant's tissues following absorption through the roots or leaves. The xylem is the principal water-conducting tissue of vascular plants: dissolved substances are transported within the transpiration stream from the roots, upwards into the stem(s) and leaves. The other long-distance transport tissue is the phloem, the principal function of which is the conduction of assimilates and food from their sites of production (leaves) to the storage organs (fruits, tubers etc.). Flows of substances within the plant can therefore take place in both upwards and downwards directions (Strasburger, 1998). Elements like K, Cs and P are readily transported in both xylem and phloem. The mobility of elements in the xylem is mainly dependent on ion valence ($K^+ > Ca^{2+} \gg$ polyvalent cations) and resulting adsorption to the cell wall (Marschner, 1986). Cobalt is translocated in phloem as a negatively charged complex. Due to physiological constraints the alkaline earth metals (Be, Mg, Ca, Ba, Sr, Ra), lead, plutonium, lanthanides and actinides are not transported in the phloem. They are therefore dependent on xylem mobility and tend to remain at the sites at which they are absorbed or are accumulated in vegetative plant parts and mostly excluded from edible plant parts (fruits, tubers) (Thiessen et al., 1999).

## 4. Quantification of soil–plant transfer

Soil-to-plant transfer is one of the major environmental pathways leading to human ingestion of radionuclides. Thus, since the 1950s and 1960s many efforts have been undertaken to predict and quantify radionuclide root uptake and to implement suitable models in radioecological/radiological models. As root uptake of radionuclides and the subsequent translocation to edible plant parts is influenced by many factors (as described in section 3), its modelling is a compromise between availability of input parameters and scientifically based mechanistic approaches. In the past, most efforts were directed to determining simple ratios of radionuclide concentration between crops and soils; more recently, radionuclide fluxes have gained importance in soil-to-plant transfer modelling. A range of different approaches is presented below.

### 4.1. *The transfer factor (TF) concept*

The transfer factor concept is the simplest model for the quantification and prediction of crop contamination with radionuclides (RN). The transfer factor (TF) is defined as:

$$\mathrm{TF} = \frac{\text{RN concentration in plant tissue (Bq kg}^{-1})}{\text{RN concentration in soil (Bq kg}^{-1})} \quad \text{(dimensionless)} \qquad (3)$$

It is usual to adhere to strict protocols when obtaining data to calculate transfer factors, as originally specified by the International Union of Radioecology (IUR, 1989). The radionuclide concentration in soil is always determined on a dry weight basis down to a depth of 20 cm for all crops except pasture grass (10 cm). When determining radionuclide concentrations in plant material the edible parts of crops (grain, tubers, fruits, edible leaves etc.) have been most often investigated although, in the case of grazed and semi-natural ecosystems, the vegetative parts of plants which provide nutrition for animals are just as important (see Chapter 5). Radionuclide concentrations are usually expressed on a dry plant matter basis. However, in

Table 3
Expected values of soil-to-plant transfer factors based on crop and soil dry weight basis for selected crops and elements (modified from IAEA, 1994)

| Element | Soil type | Cereal grain | Grass | Pea, bean (pods) | Potato tubers | Green vegetables |
|---|---|---|---|---|---|---|
| Co | Not specif. | 0.0037 | 0.054 | 0.03 | 0.06 | 0.2 |
| Zn | Not specif. | 1.2 | 0.99 | 0.71 | 35 | – |
| Sr | Clay | 0.12 | 1.1 | 1.3 | 0.15 | 2.7 |
| | Sand | 0.21 | 1.7 | 2.2 | 0.26 | 3.0 |
| | Peat | 0.020 | 0.34 | – | – | – |
| Tc | Not specif. | 0.73 | 76 | 4.3 | 0.24 | – |
| Cs | Clay | 0.01 | 0.11 | 0.017 | 0.07 | 0.18 |
| | Sand | 0.026 | 0.24 | 0.094 | 0.17 | 0.46 |
| | Peat | 0.083 | 0.53 | – | 0.27 | – |
| Ra | Not specif. | 0.0012 | 0.08 | 0.007 | 0.0011 | 0.049 |
| Th | Not specif. | 0.000034 | 0.011 | 0.00012 | 0.000056 | 0.0018 |
| U | Not specif. | 0.0013 | 0.023 | – | 0.011 | 0.0083 |
| Np | Not specif. | 0.0027 | 0.069 | 0.018 | 0.0067 | 0.037 |
| Pu | Not specif. | 0.0000086 | 0.00034 | 0.000061 | 0.00015 | 0.000073 |

Not specif.: not specified.
–: no data available or mean based on very few observations.

the older literature many TFs are calculated on a plant fresh (wet) weight basis and must be recalculated into dry-matter-based values by use of average dry matter plant contents (IAEA, 1994).

An important assumption in using the TF is that it is independent from the absolute radionuclide concentration in the soil. This assumption does not always appear to hold true under real world conditions (Bunzl et al., 2000). Other simplifications in the transfer factor concept include (i) the artificial definition of the rooting zone, (ii) the lack of discrimination between radionuclide pools of different availability in soil, (iii) the fact that the TF does not really describe the process of root uptake but merely provides a concentration ratio including, for example, mass loading of plant surfaces with contaminated soil particles (see section 4.5) and (iv) the omission of any plant physiological parameters. The influence of these and additional simplifications discussed above is reflected by the huge variability of transfer factor values obtained under field and experimental conditions which, for many radionuclides, exceeds three orders of magnitude (Coughtrey et al., 1983; IAEA, 1994). Table 3 provides literature data on soil-to-plant transfer factors for agricultural crops. Another comprehensive dataset published more recently can be found in Gerzabek et al. (1998). A careful evaluation of the available Cs and Sr transfer data was published by Nisbet et al. (1999) who investigated the effects of ageing, pH, organic matter, and exchangeable potassium (or calcium) on the soil-to-plant transfer of Cs and Sr. An extended review of pre-Chernobyl soil–plant transfer data was presented by Baes et al. (1984).

The growing awareness of the limitations of the simple empirical transfer factor concept has stimulated interest in models which better predict plant uptake following soil contamination by radionuclides. As early as 1987, Frissel et al. attempted to improve the simple TF model for radiocaesium, the most thoroughly investigated radionuclide with respect to soil-to-plant

Table 4
Soil–plant fluxes of caesium, radium and cobalt into some agricultural crops (soil inventory $^{137}Cs$: 588 $kBq\,m^{-2}$; $^{226}Ra$: 85.1–91.5 $kBq\,m^{-2}$; $^{56}Co$: 687.9 $kBq\,m^{-2}$; arithmetic means ± standard deviation; $n = 12$)

| | Flux rate ($s^{-1}$) | | |
|---|---|---|---|
| | $^{137}Cs$ ($\times 10^{-12}$) | $^{226}Ra$ ($\times 10^{-12}$) | $^{60}Co$ ($\times 10^{-12}$) |
| Mustard | 56.9 ± 53.2 | 13.1 ± 6.68 | 4.77 ± 2.43 |
| Potato tuber | 16.9 ± 15.8 | 4.39 ± 1.34 | 2.96 ± 1.93 |
| Potato leaves | 10.3 ± 7.26 | 14.4 ± 7.81 | 7.36 ± 7.19 |
| Maize grain | 11.8 ± 12.9 | 0.38 ± 0.34 | 0.28 ± 0.14 |
| Maize straw | 73.1 ± 70.7 | 23.5 ± 12.0 | 2.44 ± 0.75 |
| Bean straw | 20.6 ± 14.6 | 12.5 ± 6.72 | 10.7 ± 4.90 |
| Bean pods | 24.6 ± 22.0 | 2.35 ± 1.22 | 65.4 ± 25.2 |
| Grass | 2.03 ± 1.06 | 9.15 ± 2.93 | 1.17 ± 0.69 |
| Wheat grain | 6.15 ± 6.97 | 9.44 ± 6.56 | 0.33 ± 0.11 |
| Wheat straw | 3.44 ± 3.58 | 21.6 ± 9.87 | 0.17 ± 0.10 |
| Beet leaves | 22.4 ± 21.3 | 23.9 ± 14.4 | 4.01 ± 1.91 |
| Sugar beet | 20.3 ± 22.7 | 36.9 ± 16.8 | 10.3 ± 5.15 |

transfer, by including soil characteristics, ageing and irrigation into statistical models. The following formula, including five exponential correction terms, was proposed:

$$TF_{act} = TF_{stand} \times e^{(-0.64(pH-6))} \times e^{(0.11(OM-4))} \times e^{(-0.017(depth-20))} \times e^{(-0.082(period-2))} \times e^{(0.008(irr))}, \tag{4}$$

where $TF_{act}$ represents the actual Cs transfer factor, $TF_{stand}$ the standardised Cs transfer factor, OM the soil organic matter (%), depth the contaminated horizon thickness (cm), period the time elapsed after fallout (y) and irr the irrigation ($mm\,y^{-1}$). Based on this formula the standardised TF was calculated using the large TF database of the International Union of Radioecologists (IUR) for a soil pH of 6, OM content of 4%, a 20 cm thick contaminated horizon, two years after contamination and without irrigation.

The IUR Radflux database (Mitchell and Donelly, 2000) was designed to provide a readily available, comprehensive compilation of rate constants for the parameterisation of dynamic multi-compartmental models represented mathematically as sets of first-order linear differential equations. These rate constants (with the dimension of $time^{-1}$), representing net transfer between a donor (e.g. soil) and a receiving compartment (e.g. plant tissue), have been obtained directly or were calculated from literature data by using soil/plant TFs and available information on exposure time and biomass production (see Table 4). The consideration of these two parameters often reduces data variability in comparison with TF values. Existing transfer factor databases collated by the IUR and others were incorporated into the database after recalculation of values to a consistent format. A calculation example is given below.

$$\text{Soil–Plant Flux} = \frac{-\ln(1 - \text{lost fraction})}{\text{exposure time}} \quad (\text{time}^{-1}), \tag{5}$$

where

$$\text{Lost Fraction} = \frac{\text{RN inventory in plant (Bq m}^{-2})}{\text{total RN inventory (soil + plant, Bq m}^{-2})} \tag{6}$$

## 4.2. *Refined and alternative modelling approaches*

For the radiologically important nuclear fission products $^{137}Cs$ and $^{90}Sr$, various soil–plant modelling approaches have been suggested (e.g. Roca et al., 1997; Absalom et al., 1999; Kirchner and Ehlken, 1999). The common theme in these models is that they each focus on the influence of soil/soil solution physicochemical interactions on the concentrations of the radionuclides present in the soil solution, which are then assumed to be available to plant roots. Uptake and translocation are not modelled mechanistically, but are represented by fairly general empirical relationships. By comparing model predictions with experimental data, the developers of these models were able to demonstrate that, although they address only some of the relevant processes in soil–plant systems, their models constitute a major improvement compared with the traditional transfer factor concept.

Van Dorp et al. (1979) proposed a method for estimating radionuclide concentrations in edible plant parts based on the solubility of a given radionuclide in soil water, its ability to cross root membranes and its upward movement within the plant with the transpiration stream. This approach was chosen to represent the closely linked processes of root uptake and translocation into edible plant parts, which are often incorporated in a single soil–plant transfer factor. Measured values of $K_d$ are used together with a plant-specific fraction of absorbed radionuclide which is transferred to the edible plant part ($F$, dimensionless), root selectivity coefficient ($S$, dimensionless), total production (PT, g dry matter $cm^{-2}\,y^{-1}$), $P_{ep}$ (production of edible plant part dry matter, $g\,cm^{-2}\,y^{-1}$), transpiration coefficient (TC, $ml\,g^{-1}$ dry weight), radionuclide inventory of the soil ($C_t$, $Bq\,cm^{-2}$), rooting depth within the soil profile ($L$, cm), soil water content ($\theta$, $cm^3\,cm^{-3}$) and bulk density of the soil ($\rho$, $g\,cm^{-3}$) to estimate the radionuclide concentration in the dry matter of edible plant parts ($C_{ep}$).

$$C_{ep} = F \times S \times \left(\frac{\mathrm{PT}}{P_{ep}}\right) \times \mathrm{TC} \times \left(\frac{C_t}{L(\theta + \rho \cdot K_d)}\right) \qquad (7)$$

To our knowledge the predictive power of this approach has never been evaluated.

## 4.3. *Comparison of TF values derived from pot, lysimeter and field experiments*

In some evaluations of soil–plant transfer data for agricultural crops, pot experiments turned out to yield different average values when compared with data observed under field conditions. According to Frissel et al. (1990) the reasons for this observation are as follows.

- A more intensive exploration of the contaminated substrate by plant roots due to the limited space available in plant pots.
- In pot experiments soils are mostly contaminated artificially using radionuclides in easily soluble form (dissolved in aqueous solutions). Binding to the soil substrate can increase with time (ageing effects), until equilibrium conditions are reached.
- To avoid plant drought stress plant pots have to be irrigated regularly as the limited soil volume and high root density result in rapid water loss by evapotranspiration.

These differences can be reduced to a minimum by choosing an experimental design which allows enough space per plant for the root system and some equilibration time after artificial contamination. Table 5 compares results from pot (5 kg soil; Chernobyl contamination)

Table 5
Comparison of soil–plant transfer data derived from pot- and lysimeter experiments (TF-values, Bq kg$^{-1}$ dry matter plant/Bq kg$^{-1}$ dry matter soil; recalculated from fresh weight data for pot experiments)

| | TF-values (kg/kg) | Wheat grain | Maize grain | Wheat straw | Maize straw | Endive |
|---|---|---|---|---|---|---|
| Lysimeter ($n = 3$) | Mean | 0.0024 | 0.0020 | 0.0049 | 0.0183 | 0.0233 |
| (Gerzabek et al., 1998) | ± sd. | 0.008 | 0.0001 | 0.0009 | 0.0015 | 0.0049 |
| Average of 2 soils ($n = 36$) | Mean | 0.0008 | 0.0033 | 0.0130 | 0.0081 | 0.0280 |
| (in Horak et al., 1989) | ± sd. | 0.0005 | 0.0012 | 0.0049 | 0.0014 | 0.0112 |

and lysimeter (>1000 kg soil; artificial CsCl contamination) experiments carried out with comparable soils (Eutric cambisol, calcareous) and under the same climatic conditions. The differences between the data are within the normal range of variability and a systematic bias to higher TF values for pot experiments was not observed.

Lysimeter facilities combine the advantages of controlled experiments with almost undisturbed climatic and soil conditions. Field data on radionuclide concentrations in soil drainage waters are scarce, with the exception of Chernobyl fallout isotopes. For artificial radionuclides, which cannot be investigated under field conditions because background contamination levels are usually very low, lysimeters are often the most suitable means to obtain data for different soil types for the implementation of assessment models.

Comparison of radionuclide losses from soils via various pathways under normal agricultural conditions can be used to help quantify the contribution of mechanisms such as leaching, cropping and erosion to the long term decline of radionuclide soil inventories.

### 4.4. *Radionuclide translocation*

A parameter which is widely used in assessment models is the translocation factor (unit m$^2$ kg$^{-1}$) which, in case of direct contamination of crops, must be considered when calculating ingestion doses likely to be incurred by consumption of the vegetable matter at harvest. It is defined as the ratio of the activity concentration in the edible plant part (Bq kg$^{-1}$) to the total activity retained by the plant canopy per unit ground area (Bq m$^{-2}$). Some models relate the total activity in the edible plant parts at harvest (Bq m$^{-2}$) to the total activity initially retained by the plant canopy (Bq m$^{-2}$) in which case translocation is defined as a dimensionless parameter (Thiessen et al., 1999). For leafy vegetables such as cabbage and fodder-hay the translocation factor is, by definition, 1.0 as the entire plant shoot is consumed. An extended summary on shoot/fruit translocation factors of fruit crops was presented by Watkins and Maul (1995). A generic value of the translocation factor often used in assessment models is 0.1, representing a conservative value for most radionuclides. A collection of recommended translocation factors for agricultural crops is presented in IAEA (1994).

Although useful for application in assessment models, the translocation factor is unable to describe the complicated plant-internal transport processes, which are highly variable and depend on the developmental stage of plants at the time of deposition, seasonal effects, plant species and environmental conditions.

Table 6
Summary of soil mass loading on plant surfaces from different measurements

| Plant | Pathway | mg soil $g^{-1}$ plant DM | Number of observations | Reference |
|---|---|---|---|---|
| Grass | n.s. | $18 \pm 48$ | 26 | Arthur and Alldredge (1982) |
| Tomatoes | n.s. | 17 | | Dreicer et al. (1984) |
| Lettuce | n.s. | $260 \pm 100$ | 4 | McLeod et al. (1984a) |
| Broccoli | n.s. | $10 \pm 8.1$ | 4 | McLeod et al. (1984a) |
| Turnip | n.s. | $32 \pm 11$ | 4 | McLeod et al. (1984a) |
| Cabbage | n.s. | $1.1 \pm 1.1$ | 4 | McLeod et al. (1984a) |
| Tobacco | n.s. | $2.1 \pm 0.6$ | 12 | McLeod et al. (1984b) |
| Sunflower | n.s. | $2.6 \pm 0.9$ | 10 | Pinder and McLeod (1988) |
| Soybean | n.s. | 2.1 | 10 | Pinder and McLeod (1989) |
| Wheat | n.s. | 4.8 | 10 | Pinder and McLeod (1989) |
| Corn | n.s. | 1.4 | 10 | Pinder and McLeod (1989) |
| Broadbean | Wind erosion | 3.0 | 4 | Li et al. (1994) |
| | Rainsplash | 6.5 | 4 | Li et al. (1994) |
| | Total | 9.5 | 4 | Li et al. (1994) |
| Ryegrass | Wind erosion | 3.1 | 4 | Li et al. (1994) |
| | Rainsplash | 2.7 | 4 | Li et al. (1994) |
| | Total | 5.8 | 4 | Li et al. (1994) |

n.s.: not specified.

## 4.5. *Soil contamination of plant surfaces*

The external surfaces of plants can become contaminated by soil particles as a result of rainsplash or resuspension due to wind action. The bioavailability of radionuclides ingested in vegetable matter by humans or animals can depend strongly on whether the radionuclides are contained within the plant tissues or within soil particles associated with the plant material. Large variations in soil–plant transfer factors used in dose assessment models may also result from soil contamination of plant surfaces (Gerzabek et al. 1992a, 1992b). Several different methods have been devised to estimate the soil mass loading on the external surfaces of plants (Arthur and Alldredge, 1982; Kiriyama and Kuroda, 1982; Pinder and McLeod, 1989).

Scandium is ubiquitous in soils but scarcely absorbed by plant roots and not mobile within plant tissues. Therefore $^{46}Sc$, determined by neutron activation analysis, can serve as a tracer for soil particles associated with the external surfaces of vegetation samples. In a combined greenhouse and field experiment Li et al. (1994) investigated the effects of wind erosion and rainsplash using ryegrass and broadbean as test plants. In Table 6 the main results of this experiment, together with other complementary data from the literature, are presented.

Under field conditions 5.77 mg soil $g^{-1}$ DW (dry weight) was found to be retained on the external surfaces of ryegrass while broad bean surfaces retained 9.51 mg soil $g^{-1}$ DW. The estimated contributions of soil splash and wind erosion, respectively, were 68% and 32% for broadbean, 47% and 53% for ryegrass. The variability between the two species is probably due to differences in plant anatomy and leaf structure.

As the soil–plant transfer factor does not discriminate between external plant contamination and 'real' root uptake, the soil mass loading concept can be used for correction purposes applying the following formula:

$$A_r = A_t - A_s \times M_s \times 0.001,$$

where $A_r$ is the activity concentration in plant tissue due to root uptake (Bq kg$^{-1}$), $A_t$ is the total activity concentration measured in plant tissues (Bq kg$^{-1}$), $A_s$ is the activity concentration in the soil (Bq kg$^{-1}$) and $M_s$ is the soil mass loading value (mg soil g$^{-1}$ plant).

Applying this formula to a sample set from Austrian soil–plant transfer factor investigations under field conditions, it was shown that mass loading contributes from 3 to 23% to the overall $^{137}$Cs contamination of cereal straw samples and their respective transfer factors. This effect was due to soil mass loading values of 1.1 to 6.0 mg soil g$^{-1}$ plant (Li et al., 1994). After correction for mass loading, transfer factors ranged from 0.008 to 0.057. For other inorganic or organic pollutants with similar or even lower availability for root uptake, the soil mass loading described might be of at least similar significance.

### 4.6. *Food-chain models*

From a radiation protection viewpoint the contamination of foodstuffs for human consumption is a crucial issue. Complex assessment models have therefore been developed to predict human exposure via the different pathways by which radionuclides are transferred to food products. Examples of such models are PATHWAY (Whicker and Kirchner, 1987), ECOSYS-87 (Müller and Pröhl, 1993) and FARMLAND (Brown and Simmonds, 1995). These models address a range of important atmosphere–soil–plant interaction processes (for example deposition, interception, weathering, resuspension, root uptake, fixation and leaching in soils) and subsequent transfer to animal products. In most applications the radionuclide contamination of plant material is calculated using the transfer factor concept.

## 5. Overall mass/activity balances of radionuclides in arable systems

Between 1990 and 1996 a lysimeter experiment was carried out in Seibersdorf (Austria) to investigate radionuclide behaviour ($^{60}$Co, $^{137}$Cs and $^{226}$Ra) in agricultural soil–plant systems (Gerzabek et al., 1998). Selected data (maize) were used for the calculation of an overall activity balance and the quantification of radionuclide losses via different pathways on the basis of lysimeter soil activities, radionuclide leaching from the soils and radionuclide uptake into the maize crop. The four soil types investigated were as follows.

I Eutric Cambisol
II Dystric Cambisol (low exchangeable K content)
III Dystric Cambisol (high exchangeable K content)
IV Dystric Gleysol

Each soil was placed in triplicate lysimeters and was artificially contaminated with 588 ± 30.8 kBq m$^{-2}$ $^{137}$Cs, 79 ± 8.5 kBq m$^{-2}$ $^{226}$Ra and 688 ± 40.7 kBq m$^{-2}$ $^{60}$Co. At harvest, the radionuclide concentrations in maize straw varied considerably between different soil types and radionuclide concentrations in maize grain were always lower than those in vegetative

Table 7
Radionuclide inventories and fluxes of maize biomass grown of four different soils (soil I Eutric Cambisol, soil II Dystric Cambisol (low exchangeable K content), soil III Dystric Cambisol (high exchangeable K content), soil IV Dystric Gleysol)

| | Nuclide | Radionuclide content in harvested biomass (Bq m$^{-2}$) | | | |
|---|---|---|---|---|---|
| | | Soil I | Soil II | Soil III | Soil IV |
| Maize straw + corncobs | $^{137}$Cs | 103 ± 6.2 | 1143 ± 199 | 26.4 ± 3.8 | 1001 ± 288 |
| Maize grains | $^{137}$Cs | 5.8 ± 0.50 | 117 ± 31.7 | 1.1 ± 0.3 | 69.0 ± 22.2 |
| Maize straw + corncobs | $^{226}$Ra | 9.87±0.52 | 39.8 ± 3.7 | 39.9 ± 6.3 | 20.0 ± 3.5 |
| Maize grains | $^{226}$Ra | 0.06±0.01 | 0.13±0.01 | 0.53±0.13 | 0.18±0.05 |
| Maize straw + corncobs | $^{60}$Co | 28.3 ± 5.0 | 18.2 ± 8.0 | 18.8 ± 3.4 | 23.6 ± 7.4 |
| Maize grains | $^{60}$Co | 1.4 ± 0.45 | 1.3 ± 0.85 | 2.0 ± 0.39 | 0.64±0.01 |

Table 8
Calculated loss rates (s$^{-1}$) for different lysimeter soils and radionuclides (geometric mean ± standard deviation, $n$ = 8–12)

<table>
<tr><th>Nuclide</th><th colspan="4">Radionuclide leaching loss from soil inventory, values × 10$^{-14}$ (s$^{-1}$)</th></tr>
<tr><th></th><th>Soil I</th><th>Soil II</th><th>Soil III</th><th>Soil IV</th></tr>
<tr><td>$^{137}$Cs</td><td>1.34 ± 0.45</td><td>1.07 ± 0.15</td><td>1.65 ± 0.33</td><td>1.59 ± 0.32</td></tr>
<tr><td>$^{226}$Ra</td><td colspan="4">13.5 ± 2.4</td></tr>
<tr><td>$^{60}$Co</td><td>16.9 ± 4.2</td><td>7.5 ± 2.6</td><td>16.8 ± 3.3</td><td>11.2 ± 2.8</td></tr>
</table>

plant parts (see Table 7). From the experimental data it can be concluded that the total radionuclide activities extracted from the soils due to harvest of the maize crop were small. Despite the high biomass produced (2.37 ± 0.3 kg maize straw and 1.16 ± 0.16 kg maize grain per m$^2$) the radionuclide activities extracted by a single harvest represent on average 0.105%, 0.003% and 0.031% of the soil inventories for $^{137}$Cs, $^{60}$Co and $^{226}$Ra, respectively. Nevertheless, extraction of the radionuclides in harvested crop tissues ('phytoextraction') exceeded the annual losses due to radionuclide leaching in the soil solution, which were calculated based on radionuclide concentrations in leaching waters observed between 1990–1996 (Table 8; Gerzabek et al., 1996). Leaching losses each year amounted to approximately 0.4 × 10$^{-4}$%, 3.9 × 10$^{-4}$% and 4.3 × 10$^{-4}$% for $^{137}$Cs, $^{60}$Co and $^{226}$Ra, respectively. Therefore, radionuclide losses from the soil due to harvesting of maize were more than three orders of magnitude (2400×) higher than leaching losses in the case of $^{137}$Cs, approximately two orders of magnitude higher (70×) for $^{226}$Ra and only one order of magnitude higher (9×) for $^{60}$Co.

The time trend of leaching losses, measured half-yearly, gave no indication of increasing fixation to soil components, but remained constant. Differences between the mobilities in the soil of $^{226}$Ra and $^{60}$Co were small, while $^{137}$Cs leaching rates were lower by an order of magnitude. In general, the estimated losses during a hypothetical period of 200 years illustrate that leaching from agricultural soils leads only to marginal losses from the initial soil inventory.

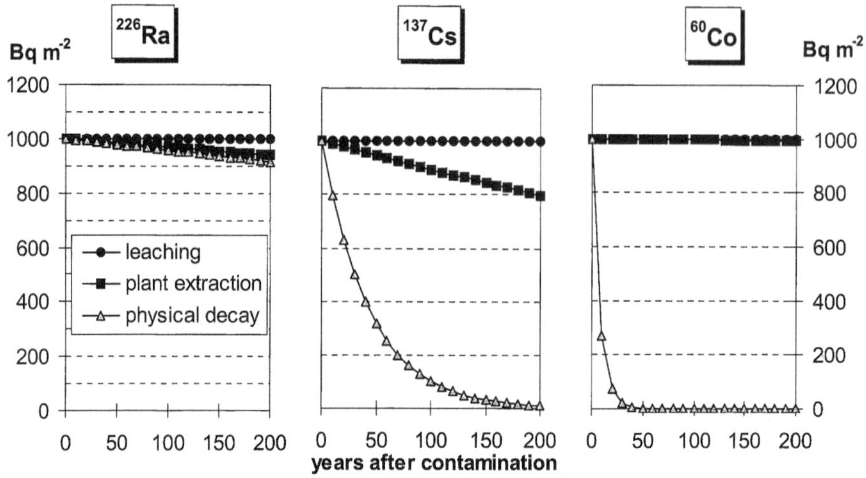


Fig. 2. Estimated radionuclide losses from soil via plant harvest, soil leaching and physical decay within a hypothetical period of 200 years.

In Figure 2 the calculated annual radionuclide losses resulting from harvesting crop biomass and soil leaching are combined with the rather more accurately known decrease in activity due to physical decay. For $^{60}Co$, with a physical half-life of 5.27 years, it can be seen very clearly that the reduction in the soil inventory due to environmental processes is negligible compared with the reduction due to physical decay. Furthermore, in comparison with the other radionuclides in this study, plant uptake of $^{60}Co$ is relatively low. As a result, contamination of the soil with $^{60}Co$ is not a long term problem and the need for soil clean-up to protect food production and groundwater is not as great compared with longer-lived radionuclides.

In contrast, radium contamination of soils is a significant long-term problem often requiring remediation (Sweeck et al., 2005). Due to the long physical half life of $^{226}Ra$ (1600 years) plant extraction represents a significant loss pathway which reduces the effective half-life of this radionuclide in the soil to approximately 600 years (Figure 2). Phytoextraction could, therefore, be a useful tool for slow soil clean-up of soils contaminated with $^{226}Ra$, if suitable plant species with high accumulation potential can be selected. From the different crops investigated sugar beet turned out to take up radium most effectively (Table 4). The topic of phytoextraction and phytoremediation is covered in more detail in Chapter 3.

## 6. Possible countermeasures

The Chernobyl accident confirmed the need for the availability of countermeasures which can be used to reduce the uptake of radionuclides by crops, but which are also cost-effective and easy to apply in routine agricultural practice. Laboratory research and experience gained in contaminated areas have shown that countermeasures applied to soils can be classified according to the following broad approaches (Desmet, 1991; Howard and Desmet, 1993; Renaud and Maubert, 1997).

(a) Dilution of radionuclides in the rooting zone by ploughing.
(b) Precipitation of radionuclides by an insoluble chemical compound(s).
(c) Binding of radionuclides to insoluble amendments.
(d) Addition of chemicals which dilute radionuclides in the soil solution and/or compete for root uptake.

The potential of ploughing to reduce root uptake of all radionuclides in all soil types is widely accepted. If practicable, deep ploughing can result in an almost complete removal of radionuclides from the rooting zone (Roed et al., 1996). The applicability of ploughing as a countermeasure, however, depends on site-specific conditions (e.g. thickness of the soil layer over bedrock, vegetation cover, soil density, slope and topography).

In contrast, the effectiveness of countermeasures (b) to (d) listed above depends on a number of factors of which the soil type and its nutrient status are of primary importance. The addition of a chemical treatment may effectively reduce the transfer of a specific radionuclide, but increase the bioavailability of others. Therefore, the effectiveness of any of the soil treatments which are based on approaches (b) to (d) may be site-specific and should be evaluated prior to their application. Fortunately, available experimental evidence combined with our present knowledge on the principles governing the behaviour of radionuclides in the soil–plant system (as outlined above) allows us to categorise the effects of chemical amendments of soils contaminated with multiple radionuclides (Ehlers et al., 1997). The results are shown in Tables 9–11, which focus on the long-lived isotopes of caesium and strontium which are known to dominate the long-term radiological consequences after a nuclear accident. In addition, information is compiled on the effect of soil based countermeasures on the bioavailability of actinides, since actinides may pose a radiological problem in areas most heavily affected by severe accidents. Only broad classifications of the effectiveness of countermeasures are given, since the actual impact of any individual countermeasure depends on a variety of additional factors, such as plant species composition, moisture regime, physico-chemical properties of the soil and the duration over which the countermeasure is applied.

In Tables 9–11 the soil types are generalised into three categories.

(i) Soils with low organic matter content ($<10\%$) and high cation exchange capacity ($> 100$ meq $kg^{-1}$). These nutrient-rich soils typically are used for intensive agriculture.
(ii) Soils with low organic matter content ($<10\%$) and low cation exchange capacity ($< 100$ meq $kg^{-1}$). In this group nutrient-poor soils are found which, nevertheless, are used agriculturally, for instance as permanent pastures.
(iii) Soils with high organic matter content ($>10\%$).

It is apparent from Tables 9–11 that a number of countermeasures suggested in the literature are of generally low effectiveness and may even increase the root uptake of caesium or strontium radioisotopes. Only very limited information is available on the impact of most of the countermeasures on the root uptake of actinides. This may reflect the fact that, because of their generally low transfer factors, actinides are not considered to contribute significantly to radiation doses due to ingestion of contaminated foodstuffs. It should be taken into account, however, that the public acceptance of a countermeasure may depend on presenting evidence

Table 9
Effectiveness of addition of chemicals on reducing the soil–plant transfer in mineral soils showing a high cation exchange capacity

| Chemical added | Soil–plant transfer[a] | | |
|---|---|---|---|
| | Cs | Sr | Actinides |
| K | (←) | (←) – (→) | |
| Phosphate | (→) | (←) – ← | |
| Nitrate | (→) | | |
| PK | (←) | (←) – (→) | |
| NK, NP, NPK | (←) | | |
| $NH_4$ | (→) – → | (←) – (→) | → |
| $NH_4 + K$ | ← | (→) | |
| Organic N (manure) | (→) | | (←) |
| Organic matter | (←) | (←) | (←) |
| Sewage sludge | (→) | | |
| Ca | 0 – (→) | (←) – ← | (←) – (→) |
| $Na_2CO_3$ | | ← | |
| Clay minerals | (←) – (→) | (←) | |
| Zeolithes | ← – → | ← | |

Key: ← : effective reduction (greater than a factor of 2), (←): slight reduction (less than a factor of 2), 0: no effect, (→): slight increase (less than a factor of 2), →: pronounced increase (greater than a factor of 2).

Table 10
Effectiveness of addition of chemicals on reducing the soil–plant transfer in mineral soils showing a low cation exchange capacity

| Chemical added | Soil–plant transfer[a] | | |
|---|---|---|---|
| | Cs | Sr | Actinides |
| K | ← | (←) – (→) | |
| Phosphate | (→) | (←) – ← | |
| Nitrate | (→) – → | | |
| PK | (←) – ← | (←) | |
| NK, NP, NPK | | | |
| $NH_4$ | (→) – → | (←) – (→) | |
| $NH_4 + K$ | ← | (→) | |
| Organic N (manure) | (→) – → | | |
| Organic matter | (←) – 0 | (←) – ← | (←) |
| Sewage sludge | | | |
| Ca | (←) | (←) – ← | 0 – ← |
| $Na_2CO_3$ | | | |
| Clay minerals | ← | (←) – ← | |
| Zeolithes | ← – (→) | ← | |

Key: ← : effective reduction (greater than a factor of 2), (←): slight reduction (less than a factor of 2), 0: no effect, (→): slight increase (less than a factor of 2), → : pronounced increase (greater than a factor of 2).

Table 11
Effectiveness of addition of chemicals on reducing the soil–plant transfer in organic soils

| Chemical added | Soil–plant transfer[a] | | |
|---|---|---|---|
| | Cs | Sr | Actinides |
| K | ← | (←) – (→) | |
| Phosphate | (→) | | |
| Nitrate | | | |
| NP | 0 | | |
| NK, PK | | | |
| NPK | ← – (→) | 0 – (→) | |
| $NH_4$ | (→) – → | 0 | |
| $NH_4 + K$ | ← | (→) | |
| Organic N (manure) | | | |
| Sewage sludge | | | |
| Ca | (←) | (←) – (→) | |
| NK/PK/NPK + Ca | (←) – ← | (←) | |
| $Na_2CO_3$ | | | |
| Clay minerals | | | |
| Zeolithes | (←) – ← | 0 | |

Key: ←: effective reduction (greater than a factor of 2), (←): slight reduction (less than a factor of 2), 0: no effect, (→): slight increase (less than a factor of 2), → : pronounced increase (greater than a factor of 2).

that it does not mobilise radionuclides which in the public perception are believed to be 'very dangerous', such as plutonium.

The general mechanisms of the most effective chemical treatments are well established.

(a) The addition of potassium or calcium has two consequences. On the one hand, the added cations compete with the radionuclides for the binding sites in the soil and, therefore, can remobilize sorbed radioisotopes, thereby increasing their root uptake. On the other hand, addition of K or Ca dilutes the concentrations of Cs and Sr in the soil solution, thereby decreasing their uptake by plants. For the dilution mechanism to dominate, potassium or calcium have to be added in quantities often well above those needed for general soil fertilisation. Their addition generally has a pronounced effect on caesium or strontium root uptake in potassium- or calcium-deficient soils only. Furthermore, it should be noted that the dilution of ions in the soil solution not only affects the radioisotopes present, but also trace nutrient ions and may result in deficiencies of essential trace minerals for the plants.
(b) Addition of calcium as lime ($CaCO_3$) to acid organic soils shifts the pH upwards, thereby increasing the cation exchange capacity of the soil. This reduces the plant availability of all cations. Calcium should not be added as $CaCl_2$ or $CaSO_4$ which both tend to reduce the soil pH.
(c) The addition of phosphate reduces the availability of strontium by precipitation of insoluble strontium phosphate.

Based on the information given in Tables 9–11 and taking into account costs, ease of application and acceptability by farmers (Nisbet, 1995), potassium fertilisation, liming and ploughing may be judged as the most promising countermeasures after a nuclear accident. Each of these is a typical agricultural treatment although, for use as countermeasures, liming and potassium additions are often required well above normal fertilisation thresholds. A number of secondary effects of these treatments on the environment have been identified, some of them being detrimental, others beneficial (Ehlers et al., 1997; Salt and Rafferty, 2001), as described below.

(a) Liming and ploughing improve soil structure and water hydraulics, whereas K addition decreases soil stability.
(b) Both chemical treatments reduce the availability of other plant nutrients due to dilution and ion competition. This may be compensated by the addition of nutrients together with the K and Ca treatments.
(c) Liming and ploughing enhance the degradation of soil organic matter resulting in decreasing soil fertility and in potential mobilisation of toxic trace substances since soluble humic acids may play a crucial role in the transport of these substances.
(d) As a consequence of ion competition, mobilisation of toxic trace substances may also be promoted by the addition of K or Ca.
(e) Changes in the ploughing regime may considerably increase erosion of topsoil.

## 7. Conclusions

Twenty years after the Chernobyl accident contaminated large areas of continental Europe, considerable progress has been made in understanding the behaviour of radionuclides, particularly those of caesium and strontium, in soil/crop systems. Data bases on soil–plant transfer have been expanded considerably and much new information on key processes such as soil sorption processes and competition for uptake between radionuclides and common soil cations has been obtained. For most radionuclides, the soil is the primary long-term reservoir in agricultural ecosystems. As a result, soil-based countermeasures to reduce food-chain contamination can be designed to reduce the degree of soil–plant transfer significantly although undesirable side effects of some of these treatments, such as heavy fertiliser treatments, may be experienced. Experimental mass/activity balance studies have shown that absolute losses from soils of key radionuclides such as $^{137}$Cs by processes such as cropping and leaching are small: radioactive decay will probably dictate the rate at which agricultural systems become decontaminated. The development of more rapid methods to clean-up soils is desirable and one of these, the use of plants to extract radionuclides from contaminated soils, is the focus of the next chapter.

## Acknowledgements

Two of the authors (Sabine Ehlken and Gerald Kirchner) gratefully acknowledge financial support of the literature review on countermeasures summarised here by the Commission of the European Communities under research contract FI4P-CT95-0021.

## References


Aarkrog, A., Botter-Jensen, L., Jiang, C.Q., Dahlgaard, H., Hansen, H., Holm, E., Lauridsen, B., Nielsen, S.P., Strandberg, M., Sogaard-Hansen, J. (1992). *Environmental Radioactivity in Denmark in 1990 and 1991.* Riso National Laboratory, Roskilde, Denmark.

Absalom, J.P., Young, S.D., Crout, N.M.J., Nisbet, A.F., Woodman, R.F.M., Smolders, E., Gillett, A.G. (1999). Predicting soil to plant transfer of radiocesium using soil characteristics. *Environmental Science & Technology* **33**, 1218–1223.

Ageets, V.Yu. (1999). *Belarus and Chernobyl: The Second Decade.* Gomel, Belorus: Scientific-Research Institute of Radioecology of the Ministry of Emergencies Republic of Belorus, 89 pp.

Arthur, W.J. III, Alldredge, A.W. (1982). Importance of plutonium contamination on vegetation surfaces at Rocky Flats, CO. *Environmental and Experimental Botany* **22**, 33–38.

Baes, C.F. III, Sharp, R.D., Sjoreen, A.L., Shor, R.W. (1984). A Review and Analysis of Parameters for Assessing Transport of Environmentally Released Radionuclides through Agriculture. ORNL-5786. Oak Ridge National Laboratory, Oak Ridge, Tennessee (available via Internet: http://homer.ornl.gov/documents/ornl5786.html; 12.09.2001).

Brown, J., Simmonds, J.R. (1995). *FARMLAND: A Dynamic Model for the Transfer of Radionuclides through Terrestrial Foodchains.* National Radiation Protection Board, London, UK, ISBN 0-85951-380-7. NRPB-Report No. R273.

Brückmann, A., Wolters, V. (1994). Microbial immobilization and recycling of $^{137}$Cs in the organic layers of forest ecosystems: relationship to environmental conditions, humification and invertebrate activity. *The Science of the Total Environment* **157**, 249–256.

Bunzl, K. (1987). Das Verhalten von Radionukliden im Boden. *Deutsche Tieraerztliche Wochenschrift* **94**, 357–359.

Bunzl, K. (1997). Radionuklide. In: Blume, H.-P., Felix-Henningsen, P., Fischer, W.R., Frede, H.-G., Horn, R., Stahr, K. (Eds.), *Handbuch der Bodenkunde.* Ecomed, Landberg/Lech, Germany, pp. 1–18, chapter 6.5.2.6.

Bunzl, K., Schimmack, W. (1988). Distribution coefficients of radionuclides in the soil: analysis of the field variability. *Radiochimica Acta* **44/45**, 355–360.

Bunzl, K., Kracke, W., Schimmack, W. (1995). Migration of fallout $^{239,240}$Pu, $^{241}$Am and $^{137}$Cs in the various horizons of a forest soil under pine. *Journal of Environmental Radioactivity* **28**, 17–34.

Bunzl, K., Albers, B.P., Schimmack, W., Belli, M., Ciuffo, L., Menegon, S. (2000). Examination of a relationship between Cs-137 concentrations in soils and plants from alpine pastures. *Journal of Environmental Radioactivity* **48**, 145–158.

Coughtrey, P.J., Jackson, D., Thorne, M.C. (1983). Radionuclide Distribution and Transport in Terrestrial and Aquatic Ecosystems. *A Critical Review of Data*, vol. 1. A.A. Balkema, Rotterdam, Netherlands.

Cremers, A., Elsen, A., De Preter, P., Maes, A. (1988). Quantitative analysis of radiocesium retention in soils. *Nature* **335**, 247–249.

Dahlman, R.C., Bondietti, E.A., Eyman, L.D. (1976). Biological pathways and chemical behaviour of plutonium and other actinides in the environment. In: Friedman, A.M. (Ed.), *Actinides in the Environment*, American Chemical Society Symposium Series, No. 35.

Delvaux, B., Kruyts, N., Cremers, A. (2000). Rhizosphere mobilization of radiocesium in soils. *Environmental Science & Technology* **34**, 1489–1493.

Dertinger, H., Müller, A., Nagel, K., Riedl, A., Strack, S. (1986). Fixierung von radioaktivem Jod im Boden. Kernforschungszentrum Karlsruhe KFK-Report 3916, Karlsruhe, Germany, 78 pp. (in German).

Desmet, G. (1991). Improvement of practical countermeasures: the agricultural environment. Commission of the European Communities, Report EUR-12554.

Dighton, J., Clint, G.M., Poskitt, J. (1991). Uptake and accumulation of $^{137}$Cs by upland grassland soil fungi: a potential pool of Cs immobilization. *Mycological Research* **95**, 1052–1056.

Dreicer, M., Hakonson, T.E., White, G.C., Whicker, F.W. (1984). Rainsplash as a mechanism for soil contamination of plant surfaces. *Health Physics* **46**, 177–187.

D'Souza, T.J., Mistry, K.B. (1980). Absorption of gamma-emitting fission products and activation products by rice under flooded and unflooded conditions from two typical soils. *Plant and Soil* **55**, 189–198.

Ehlers, H., Ehlken, S., Kirchner, G. (1997). Countermeasures applied at the soil–plant interface: effectiveness and side-effects—A literature review. CEC Research Project 'Countermeasures: Environmental and Socio-Economic Responses—A Long-Term Evaluation'. Internal Report, University of Bremen (available on request).

Ehlken, S., Kirchner, G. (1996). Seasonal variations in soil-to-grass transfer of fallout strontium and cesium and of potassium in North German soils. *Journal of Environmental Radioactivity* **33**, 147–181.

Epstein, E., Hagen, C.E. (1952). A kinetic study of the absorption of alkali cations by barley roots. *Plant Physiology* **27**, 457–474.

Epstein, E., Leggett, J.E. (1954). The absorption of alkaline earth cations by barley roots: kinetics and mechanisms. *American Journal of Botany* **41**, 785–791.

Frissel, M.J., Soutjesdijk, J.R., Koolwijk, A.C., Köster, H.W. (1987). The Cs-137 contamination of soils in the Netherlands and its consequences for the contamination of crop products. *Netherlands Journal of Agricultural Science* **35**, 339–346.

Frissel, M., Noordijk, H., Van Bergeijk, K.E. (1990). The impact of extreme environmental conditions, as occuring in natural ecosystems, on the soil-to-plant transfer of radionuclides. In: Desmet, G., Nasimbeni, P., Belli, M. (Eds.), *Transfer of Radionuclides in Natural and Semi-Natural Environments*. Elsevier, London, UK, pp. 40–47.

Gastberger, M., Steinhäusler, F., Gerzabek, M.H., Hubmer, A., Lettner, H. (2000a). $^{90}$Sr and $^{137}$Cs in environmental samples from Dolon near the Semipalatinsk nuclear test site. *Health Physics* **79**, 257–265.

Gastberger, M., Steinhäusler, F., Gerzabek, M.H., Lettner, H., Hubmer, A. (2000b). Soil-to-plant transfer of fallout caesium and strontium in Austrian lowland and Alpine pastures. *Journal of Environmental Radioactivity* **49**, 217–233.

Gerzabek, M.H., Artner, C., Horak, O., Mück, K. (1992a). Results of field studies on $^{90}$Sr and stable Sr soil-to-plant transfer. Austrian Research Center Seibersdorf, Seibersdorf, Austria, Report 4617, 25 pp.

Gerzabek, M.H., Mohamad, S.A., Mück, K. (1992b). $^{137}$Cs in soil texture fractions and its impact on $^{137}$Cs soil-to-plant transfer. *Communications of Soil Science and Plant Analysis* **23**, 321–330.

Gerzabek, M.H., Mück, K., Steger, F., Algader, S.M. (1996). Die Auswaschung von $^{60}$Co, $^{137}$Cs und $^{226}$Ra im Lysimeterversuch. *Die Bodenkultur* **47**, 71–80.

Gerzabek, M.H., Strebl, F., Temmel, B. (1998). Plant uptake of radionuclides in lysimeter experiments. *Environmental Pollution* **99**, 93–103.

Gerzabek, M.H., Muramatsu, Y., Strebl, F., Yoshida, S. (1999). Iodine and bromine contents of some Austrian soils and relations to soil characteristics. *Journal of Plant Nutrition and Soil Science* **162**, 415–419.

Guivarch, A., Hinsinger, P., Staunton, S. (1999). Root uptake and distribution of radiocesium from contaminated soils and the enhancement of Cs adsorption in the rhizosphere. *Plant and Soil* **211**, 131–138.

Haunold, E., Horak, O., Gerzabek, M.H. (1987). Umweltradioaktivität und ihre Auswirkung auf die Landwirtschaft. I. Das Verhalten von Radionukliden in Boden und Pflanze. *Die Bodenkultur* **38**, 95–118.

Hilton, J., Comans, R.N.J. (2001). Chemical forms of radionuclides their quantification in environmental samples. In: *Radioecology: Radioactivity and Ecosystems*. BVG Press, Chapter 3.3, pp. 99–111.

Horak, O., Mück, K., Gerzabek, M.H. (1989). Cs-137 Soil to Plant Transfer Factors Derived from Pot Experiments and Field Studies. Austrian Research Center Seibersdorf Report OEFZS-4515, pp. 1–11.

Howard, B.J., Desmet, G. (Eds.) (1993). Relative effectiveness of agricultural countermeasure techniques. *Science of the Total Environment (Special Issue)* **137**, 1–315.

IAEA (International Atomic Energy Agency) (1994). Handbook of Parameter Values for the Prediction of Radionuclide Transfer in Temperate Environments. IAEA Technical Report Series 364, Vienna, Austria, 74 pp.

IAEA (International Atomic Energy Agency) (1996). International basic safety standards for protection against ionizing radiation and for the safety of radioactive sources. IAEA Safety Series No. 115, Vienna, Austria.

IUR (1989). International Union of Radioecologists: VIth annual report of the working group on soil-to-plant transfer. Published by RIVM, Bilthoven, the Netherlands.

Juo, A.S.R., Barber, S.A. (1970). The retention of strontium by soils as influenced by pH, organic matter and saturation cations. *Soil Science* **99**, 143–148.

Kawasaki, T., Moritsugu, M. (1987). Effect of calcium on the absorption and translocation of heavy metals in excised barley roots: multi-compartment transport box experiment. *Plant and Soil* **100**, 21–34.

Kirchner, G., Ehlken, S. (1999). Soil-to-plant transfer of strontium: another test of the observed ratio model using data from field experiments. In: Gerzabek, M.H. (Ed.), *Soil-Plant-Relationships*, pp. 194–200. Proc. XXIXth Annual Meeting of the European Society for Nuclear Application and Intenational Union of Radioecologists (ESNA/IUR) Austrian Research Center Seibersdorf Report L-209.

Kiriyama, T., Kuroda, R. (1982). Ion exchange separation and spectrophotometric determination of titanium in biological materials. *Fresenius Journal of Analytical Chemistry* **313**, 328–330.

Li, G., Gerzabek, M.H., Mück, K. (1994). An experimental study on mass loading of soil particles on plant surfaces. *Die Bodenkultur* **45**, 15–24.

Livens, F.R., Loveland, P.J. (1988). The influence of soil properties on the environmental mobility of caesium in Cumbria. *Soil Use and Management* **4**, 69–75.

Maathuis, F.J.M., Sanders, D. (1995). Contrasting roles in ion transport of two $K^+$-channel types in root cells of *Arabidopsis thaliana*. *Planta* **197**, 456–464.

Maathuis, F.J.M., Sanders, D. (1996). Mechanisms of potassium absorption by higher plant roots. *Physiologia Plantarum* **96**, 158–168.

Marschner, H. (1986). *Mineral Nutrition of Higher Plants*. Academic Press Inc., San Diego, USA, 674 pp.

Marschner, H., Römheld, V., Horst, W.J., Martin, P. (1986). Root-induced changes in the rhizosphere: Importance for the mineral nutrition of plants. *Zeitschrift für Pflanzenernährung und Bodenkunde* **149**, 441–456.

McLeod, K.W., Alberts, J., Adriano, D.C., Pinder, J.E. III (1984a). Plutonium contents of broadleaf vegetable crops grown near a nuclear fuel chemical separation facility. *Health Physics* **47**, 261–267.

McLeod, K.W., Pinder, J.E. III, Watts, J.R. (1984b). Contribution of a nuclear fuel chemical separations facility to the plutonium content of a tobacco crop. *Health Physics* **46**, 1205–1211.

Meisel, S., Gerzabek, M.H., Müller, H.K. (1991). Influence of plowing on the depth distribution of various radionuclides in the soil. *Journal of Plant Nutrition and Soil Science* **154**, 211–215.

Merckx, R., Sinnaeve, J., Van Ginkel, J.H., Cremers, A. (1983). The effect of growing plant roots on the speciation of Co-60, Zn-65, Mn-54 and Fe-59 in the rhizosphere. In: *Proceedings of the Seminar on the Transfer of Radioactive Materials in the Terrestrial Environment Subsequent to an Accidental Release to Atmosphere*, vol. 1. Commission of the European Communities, Luxembourg, pp. 275–288.

Michelet, B., Boutry, M. (1995). The plasma membrane $H^+$-ATPase. *Plant Physiology* **108**, 1–6.

Mitchell, N.G., Donelly, C.E. (2000). Online publication of the International Union of Radioecologists, http://www.iur-uir.com/Publications/iur_scientific_publications.pdf (30.08.2001).

Mortvedt, J.J. (1994). Plant and soil relationships of uranium and thorium decay series radionuclides—A review. *Journal of Environmental Quality* **23**, 643–650.

Müller, H., Pröhl, G. (1993). ECOSYS-87: A dynamic model for assessing radiological consequences of nuclear accidents. *Health Physics* **64**, 232–252.

Nikolova, I., Johanson, K.J., Clegg, S. (2000). The accumulation of $^{137}Cs$ in the biological compartment of forest soils. *Journal of Environmental Radioactivity* **47**, 319–326.

Nisbet, A.F. (1995). An evaluation of the applicability of agricultural countermeasures for use in the UK. National Radiological Protection Board, Chilton, Report NRPB-M 551.

Nisbet, A.F., Woodman, R.F.M., Haylock, R.G.E. (1999). Recommended soil-to-plant transfer factors for radiocaesium and radiostrontium for use in arable systems. National Radiological Protection Board, Chilton, UK, Report NRPB-R304, 50 pp.

Pendleton, R.C., Uhler, R.L. (1960). Accumulation of $^{137}Cs$ by plants grown in a simulated pond, wet meadow and irrigated field environments. *Nature* **185**, 707–708.

Pinder, J.E. III, McLeod, K.W. (1988). Contaminant transport in agroecosystems though retention of soil particles on plant surfaces. *Journal of Environmental Quality* **17**, 397–407.

Pinder, J.E. III, McLeod, K.W. (1989). Mass loading of soil particles on plant surfaces. *Health Physics* **57**, 935–942.

Renaud, P., Maubert, H. (1997). Agricultural countermeasures in the management of a post-accidental situation. *Journal of Environmental Radioactivity* **35**, 53–69.

Robson, A.D., Pitman, M.G. (1983). Interaction between nutrients in higher plants. In: Läuchli, A., Bieleski, R.L. (Eds.), *Inorganic Plant Nutrition*. In: *Encyclopedia of Plant Physiology, New Series*, vol. 15A, pp. 146–180.

Roca, M.C., Vallejo, V.R., Roig, M., Tent, J., Vidal, M., Rauret, G. (1997). Prediction of cesium-134 and strontium-85 crop uptake based on soil properties. *Journal of Environmental Quality* **26**, 1354–1362.

Roed, J., Andersson, K.G., Prip, H. (1996). The skim and burial plough: a new implement for reclamation of radioactively contaminated land. *Journal of Environmental Radioactivity* **33**, 117–128.

Russell, R.S., Newbould, P. (1966). Entry of strontium-90 into plants from the soil. In: Russell, R.S. (Ed.), *Radioactivity and Human Diet*. Pergamon Press, Oxford, UK, pp. 215–245.

Salt, C.A., Rafferty, B. (2001). Assessing potential secondary effects of countermeasures in agricultural systems: a review. *Journal of Environmental Radioactivity* **56**, 99–114.

Scheffer, F., Schachtschabel, P. (1989). *Lehrbuch der Bodenkunde*, 12th edition. Ferdinand Enke Verlag, Stuttgart, Germany.

Shalhevet, J. (1973). Effect of mineral type and soil moisture content on plant uptake of $^{137}Cs$. *Radiation Botany* **13**, 165–171.

Shaw, G., Bell, J.N.B. (1991). Competitive effects of potassium and ammonium on caesium uptake kinetics in wheat. *Journal of Environmental Radioactivity* **13**, 283–296.

Shaw, G., Wang, X. (1996). Caesium and plutonium migration in forest soils of the Chernobyl 30 km zone. *Mitteilungen der Österreichischen Bodenkundlichen Gesellschaft* **53**, 27–34.

Smolders, E., Sweeck, L., Merckx, R., Cremers, A. (1997). Cationic interactions in radiocaesium uptake from solution by spinach. *Journal of Environmental Radioactivity* **34**, 161–170.

Strasburger, E. (1998). *Lehrbuch der Botanik* (Textbook Botany), 34th edition. Gustav Fischer Publishers, Stuttgart, Germany, 1007 pp.

Sweeck, L., Wauters, J., Valcke, E., Cremers, A. (1990). The specific interception potential of soils for radiocesium. In: Desmet, G., Nassimbeni, P., Belli, M. (Eds.), *Transfer of Radionuclides in Natural and Semi-Natural Environments.* Elsevier, London, UK, pp. 249–266.

Sweeck, L., Kanyar, B., Krajewski, P., Kryshev, A., Lietava, P., Nenyei, A. (2005). Model testing for the remediation assessment of a radium contaminated site in Olen, Belgium. *Journal of Environmental Radioactivity* **84**, 245–258.

Tanner, W., Caspari, T. (1996). Membrane transport carriers. *Annual Review of Plant Physiology and Plant Molecular Biology* **47**, 595–626.

Thiessen, K.M., Thorne, M.C., Maul, P.R., Pröhl, G., Wheater, H.S. (1999). Modeling radionuclide distribution and transport in the environment. *Environmental Pollution* **100**, 151–177.

Thiry, Y. (1997). Etude du Cycle du Radiocesium en Ecosysteme Forestier: Distribution et Facteurs de Mobilité. Thesis, Université Catholique de Louvain, Louvain-la-Neuve, Belgium.

Tikhomirov, F.A. (1988). Long lived radionuclides in soil plant systems. In: Carter, M.W. (Ed.), *Radionuclides in the Food Chain.* Springer Publishers, Berlin, Germany, pp. 136–144.

Tinker, P.B., Nye, P. (2000). *Solute Movement in the Rhizosphere*, 2nd edition. Oxford University Press, Oxford, UK.

UNSCEAR (United Nations Committee on the Effects of Atomic Radiation) (1988). *Sources, Effects, and Risks of Ionizing Radiation.* United Nations, New York, USA, pp. 647.

Van Dorp, F., Eleveld, R., Frissel, M.J. (1979). A new approach for soil–plant transfer calculations. In: International Symposium on Biological Implications of Radionuclides Released from Nuclear Industries. International Atomic Energy Agency, Vienna, Austria, Report IAEA-SM-237/134, 26–30 March 1979.

Watkins, B.M., Maul, P.R. (1995). Transfer of deposited radioactivity from foliage to edible crop parts. Report of the work undertaken by Intera Information Technologies under Ministry of Agriculture, Fisheries and Food project reference 1B058.

Whicker, W.F., Kirchner, T.B. (1987). Pathway: a dynamic food-chain model to predict radionuclide ingestion after fallout deposition. *Health Physics* **52**, 717.2.

White, P.J. (1997). Cation channels in the plasma membrane of rye roots. *Journal of Experimental Botany* **48**, 499–514.

White, P.J. (1998). Calcium channels in the plasma membrane of root cells. *Annals of Botany* **81**, 173–183.

Whitehead, D.C. (1984). The distribution and transformations of iodine in the environment. *Environment International* **10**, 321–339.

# Phytoremediation of soils contaminated with radionuclides

Neil Willey[a], Chris Collins[b]

[a]*Centre for Research in Plant Science, Faculty of Applied Sciences, University of the West of England, Frenchay, Bristol, BS16 1QY, UK*
[b]*Department of Soil Science, School of Human and Environmental Sciences, The University of Reading Whiteknights, Reading, RG6 6DW, UK*

## 1. Phytoremediation—a 21$^{st}$ century challenge for radioecology

### 1.1. *Past, present and future contamination of soils with radionuclides*

Contamination of the environment has been a frequent legacy of industrialisation. Recognition of the adverse health, environmental and economic effects of this contamination resulted in legislation for its minimisation and monitoring. There is now increasing pressure to develop effective technologies not just to minimise and monitor but also to decontaminate ecosystem compartments such as soils that have become contaminated; the nuclear industry is similar to other industries in these respects. There have been a variety of estimates of radioactively contaminated soil volumes at nuclear sites around the world and thorough assessments for some locations (e.g. Haywood and Smith, 1992). In a detailed comparison of decontamination options, it was noted that for many of these radionuclide-contaminated soils in the USA there were simply no established options other than removal and burial in waste repositories (USEPA, 1997). This classic 'dig and dump' option is the only method used for a number of non-radioactive contaminants and the only one considered at many radioactively contaminated sites. However, in the UK, and probably in many other countries, in addition to the expense and disruption that 'dig and dump' entails, there is probably not sufficient capacity in approved waste repositories for existing radionuclide-contaminated soils (Hall and Watt, 2002).

In the absence of practical decontamination methods, estimates of decontamination costs for radionuclide-contaminated soil must be general but are certainly high because, although the geographic spread of radionuclide contaminated soils is much less than for many other contaminants, radionuclides can be uniquely toxic and expensive to clean up. The United States Department of Energy estimated in 1995 that it is responsible for 67 million m$^3$ of soil contaminated with radionuclides with associated clean up costs of about $13 billion, primarily

RADIOACTIVITY IN THE ENVIRONMENT
VOLUME 10 ISSN 1569-4860/DOI: 10.1016/S1569-4860(06)10003-0

at nuclear weapon facilities (USDoE, 1995, Glass, 2000). At nuclear weapons test sites around the world, often despite relatively low radiation dose rates, it is possible that clean-up costs will be of similar magnitude (Simon and Bouville, 2002). The necessity for environmental clean-up has long been recognised at weapons test sites and a variety of decontamination trials initiated. These have emphasised that clean-up costs to national governments of contaminated soils from nuclear weapons production may total many tens of billions of dollars worldwide. There are few estimates of total clean-up costs for contaminated soils from accidents such as those at Khyshtym or Chernobyl, probably because there is little real possibility that there ever will be complete clean up. There is certainly no prospect of using the only established soil decontamination method—burial in waste repositories. In the Chernobyl exclusion zone research has had to focus on reclaiming contaminated land through, for example, ploughing regimes (Roed et al., 1996) and alternative crops such as willows (Victorova et al., 2000) or other rehabilitation strategies (Firsakova et al., 2000). There is, therefore, a problematic legacy of radioactively contaminated soil from past nuclear activities. The continuing presence of radionuclide contaminated soils, which evokes much public concern, indicates that it is not a problem that currently has a cheap and easy solution.

In addition to the high profile radionuclide decontamination challenges outlined above, the first nuclear power stations constructed for generating electricity have begun to be decommissioned. At nuclear power stations presently being dismantled there are some areas of radioactive contamination that, despite being of small scale compared to those at nuclear weapons sites, require remediation before delicensing (e.g. Willey et al., 2001). The decommissioning of nuclear power stations is an ongoing source of potential soil decontamination challenges and with more than 400 nuclear power stations now in operation world-wide will be so in the future. Furthermore, much of the radioactive waste generated by nuclear industries is not yet in permanent waste repositories. Restrictions on discharges to the sea tightened considerably towards the end of the 20$^{th}$ century (e.g. Brown et al., 1999) and are becoming more restrictive, e.g. OSPAR (2003), whilst many terrestrial waste repositories currently being used are filling up rapidly. These factors have increased the pressure to develop permanent terrestrial nuclear waste repositories in the 21$^{st}$ century. Possessing techniques to decontaminate soil can be a useful part of approval for filling and operating waste disposal sites. It can reduce volumes but also help deal with any leakages of radioactivity that may occur. At the start of the 21$^{st}$ century there is also considerable interest in contamination from depleted U munitions (Durante and Pugliese, 2002) and potential terrorist use of 'dirty' nuclear devices. These are circumstances in which possessing techniques to decontaminate soil of radioactivity might be necessary in the future. There is, therefore, significant pressure to develop decontamination technologies not only to deal with soil contaminated with radionuclides in the past but also because of present and future decontamination challenges.

### **1.2.** *The allure and challenge of phytoremediation for radionuclides*

The allure of bioremediation is great. Using living organisms to degrade or remove contaminants resonates with a 21$^{st}$ century desire for cheap and sustainable solutions to contamination. Higher plants, which have evolved as solar powered extractors of inorganics from soils, are particularly seductive (Willey, 2001). Phytoremediation seems viable where decontamination times can be long and the pollution is dispersed at low concentration, as is often the case in radionuclide contaminated soils. Romney et al. (1957) and Nishita et al. (1958) reported the first direct calculations of the potential of plants to decontaminate soil of radionuclides. A single crop of *Trifolium repens* (clover) was able to remove 4.42% of added Sr and 9 crops over 520 days removed 24% (Nishita et al., 1958). There is now much experience of bioremediation using microbes (Hinchee et al., 1995). It has probably not lived up to initial expectations of researchers or end-users (Watanabe, 2001) but has had some notable successes and is now part of the toolkit of environmental engineers. However, the life sciences in particular, but also the environmental sciences, have been transformed since the inception of bioremediation. With developments in molecular biology the manipulation of plants for phytoremediation has been a reality for a number of years (e.g. Raskin, 1996) and there have been some notable phytoremediation successes. It is, therefore, an appropriate time to assess the potential of phytoremediation for decontaminating soils of radionuclides.

The urgencies of the Cold War hamstrung the nuclear industries and radioecology, and it is unfortunate that the sums of money, the person hours expended and scientific breakthroughs of the nuclear age have been so little focused on environmental clean-up until recently. Decontamination of soils with radionuclides will be a significant challenge in the 21$^{st}$ century and radioecologists have much expertise to aid this effort. Developing decontamination techniques may even affect variables that are almost impossible to cost such as general perceptions of the viability of nuclear energy sources. With increasing pressure to develop low carbon economies, it seems likely that nuclear power will be used as a significant, even increasing, source of electricity in the 21$^{st}$ century (Starr, 2000), and that all aspects of soil decontamination for radionuclides will then receive great attention. In this chapter, we suggest that knowledge of radionuclide behaviour in the environment, plant science and environmental modelling might allow phytoremediation to meet some of the challenges of cleaning up soil contaminated with radionuclides. We also suggest that focusing on this challenge might provide new directions for radioecology in the 21$^{st}$ century.

## 2. A phytoremediation perspective on the biogeochemistry of radionuclides

### 2.1. *Partitioning of radionuclides between ecosystem compartments*

An important aspect of radionuclide behaviour in terrestrial environments is partitioning between ecosystem compartments (Avery, 1996). Radionuclides released to the atmosphere may, in some instances, remain there for extended periods, but will eventually deposit on the earth's surface. Radionuclides can also be released directly into the regolith, a pathway of great importance to assessments of nuclear waste repositories. Once radionuclides have contaminated the soil, a primary concern of radioecology is their availability for biological uptake and subsequent partitioning between hydrosphere and biosphere. At its most fundamental level, bioremediation involves managing and manipulating partitioning of contaminating substances, such as radionuclides, to the biosphere for the purposes of soil decontamination. Soil binding characteristics of radionuclides and uptake patterns by organisms, both of which have been much investigated by radioecologists, therefore provide the keys to phytoremediation. In particular, radioecological studies carried out in different ecosystems, such as those described in other chapters of this book, provide interesting insights into the processes that might underpin phytoremediation.

Radioactive contaminants exhibit a great diversity of chemical characteristics and soil behaviour. Results from some compilations of $K_d$ values (see section 2 of Chapter 2 for a definition of $K_d$) for radionuclides are shown in Table 1, which also shows $K_d$ values for the nutrient ions that plants take up in order to carry out their autotrophic nutrition. It is clear that many radionuclides in many soils have large $K_d$ values and it is easy to dismiss the potential of phytoremediation on this basis. However, the $K_d$ values for Sr and technetium (Tc) are consistently low and there is great variation for other radionuclides. The $K_d$ values for macronutrient ions, which it is quite clear plants can remove in significant quantities from most soils, are somewhat lower than for many radionuclides, but those for micronutrient ions are similar. It is widely accepted that for fertiliser ions with low $K_d$ values, such as $NO_3^-$ and $K^+$, 50–60% and 75–90% respectively can be removed in a single cropping after fertiliser addition (Fink, 1982). Lower values of 5–15% removal are reported for single croppings for ions with higher $K_d$s such as $PO_4^{2-}$ but long term field experiments indicate 50–100% removal over 10–20 years (Fink, 1982). High $K_d$ values do at least keep surface deposited contaminants in the rooting zone. Smith et al. (2000) have shown how significant quantities of Cs can be removed by plants over a number of years even when $K_d$ values in the short term are high. It must be also remembered that radioecologists have much knowledge that might be used to decrease $K_d$ and improve the bioavailability of radionuclides for phytoremediation.

There are numerous studies of the effect of soil variables such as pH and organic matter on radionuclide bioavailability (e.g. Echevarria et al., 2001; Entry et al., 2001). These are variables that can be manipulated in farming systems and, perhaps, phytoremediation systems. Furthermore, in assessments of radionuclide binding to soil, a variety of soil extraction methods have been developed for radionuclides (e.g. Howe et al., 2002). A number of chemicals are, therefore, already known on which to base research into increasing the bioavailability of radionuclides in soil. For example, ammonium-based extractions have long been used by soil scientists to estimate available concentrations of monovalent cations in soils. $NH_4^+$ has proved effective at displacing $Cs^+$ from some soils (Lasat et al., 1997; Dushenkov et al., 1999), although there are others in which the Cs has proved difficult to extract using any current method. The desorption of radiocaesium by stable Cs is also significant, with predictable isotopic dilution effects. An increase in stable Cs carrier can increase radiocaesium uptake by plants (Nishita et al., 1960; Wallace et al., 1982), an effect that might be useful in phytoremediation (Watt, 2004). The detailed understanding of processes controlling radiocaesium mobilisation in soils recently reported (Delvaux et al., 2000, 2001) might help to manipulate Cs bioavailability. There is also a variety of compounds that have been used in experiments on radionuclide bioavailability or in efforts to decrease the transfer of radionuclides to living organisms. For example, silver thiourea has been used to demonstrate the role of Cs binding to frayed edge sites on 2:1 clay minerals (Cremers et al., 1988), and sodium tetraphenylborate has been used to precipitate free $K^+$ in an effort to extract K from interlayer sites on clays (Cox and Joern, 1997) and can affect Cs bioavailability (Delvaux et al., 2000). ACFC (ammonium ferric hexacyano ferrate) has been used to decrease soil to plant transfer of Cs (Vandenhove et al., 1996) and it has been proposed that a fungal pigment, norbadione A, has binding properties that might aid soil clean up of Cs (Garaudée et al., 2002). Such knowledge is less extensive for other radionuclides and has seldom been focused on increasing soil bioavailability but might provide a foundation for manipulating availability for phytoremediation.

A few years ago, ecosystem compartmentation of Hg and As did not indicate that these elements had any greater phytoremediation potential than is currently the case for some radionuclides. Significant progress has now been made in realising the potential of phytoremediation for both Hg (Rugh et al., 1998) and As (Dhankher et al., 2002; Nie et al., 2002). From the perspective of recent advances in phytoremediation, some radionuclides are available in a wide variety of soils (Sr, Tc, Cl) and some radionuclides are available in at least some soils (e.g. Cs). For most radionuclides their behaviour in at least some soil types has been well investigated and there are some established methods for investigating and manipulating their bioavailability. There are radionuclides and soil types for which phytoremediation is unlikely ever to be possible but an agronomic perspective on $K_d$, the potential for manipulating radionuclide bioavailability and experience with other elements indicate that, if plants can be found that take up specific radionuclides in sufficient quantities, phytoremediation might have real potential.

## 2.2. *Plant uptake of radionuclides*

Plants actively mine the soil for simple inorganic ions and remove applied fertilisers, some of which (e.g. triple superphosphate) are relatively unavailable in the soil, in large quantities. Ion uptake is primarily energised by $H^+$ efflux through a P-type ATPase that creates a transmembrane proton gradient. About 13% of the *Arabidopsis* genome is related to transport activity, i.e. about 2,300 proteins (Arabidopsis Genome Initiative, 2000). Extracting inorganic ions from the soil and translocating them, and products including them, is a significant part of what plants have evolved to do. A useful insight into the potential of plants to take up radionuclides instead of nutrients is provided by experiments on uptake in hydroponics. Those radionuclides that are most similar to plant nutrients have the highest potential uptake rates. Cl and S are plant nutrients and Cs, Sr and Co similar to the plant nutrients K, Ca and Ni respectively. For radioisotopes of these elements it is clear that if they are available in the soil, then plants should be able to extract them in significant quantities.

The uptake of elements like U and Pu, which are dissimilar to plant nutrients, is clearly much lower than the uptake of radionuclides with greater similarities to plant nutrient ions, but can potentially be significant (Table 1). The chemistry of radionuclides that are dissimilar to nutrient ions is often complex but nevertheless simple inorganic ions of most of these radionuclides are possible. Although these simple inorganic ions are not directly analogous to nutrient ions, they can be taken up by plants. Plants have already been engineered with dramatically altered Fe and As uptake and that can change conditions in the rhizosphere to alter chemical form and bioavailability of ions (Takahashi et al., 2001; Dhankher et al., 2002). A phytoremediation perspective suggests, therefore, that for many radionuclides there is great plant uptake capacity and for others there is now the real possibility that it can be engineered.

## 2.3. *Extremes of soil to plant transfer of radionuclides*

Many compilations of soil–plant transfer factors (TFs—for a definition see equation (3) of Chapter 2) report low values that might discourage investigations into the potential of phytoremediation for radionuclides (Table 1). For example, the IUR database and values from the former Soviet Union (Frissel, 1992) have few TF values greater than 1 and extended datasets for Cs and Sr have reported similarly low values (Nisbet and Woodman, 2000). Such compilations aimed to quantify mean TF values for radiological protection purposes, and are primarily based on temperate agricultural soils. However, extreme values for soil to plant transfer are also relevant to assessing the possibilities of phytoremediation. They demonstrate the boundaries of the possible and provide clues as to how it might be brought about.

There are a number of reports of plants completely removing significant quantities of $^{99}$Tc from soils in a single cropping (Bennett and Willey, 2002). There are also reports of high TFs and removal rates (25%) for Sr (e.g. Nishita and Haug, 1972). However, even with radionuclides of more limited bioavailability there are reports which demonstrate the significant potential of phytoremediation. Fredrikkson (1970) reported experiments carried out on the removal of Cs from experimentally contaminated alluvial and lateritic soil from South America. White mustard removed 22.9% of added Cs in a single cropping after treatment with $NH_4NO_3$. This is the highest percentage removal of Cs reported from a contaminated soil in a single cropping, but there are other reports of significant removal by plants from laterites (Fredrikkson, 1970; D'Souza and Mistry, 1980; Delmas et al., 1987). High uptake of Cs has also long been reported on soils in which sands or kaolinites dominate the ion exchange complex (e.g. Cummings et al., 1969) and some experiments have removed approaching 10% of added Cs from such soils (Andersen, 1967; Abbazov et al., 1978). High TFs for Cs uptake from organic soils have long been reported (e.g. van Bergeijk et al., 1992) and indicate that significant Cs removal from them is possible. Coughtrey et al. (1990) reported 15–25% removal of Cs by plants in upland Britain from organic soils. Organic soil amendments have reportedly increased Cs removal by plants up to 60% in pot experiments (Entry et al., 2001). Lateritic, sandy, kaolinitic and organic soils together cover approaching 40% of the earth's surface (USDA, 1999). Although data from them is sparse they offer encouragement for the development of Cs phytoremediation to weigh against the obstacles presented by clay soils. Even for the generally unavailable U very significant removal rates have been found under specific conditions (Huang et al., 1998).

Although the concept of phytoremediation developed after the establishment of the nuclear industry, much information collected on the behaviour of radionuclides in the soil–plant system since the dawn of the nuclear age is useful in developing phytoremediation systems. In fact, the extended time over which data on radionuclide behaviour in terrestrial ecosystems have been collected provides a resource for phytoremediation design that does not exist for many other types of contaminants. Many early researchers recorded variables relating to soil to plant transfer that are of direct use in discussing phytoremediation of radionuclide contaminated soils and Romney et al. (1957) and Nishita et al. (1958) directly tested the concept. In comparison with nutrient ions and other contaminants, for some of which there has been significant progress in phytoremediation, many radionuclides are reasonably available in soils, taken up by plants quite rapidly and there are recorded examples of high soil to plant transfer. It is, perhaps, this perspective that has encouraged the field trials of phytoremediation for radionuclides carried out to date.

## 3. Phytoremediation of soil contaminated with radionuclides

### 3.1. *Phytoremediation of Cs contaminated soils*

Following Romney et al. (1957) and Nishita et al. (1958) the next investigation of phytoremediation for a radionuclide-contaminated soil was at Brookhaven National Laboratory, NY, USA. It focused on $^{137}Cs$, the radionuclide most commonly of concern in radioactively contaminated soils and which, at Brookhaven, has contaminated the soil since the 1950s. Pot experiments established that plant taxa differed in their Cs uptake, that ammonium salts could mobilise Cs from the soil (Lasat et al., 1997), that *Brassica oleracea* var. *capitata* (cabbage) removed the greatest amount of Cs and that additions of ammonium nitrate could increase Cs uptake by plants. Experiments in the field showed that *Amaranthus cruentus* had higher uptake than cabbage and was the taxon that eventually removed the greatest amounts of Cs from radiocaesium contaminated soil. The additions of ammonium salts to soils in the field were not as effective as in pot experiments (Lasat et al., 1998). Amongst the reasons for this might be leaching out of the root zone, the microbial transformation of ammonium salts in the field and the dislike of plants such as *Brassica oleracea* and *Amaranthus cruentus* for ammonium as a nitrogen source. The relationship between $^{137}Cs$ in soil and $^{137}Cs$ in shoot tissues of *A. cruentus* was linear, with a calculated concentration factor of 0.5. Fuhrman et al. (2002) used the equation $C_t = C_0 e^{-kt}$ (where $k$ is plant mass/y/m$^2$/soilmass $\times$ TF) to show that plant uptake would take about 18 years to reduce the $^{137}Cs$ in soil to 50% of its initial value at Brookhaven, assuming that TF did not change as the soil concentration decreased. The timescales suggested by the studies at Brookhaven were the first estimates of phytoremediation times for radiocaesium contaminated soils under field conditions. As the authors note, their equation is unlikely to be the most refined for estimating phytoremediation times but it, and the values from studies at Brookhaven, provided a guide for further research.

The potential of phytoremediation for Cs has also been investigated in the Chernobyl exclusion zone by Dushenkov et al. (1999) who investigated a variety of *Amaranthus* species together with soil amendments. Amaranth species were chosen because of previous reports of high uptake by plants in this genus. Ammonium salts were shown to have the greatest desorbing power in the soils used. In field trials *A. caudatus* removed the greatest amount of Cs, but there was no increased Cs uptake into the plants with soil amendments. The highest TF recorded was 7.5 and it was estimated that, with invariant TF, 'minimum possible' decontamination time would be about 7 years, considerably shorter than those reported for the soils at Brookhaven. This was related to the higher uptake of the plant taxa used and to the somewhat higher availability of Cs from the soils. British Nuclear Fuels investigated the potential of phytoremediation for Cs contaminated soils at the Bradwell Nuclear Power Station, Essex, UK, a site which has great variation in Cs concentration. *Helianthus annuus*, *Brassica juncea* and two varieties of *Beta vulgaris* were sown and TFs measured for individual plants. A first series of experiments showed that *B. vulgaris* had the greatest phytoremediation potential and that TFs and biomass production suggested a minimum decontamination time of about 20 years was possible (Willey et al., 2001). Interestingly, for each species the 50 plants grown on a variety of Cs concentrations took up about the same concentration of Cs irrespective of the Cs concentrations on which they were growing. The TF was significantly higher at the low Cs concentrations (Willey et al., 2001). The TFs reported in the first experiments at this site are not as high as those reported by Dushenkov et al. (1999) and the phytoremediation times are proportionately longer. They do, however, provide the first evidence that, given sufficient desorption from soil, TFs might not change, and might even increase, as soil concentration decreases during phytoremediation. Investigations of phytoremediation at the Bradwell site are on-going (Watt, 2004). Bench scale studies at the Argonne National Laboratory-West (Idaho, USA) suggested yearly removal rates of 2–4% of radiocaesium and led to phytoremediation field trials currently using *Kochia scoparia* on $^{137}$Cs contaminated soils (Negri and Hinchman, 2000). Other institutes where phytoremediation of radiocaesium is being investigated include the Kazakh National University (K.N. Evstyukhin, pers. comm.), the University of Montana (M. Rillig, pers. comm.) and Idaho State University (R. Inouye, pers. comm.).

These first trials of phytoremediation for radiocaesium contaminated land provide a number of clear research directions. They have not suggested that phytoremediation of Cs will inevitably take so long as to always be unworthwhile, but that TFs and/or biomass production need to be increased and sustained year on year. Root to shoot translocation of Cs is rapid so availability and uptake are the keys to achieving this. It is quite likely that, as is the case with K, manipulations at agronomic, physiological and molecular scales can be used to increase Cs uptake rates significantly. After relatively few field trials this represents encouraging progress of a nascent technology.

### 3.2. *Phytoremediation of U contaminated soils*

The potential for phytoremediation of U from contaminated substrates has also been investigated. This is primarily because of the existence of U contaminated effluents and mine tailings but environmental contamination with depleted U from munitions is providing further impetus. Although U is not chemically similar to any ion which is metabolically essential to plants, there is some capacity for uptake of U by plants (Ebbs et al., 1998) and elevated concentrations of U have been reported in plants growing over uraniferous rocks. Investigations using plants to filter U from effluents have shown that plant roots have a significant capacity to adsorb it. For example, sunflowers have been used to remove U from contaminated effluents (Dushenkov et al., 1997). However, almost all the U remains on the external surfaces of plant roots and transfer to above ground parts is slow. Furthermore, in contaminated soils the form of U is frequently not that which plants can absorb. Huang et al. (1998) overcame this problem by the addition of citric acid to the soil, which increased soil availability and U uptake rates 1000 fold in a few days. Citric acid can be toxic to plants but further work has suggested that enhanced phytoaccumulation of U can be effected using it (Shahandeh and Hossner, 2002). Phytoextraction of U has, therefore, emerged as a possible alternative to the more widely used *ex-situ* leaching techniques (Mason et al., 1997) but recent reviews have concluded that further research is necessary to maximise the absorption and, in particular, translocation of U by plants (Shahandeh et al., 2001). The required improvement in absorption and translocation is of a greater magnitude than that for $^{137}$Cs but there remains significant potential to increase U availability in soils.

### 3.3. *Phytoremediation of Sr, Tc and Cl contaminated soils*

$K_d$s are lower and TFs higher for Sr, Tc and Cl than for Cs and U but they have been less extensively investigated for phytoremediation. As early as 1957, Romney et al. noted that there was some potential to remove $^{90}$Sr from some soils using plants. There have been few subsequent estimates of phytoremediation times for soil contaminated with $^{90}$Sr but comparisons of TFs (e.g. Nisbet and Woodman, 2000) consistently indicate that they are likely to be less than half those for $^{137}$Cs. Fuhrman et al. (2002), using field data, calculated that Sr could be phytoextracted from soils in less than half the time of $^{137}$Cs at Brookhaven, NY. In some comparisons of $K_d$s, values for Sr can be a tenth or less than those of Cs (e.g. Sheppard and Thibault, 1990; USEPA, 1999), a difference sometimes reflected in plant uptake in the field (Frissel, 1992). Other studies have confirmed these patterns (e.g. Nishita et al., 1958; Entry and Watrud, 1998) and shown that taxon selection and agronomic practice can increase soil to plant transfer of $^{90}$Sr (Entry et al., 1999, 2001). Given agricultural experience with Ca, there is probably room for improvement in soil to plant transfer of the radiologically significant $^{90}$Sr and hence a real possibility that phytoremediation times of a few years are attainable.

Tc is generally thought to be of little radioecological significance in terrestrial ecosystems. However, reprocessing of nuclear fuel and restrictions on discharge to aquatic systems mean that much Tc will have to be put into terrestrial waste repositories. Methods for dealing with any terrestrial contamination arising from such activities are desirable. The availability of Tc is great in most aerobic soils, plants have a great uptake capacity for Tc and there is thus great potential for removing any Tc contamination from soils using phytoremediation (Bennett and Willey, 2002). A number of studies have shown that plants can remove Tc almost entirely from soils in a short time (e.g. Bell et al., 1988). There is similarly great potential for remediating soils contaminated with $^{36}$Cl using phytoremediation. As with $^{99}$Tc, much $^{36}$Cl will have to be stored in terrestrial waste repositories and there are well demonstrated pathways for $^{36}$Cl movement to soil surface in evapotranspiration flows. This means that Cl can move freely from depth to the soil surface (Shaw et al., 2004). Plants have a large capacity for Cl uptake (White and Broadley, 2001) and there is therefore clearly the potential for removing $^{36}$Cl contamination from soils using plants. In fact, $^{99}$Tc and $^{36}$Cl probably have as much potential for phytoremediation as any other environmental contaminants. They are currently of less radioecological concern than other radionuclides but provide an opportunity for radioecologists to demonstrate that the technology can work.

### **3.4.** *Phytoremediation of other radionuclides*

There have been few investigations of the potential of plants to decontaminate soils of other radionuclides. Hydroponic experiments with plutonium (Pu) have shown that plants can absorb this element (Lee et al., 2002) and there is a great deal known about the behaviour of Pu in soil and how to manipulate it (Shahandeh et al., 2001). Soil availability of Pu, with the exception of sandy soils, is low (Shahandeh et al., 2001), as is plant uptake. However, it is a good example of a radionuclide for which knowledge of its behaviour in soil is extensive enough to envisage the design of mobilising agents or the engineering of uptake. Large increases in plant uptake will, however, be necessary if phytoremediation is to be a useful decontamination option for Pu.

Some plants take up cobalt (Co) to satisfy metabolic demands, in particular plants in the Fabaceae (legumes) (Marschner, 1995). About 30 species of plant occur that hyperaccumulate Co from soils in which it occurs naturally (Reeves and Baker, 2000). Cobalt hyperaccumulators can take up Co to in excess of 1000 $mg\,kg^{-1}$ and suffer no detrimental effects. $^{60}Co$, with its relatively short physical half-life (5.26 years), is infrequently a serious radioecological problem but it does occur in a number of effluents and contaminates a number of sites worldwide. If removal of $^{60}Co$ from a soil is necessary it is possible that phytoremediation might help in the future. Hyperaccumulator traits are currently the subject of a number of detailed investigations (Clemens et al., 2002) with a view to removing significant concentrations of stable Co from soils. Any advances in phytoremediation of stable Co are likely to have a great impact on the ability of plants to remove the relatively low concentrations of $^{60}Co$ that cause contamination problems. Ruthenium ($^{106}Ru$), cerium ($^{144}Ce$) and cadmium ($^{109}Cd$) are often reported to be quite available in soils whereas zinc ($^{65}Zn$), iron ($^{59}Fe$) and manganese ($^{54}Mn$) are less so (e.g. D'Souza and Mistry, 1980; Sheppard and Thibault, 1990) but advances in manipulating heavy metal uptake by plants could make them all suitable targets for phytoremediation.

### 3.5. *State of the art*

A first phase of phytoremediation trials for radionuclides has now been completed. Only in the case of Tc has contaminated soil been completely cleaned of radionuclide, but this was not the specific intention of any of the trials reviewed here. The trials that have been carried out have mostly been focused on Cs and U but lessons have been learned that can be applied to a range of radionuclides. On the basis of the first trials, phytoremediation should not be dismissed as a potential decontamination technology, although it is still currently a technology with potential rather than proven ability. It seems likely that, even after significant refinement, decontamination times will be measured in years. If this seems slow it is important to remember that many other decontamination methods do not provide instant fixes and can also take substantial periods of time. Phytoremediation also has the advantage that it should be cheaper and less damaging to the environment than engineering based methods. Field trials of phytoremediation for radionuclides have identified the obstacles that need to be overcome for it to become a viable decontamination method. Depending on the individual radionuclide being considered these obstacles are phyto-availability and/or uptake rate.

There have been experiments to increase Cs availability in soils (Lasat et al., 1997; Dushenkov et al., 1999) and U availability in soils (Huang et al., 1998). We suggest that, although it is not necessary in all cases, it is largely a question of resources as to whether mechanisms of mobilising recalcitrant radionuclides from soils can be found. Ammonium, which can mobilise Cs successfully, and citrate, which mobilises U, rely on effects that have long been known and were not the result of sustained, well-resourced research programmes specifically targeted at phytoremediation. Given that, over a period of years, isotopes of Cs, for example, are not as irreversibly fixed as often assumed (Smith et al., 2000) it is possible that appropriate mobilising agents can be found for these and other isotopes. Rapid changes in the ability to genetically engineer ion uptake mechanisms also mean that, given resources, plant uptake and translocation rates might also be increased (see section 5.3). Such resources depend on the priority given to decontamination and the performance of phytoremediation relative to other decontamination options. Although there is extensive modelling expertise amongst radioecologists, models have been little used to assess the performance of phytoremediation and the magnitude of increases in soil to plant transfer necessary for phytoremediation to become viable. Modelling of soil to plant transfer of radionuclides over time can enable the magnitude of increases in transfer necessary to be estimated, the potential performance of phytoremediation to be gauged and hence the magnitude of resources worth investing to be calculated.

## 4. Assessing the feasibility of phytoremediation of radionuclides using a simple model

### 4.1. *Background*

The behaviour of radionuclides in terrestrial ecosystems has been modelled extensively over a variety of time scales. In most models the uptake of radionuclides from soils into plant tissues has been represented using simple soil–plant transfer factors, as described in section 4 of Chapter 2. The existence of extensive data bases of soil–plant TFs (as summarised in Table 1) is a major advantage for phytoremediation research and can be used to guide its development in a number of ways. One of these is in the development of relatively simple scoping models which can be used to assess the comparative feasibility of phytoextraction of a wide variety of radionuclides.

The feasibility of phytoremediation depends on a variety of considerations, but principal amongst these is the fundamental efficiency of the clean-up process. In the case of phytoextraction of radionuclides, this efficiency depends on the total radioactivity of a specific radionuclide taken up and retained by the harvestable tissues of a particular plant species. These tissues can then be harvested in the conventional way and treated or disposed of according to their degree of contamination. Another consideration is the degree of clean-up required at a particular site. Absolute clean-up of a soil might not be necessary. In many situations the reduction of the soil radionuclide activity concentration below an arbitrary threshold might be deemed sufficient. Legal definitions of radioactive substances are generally based on naturally occurring radionuclide activities and to reduce artificial radionuclides substantially below these activities might be desirable but unnecessary. Any remediation exercise not involving complete removal of contaminated soil inevitably follows a trend of diminishing returns. In other words, it is usually relatively easy (and cheap) to remove the first 50% of contamination present, but removal of 50% of what remains is often more difficult, more time consuming and more expensive. A useful measure of clean-up efficiency, then, is the decontamination half time. In the case of phytoremediation this is the period during which it might be expected that 50% of the initial contamination can be removed.

Thus far, little consideration has been given to the use of models as design tools for phytoremediation, but in almost all other instances of environmental management, such as fertiliser and pesticide applications, effective monitoring and management are aided enormously by systems decision support tools. Existing models of radionuclide behaviour in terrestrial ecosystems clearly provide a basis on which to design phytoremediation systems and build decision support systems for them. This is a useful foundation from which to develop specific decontamination strategies. The first step in this process, however, might be the development of simple models to demonstrate ‘proof of concept’ for phytoremediation of radionuclides, based on existing data within the literature.

Table 1
A compilation of $K_d$ values and soil–plant transfer factors (TFs) for radionuclides and plant nutrients

| | $K_d$ | | | TF | | |
|---|---|---|---|---|---|---|
| Element | Min | Max | Mean/Median | Max | Min | Mean/Median |
| Cs | 230[e] | 42,610[e] | 1896[e] | 0.9[a] | 0.00045[a] | 0.07[a] |
| | 10[h] | 52,000[h] | | 0.631[b] | 0.001[b] | 0.043[b] |
| | | | | 0.76[f] | 0.05[f] | |
| K | 2[h] | 9[h] | | 0.626[g] | 0.0095[g] | |
| | | | 87[i] | | | 1142[k] |
| Rb | | | 16,950[i] | | | |
| Sr | 42[e] | 498[e] | 163[e] | 9.6[a] | 0.0051[a] | 1.01[a] |
| | | | | 0.9[f] | 0.11[f] | |
| Ca | 1.2[h] | 9.8[h] | | | | 1.5[k] |
| | | | 44[i] | | | |
| Mg | 1.6[h] | 13.5[h] | | | | |
| S | | | | | | 23[k] |
| U | 10.4[h] | 4400[h] | | | | |
| | | | 515[i] | | | |
| Pu | 11[h] | 300,000[h] | | 0.65[c] | 0.0000035[c] | 0.00055[c] |
| | | | 2187[i] | 0.0001[d] | 0.000001[d] | |
| Np | 0.16[h] | 929[h] | | 0.57[c] | 0.00023[c] | 0.0032[c] |
| | | | | 0.1[d] | 0.0001[d] | |
| Am | 1[h] | 47,230[h] | | 0.77[c] | 0.0000015[c] | 0.00067[c] |
| | | | | 0.01[d] | 0.00001[d] | |
| Cm | 93[h] | 51,900[h] | | 0.012[c] | 0.0000014[c] | 0.00061[c] |
| | | | | 0.01[d] | 0.00001[d] | |
| Tc | 0.0029[h] | 0.28[h] | | 76[j] | 0.73[j] | 8.1[j] |
| Ra | | | 12,000[i] | 0.7[b] | 0.001[b] | 0.068[b] |
| Co | 130[e] | 103,600[e] | 1735[e] | 0.39[b] | 0.001[b] | 0.018[b] |
| Ru | | | | 0.026[g] | 0.0088[g] | |
| Ce | 58[h] | 6000[h] | | 0.045[g] | 0.005[g] | |
| As | 1.9[h] | 18[h] | | | | |
| P | | | 39[i] | | | |
| Mn | 39.7[e] | 79,040[e] | 271[e] | | | |
| Ni | | | 612[i] | | | |
| Mo | | | 62.5[i] | | | |
| Fe | | | 446[i] | | | |
| Zn | 185[e] | 153,100[e] | 1756[e] | | | |
| | | | 1375[i] | | | |

[a]Nisbet and Woodman (2000).
[b]Gerzabek et al. (1998).
[c]Frissel (1992).
[d]Romney et al. (1981).
[e]Yasuda et al. (1995).
[f]Askbrant and Sandalls (1998).
[g]D'Souza and Mistry (1980).
[h]Baes and Sharp (1983).
[i]Sheppard and Thibault (1990).
[k]Hopkins (1999).
[j]Frissel and van Bergeijk (1989).

### 4.2. *Calculation of phytoextraction and effective phytoremediation half times*

In order to estimate a 'half time' for the removal of radionuclides from a soil by phytoextraction it is necessary to calculate the fraction of the total radionuclide inventory which can be removed in a specified period of time. This can be achieved using a simple activity balance calculation based on soil-plant transfer factors and other fundamental data widely available within the literature. A schematic diagram of the calculation is shown in Figure 1. Using the parameters described in Table 2, the individual steps in the calculation are carried out as follows. As described in section 4.1 of Chapter 2, the transfer factor is defined as:

$$\mathrm{TF} = \frac{C_\mathrm{p}}{C_\mathrm{s}} \tag{1}$$

Therefore, the activity concentration in plant tissues can be calculated as the product of the soil activity concentration and the appropriate TF:

$$C_\mathrm{p} = \mathrm{TF} \times C_\mathrm{s} \tag{2}$$

The total activity removed when these plant tissues are harvested is simply the activity concentration at harvest multiplied by the biomass of plant tissue removed at plant harvest:

$$A_\mathrm{p} = \mathrm{TF} \times C_\mathrm{s} \times W \tag{3}$$

One or more harvests may be made per year or on a longer term, depending on the crop. The frequency of harvesting can be introduced to calculate the activity removed per year:

$$A_\mathrm{p}/\mathrm{y} = \mathrm{TF} \times C_\mathrm{s} \times W \times H \tag{4}$$

Equation (4) can be used to calculate the absolute activity expected to be removed by phytoextraction in one year. However, the efficiency of this removal process can only be judged in relation to the absolute activity present within the soil depth of interest. This is given by:

$$A_\mathrm{s} = C_\mathrm{s} \times \rho \times \Delta \tag{5}$$

Equations (4) and (5) can now be combined to give a measure of the efficiency of phytoextraction of a radionuclide within one year.

$$k_\mathrm{p} = \frac{\mathrm{TF} \times C_\mathrm{s} \times W \times H}{C_\mathrm{s} \times \rho \times \Delta} \tag{6}$$

The parameter $k_\mathrm{p}$ is a rate coefficient which quantifies the fractional removal of radionuclide per year. This can now be converted to a phytoextraction half time, as follows.

$$\frac{T_\mathrm{p}}{2} = \frac{\ln(2)}{k_\mathrm{p}} \tag{7}$$

$T_\mathrm{p}/2$ indicates the time period required for plant uptake and harvesting to reduce the initial radionuclide inventory of a soil by 50% in the absence of physical (radioactive) decay of the radionuclide concerned. For long-lived radionuclides such as $^{239}$Pu and $^{99}$Tc, with physical half lives of 24,110 and 211,100 years, respectively, radioactive decay is irrelevant to the time scales over which phytoremediation schemes would be expected to operate. However, for important radionuclides such as $^{137}$Cs and $^{90}$Sr, with physical half lives of 30.1 and 28.8 years,

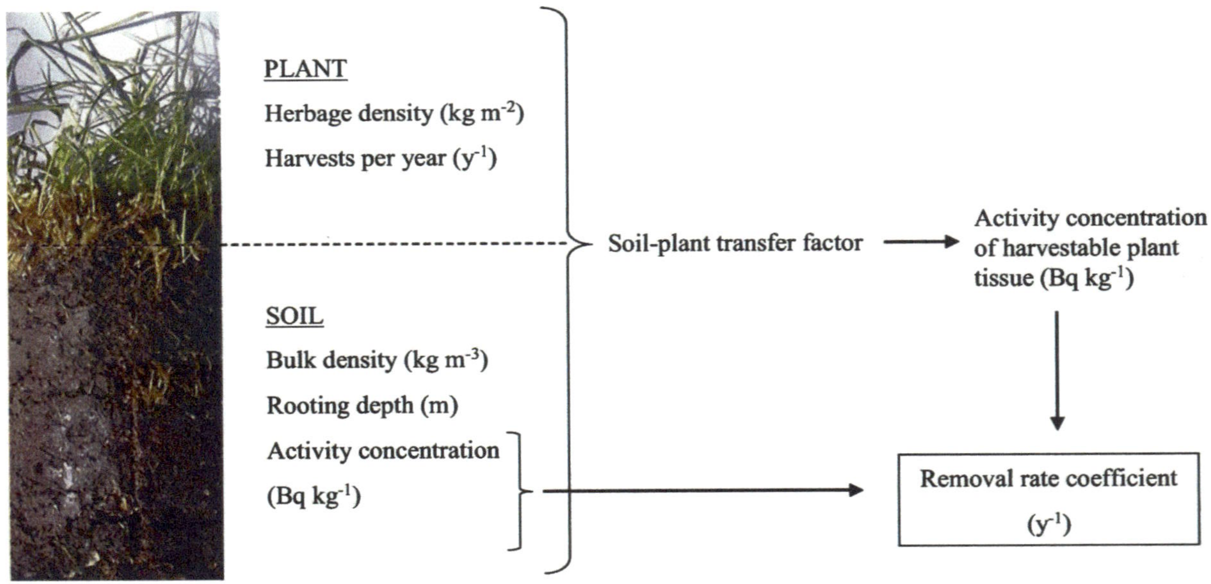


Fig. 1. Schematic description of the simple calculation used to estimate the phytoextraction rate coefficient.

Table 2
Symbols and units used in the calculation of the effective phytoremediation half time. († values are calculated from the other parameters as described in section 4.2)

| Symbol | Description | Typical value | Unit |
|---|---|---|---|
| TF | Soil–plant transfer factor | See Table 1 | Dimensionless |
| $C_s$ | Activity concentration of radionuclide in soil (rooting zone) | See Table 2 of Chapter 2 | $Bq\,kg^{-1}$ |
| $C_p$ | Activity concentration of radionuclide in harvestable plant tissue | † | $Bq\,kg^{-1}$ |
| $A_p$ | Activity inventory in plant | † | $Bq\,m^{-2}$ |
| $A_s$ | Activity inventory in soil (rooting zone) | † | $Bq\,m^{-2}$ |
| $W$ | Herbage density | 0.5–10 | $kg\,m^{-2}$ |
| $H$ | Number of harvests per year | 1–2 | $y^{-1}$ |
| $\rho$ | Soil bulk density | 800–1200 | $kg\,m^{-3}$ |
| $\Delta$ | Depth of soil exploited by plant roots (rooting zone) | 0.1–0.2 | m |
| $T_p/2$ | Phytoextraction half time | † | y |
| $T_{eff}/2$ | Effective phytoremediation half time | † | y |
| $k_p$ | Rate coefficient of phytoextraction | † | $y^{-1}$ |
| $k_r$ | Rate coefficient of physical (radioactive) decay | † | $y^{-1}$ |

respectively, radioactive decay will contribute to their gradual removal from soil by phytoextraction over a period of decades, which is consistent with current and proposed clean-up schedules for major nuclear installations. The effective phytoremediation rate coefficient is a combination of the phytoextraction and the physical decay rate coefficients ($k_p$ and $k_r$). The effective phytoremediation half time is given by:

$$\frac{T_{eff}}{2} = \frac{\ln(2)}{k_p + k_r} \tag{8}$$

Figure 2 shows the relationship between the soil–plant transfer factor and the effective phytoremediation half time calculated for a radionuclide with a 30 year physical half life. Two curves are shown which indicate the effect which harvestable biomass is expected to have on phytoextraction efficiency. A biomass of 1 $kg\,m^{-2}\,y^{-1}$ is typical of an intensively grown crop such as a cereal while 10 $kg\,m^{-2}\,y^{-1}$ is achievable with 'biomass' crops such as willow (*Salix* spp.) or Elephant grass (*Miscanthus giganteus*). Superimposed on these curves are the ranges of TFs for $^{137}Cs$, $^{90}Sr$ and $^{99}Tc$ obtained from Table 1. These indicate that at relatively low harvestable yields (1 $kg\,m^{-2}\,y^{-1}$) TFs would need to be maximal to achieve phytoremediation half times of 25, 10 and 2 years for $^{137}Cs$, $^{90}Sr$ and $^{99}Tc$, respectively. However, the same phytoremediation half times could be achieved using a high biomass crop with average soil–plant TFs for each radionuclide. If maximal TFs are combined with high biomass the phytoremediation half times are reduced to approximately 10, 2 and 0.5 years for $^{137}Cs$, $^{90}Sr$ and $^{99}Tc$, respectively. For $^{90}Sr$ and $^{99}Tc$ the combination of 'average' soil–plant TFs and crops yielding high annual harvestable biomass is probably sufficient to provide an adequate phytoremediation rate without further enhancement, although the rates could doubtless be accelerated. For

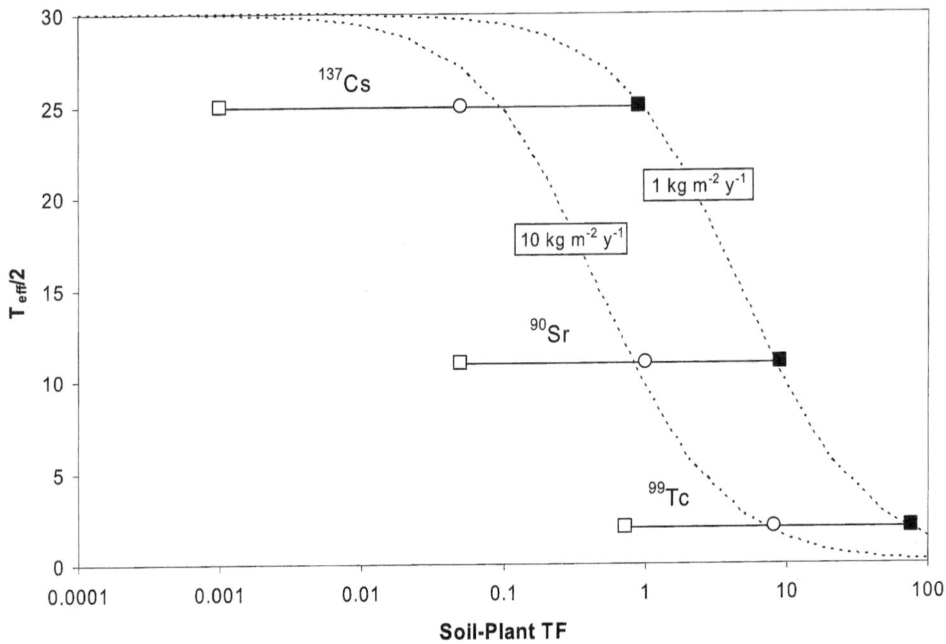


Fig. 2. The curves show the relationships between the soil–plant transfer factor (TF) and the effective phytoremediation half time ($T_{eff}/2$) calculated for two assumed herbage densities using the simple equations described in section 4.2. Both curves incorporate a 30 year physical half life which is approximately applicable to $^{137}$Cs and $^{90}$Sr and only marginally affects the effective phytoextraction half time for $^{99}$Tc. The horizontal lines show the ranges of TFs obtained from the literature for $^{137}$Cs, $^{90}$Sr and $^{99}$Tc with each point representing the minimum (□), mean (○) and maximum (■) TF, respectively.

$^{137}$Cs a phytoremediation rate which is substantially greater than the rate of physical decay of the radionuclide is only likely to be achievable by increasing the degree of soil–plant transfer. The potential for the manipulation of agronomic practices, taxonomic selection and/or genetic traits to increase the radionuclide uptake efficiencies of plants is discussed in the following section (5). However, it can be seen from Figure 2 and Table 1 that, for radionuclides such as $^{239}$Pu with soil–plant TFs generally lower than $10^{-4}$, any increase in plant uptake efficiency would need to be dramatic to improve the prospects of using phytoremediation as a clean-up technology.

## 5. The future of phytoremediation for radionuclides

### 5.1. *Agronomic regime and phytoremediation*

There is a significant reservoir of knowledge on how to grow the plant species widely used in agricultural systems. Few agronomic lessons have been applied to phytoremediation in general, but especially not to phytoremediation of land contaminated with radionuclides. Although these lessons have yet to be fully explored they indicate some ways to bring about the changes in soil to plant transfer shown to be necessary for phytoremediation of radionuclides in section 4.

Agronomists have extensive experience of growing crops in order to maximise nutrient use efficiency and biomass production, and to minimise soil erosion and nutrient leaching. Phytoremediation has already been used to manage Mo contamination of agricultural systems (Neunhauserer et al., 2001). This experience might be invaluable in managing phytoremediation systems, especially for radionuclides which are similar to nutrient ions. Currently, there is much research that has shown that significant improvements in nutrient uptake efficiency are achievable through agronomic practice (Tilman et al., 2002). Such practices might be directly relevant to phytoremediation systems for radionuclides which are similar to nutrient ions, but they are also likely to benefit the design of systems for other radionuclides. Many of these agronomic advances have been developed in conjunction with models of nutrient movements in agricultural systems, demonstrating how important models can be for optimising such management systems. The importance of sustainable agricultural ecosystems to long-term global food security means that significant funds are focused on related research and that significant advances in efficiency of nutrient uptake are likely in the future (Tilman et al., 2002).

Agricultural scientists also have extensive experience in the use of agrochemicals. Much research on agrochemicals is focused on pest control and plant growth regulation, both of which might be required in phytoremediation systems. For example, continuous cropping of the same species on the same piece of land over more than a couple of years is rarely without problems. If species being used in phytoremediation programmes are agricultural crops, or their close relatives, this should present no serious problems because of the extensive experience with pest control in agriculture. However, the use of species which are not agricultural crops or their close relatives might provide some challenges for phytoremediation. Agrochemicals have long been used as soil amendments; for example, nitrification inhibitors that limit the microbial transformation of $NH_4^+$ to $NO_3^-$ in aerobic soils have been widely used in agriculture (Zerulla et al., 2001). Thus far the experience of agronomists who have designed and applied soil amendments has not been focused on mobilising radionuclides in soil. However, as noted in section 2.1, radioecologists have quite extensive knowledge of chemicals that can alter Cs availability in soils and that have been previously used as amendments to reduce Cs uptake by crops. A focused attempt by agricultural chemists and radioecologists to design amendments that increase the availability of radionuclides in soils for phytoremediation might be a fruitful area for future research. It is certainly required if phytoremediation is to work and it seems likely that agronomic practice could be used to bring about a significant proportion of the changes shown to be necessary for phytoremediation in section 4. It will, however, have to overcome problems not just of availability, but also the potential production of radioactively contaminated leachates from treated soils.

### 5.2. *Taxon selection for phytoremediation*

In bioremediation systems a fundamental resource is biodiversity. In phytoremediation systems in general, and for phytoremediation of radionuclides in particular, there has been little systematic attempt to maximise the exploitation of this diversity. Plant species and varieties differ not only in their ability to take up radionuclides but also in their ability to grow in contrasting environments. These phenotypic variations might be vital to the development of phytoremediation as a decontamination option for soil contaminated with radionuclides, for two reasons. First, it provides the raw material (TFs) for phytoremediation systems, exploitation of which has provided great advances for the phytoremediation of As (Ma et al., 2001). Secondly, it aids the identification of the controls on radionuclide accumulation by plants that might guide genetic manipulation.

As the threat to the planet's biodiversity increases so do scientific assessments of biodiversity. These include studies into how biodiversity might be quantified and how it has evolved (e.g. Pachepsky et al., 2001). Studies of the evolution of biodiversity are proving particularly fruitful to understanding the relationship of biodiversity to variation in biological phenomena. These studies are useful to understanding inter-specific differences in radionuclide uptake by plants and how they might be exploited for phytoremediation. Gene sequences have provided a powerful measure, in addition to the biochemical, anatomical and morphological characters traditionally used, to chart the evolution of groups of organisms. Increases in computing power and novel algorithms have provided the means to reconstruct the evolutionary history (phylogeny) of numerous groups of organisms using a suite of characters including gene sequences (Harvey et al., 1996). There are new phylogenies for plants that can provide a framework for explaining and exploiting biodiversity in phenotypes such as radionuclide uptake. They have already proved useful for other studies in comparative plant ecology (Ackerley, 2001). Figure 3 shows a recent phylogeny for the major orders of angiosperms (flowering plants). Mapping biodiversity in a phenotype such as radionuclide uptake onto phylogenies such as Figure 3 can be helpful in describing the distribution of biodiversity in phenotype across the flowering plant phylum. If there are phylogenetic constraints they can be very useful in identifying plants with unusual uptake characteristics and the features associated with them.

There have been few direct assessments of the biodiversity in radionuclide uptake in plants that might be exploited for phytoremediation, but the radioecological literature is rich in data that are useful in such assessments. Broadley et al. (1999) first compiled data on Cs uptake by plants and suggested that for Cs the Caryophyllid clade (Figure 3) might include the plants with the highest Cs uptake and phytoremediation potential. Willey et al. (2005), in the largest data set of species inter-comparisons of uptake yet published for radionuclides, have extended this compilation to 265 species and confirmed the existence of high uptake on the Caryophyllid clade. It has been suggested that the lack of discrimination between Cs and K during uptake by plants on this clade might underlie this phenomenon (Broadley and Willey, 1997) and that voltage independent channels (VIC) for cations in these plants might be worth investigating with a view to manipulating Cs uptake by plants (White and Broadley, 2000). The phylogenetic constraints identified by these studies only explain a proportion of the inter-specific differences in Cs uptake by plants, but they demonstrate a method through which plants, and ultimately genes, that might be useful for phytoremediation can be identified from the myriad possibilities in the angiosperms. There are similar published studies of the biodiversity of plant uptake of other radionuclides which suggest that Sr uptake is high in the Cucurbitales (Willey and Fawcett, 2006a), that S uptake is high in the Malphigiales and Brassicales (Willey and Wilkins, 2006), that Ru uptake is high in the Geraniales and Asterales (Willey and Fawcett, 2006b) and I uptake is high in the Poales and Fabales (Willey et al., 2002). It will be very useful to compare phylogenetic effects and the biodiversity 'landscape'

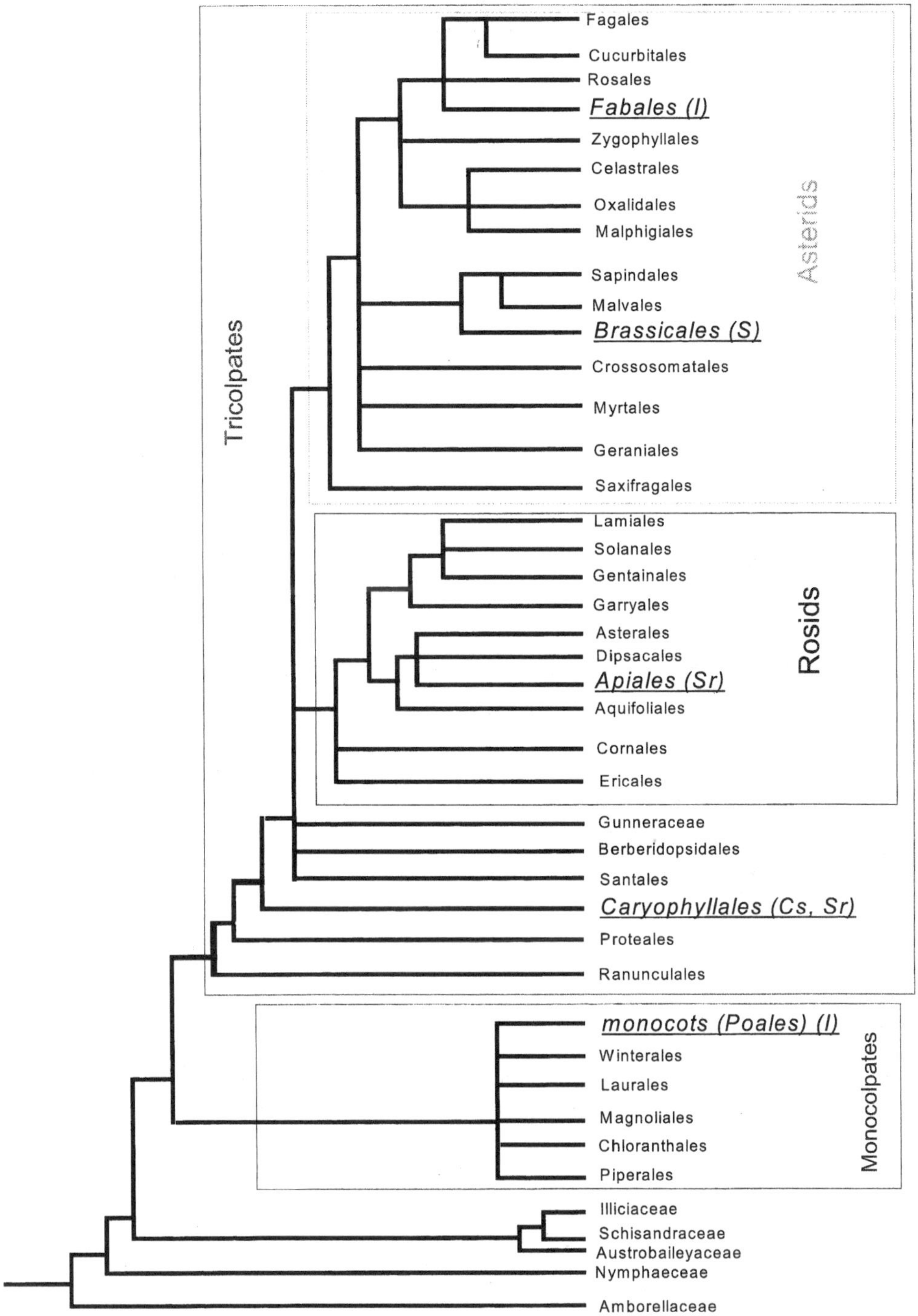


Fig. 3. The phylogeny of flowering plants showing some groups with high radionuclide uptake.

of plant uptake of different radionuclides, especially if plants that might phytoremediate more than one radionuclide simultaneously are being sought.

Previous studies of factors such as rooting depth (Van Rees and Comerford, 1986), mycorrhizae (Drissner et al., 1998) and ecological type (Willey and Martin, 1997) have suggested that they might contribute to inter-species diversity of radionuclide uptake by plants. In addition to studies of phylogenetic constraints, the quantification of the impact of these other factors will help to identify those plants with the greatest potential for phytoremediation. Studies of phytoremediation for Cs have already used plants on the Caryophyllid clade, but it seems unlikely that the plants with the very highest uptake have yet been found. Charting biodiversity and exploiting it directly through taxon selection or indirectly through genetic manipulation has the potential to increase significantly the uptake of radionuclides by plants used for phytoremediation.

### 5.3. *Genetic manipulation of plants and phytoremediation*

Discovering the structure of DNA was as useful to biology as discovering the structure of atoms was to chemistry, gene sequencing to describing the constituents of atoms, and genetic manipulation to manipulating fission reactions. Biology is currently in the phase not only of a paradigm shift in understanding living things but, in much the same way that manipulating fission reactions did, also providing previously undreamed of possibilities. The draft sequences of the genomes of *Arabidopsis thaliana* (Thalecress) (Arabidopsis Genome Initiative, 2000) and *Oryza sativa* (rice) (Yu et al., 2002) open up a new world of possibilities in plant science. Post-genomic successes with micro-organisms indicate that proteomics and metabolomics will elucidate plant functioning in unprecedented detail and enable detailed genetic engineering.

The primary constraints of using phytoremediation to extract radionuclides from soils, if they are in the rooting zone, are soil availability and plant uptake rates (section 3.5). The genetic manipulation of plants could help to overcome both these obstacles. This is possible because such manipulations have already been successfully carried out for the availability and uptake of some nutrients and contaminants. The availability of Fe is generally very low in alkaline soils (which cover up to 30% of the world's surface). Rice plants, which are grown in alkaline soils and can suffer Fe deficiency, have been genetically manipulated to over-express phytosiderophores (mugineic acids) which are exuded from the roots and increase Fe availability to rice in alkaline soils (Takahashi et al., 2001). Transgenic plants which increase the availability of recalcitrant phosphates have also been reported (López-Bucio et al., 2000). Plants that detoxify Hg (Rugh et al., 1998), volatilise Se (Pilon-Smits et al., 1998) and degrade explosives (French et al., 1999) through engineered enzyme production have been created. Pigeon peas are well known to exude piscidic acid from their roots which mobilises the almost insoluble iron phosphate (Ae et al., 1990). The use of plants as producers and exuders of compounds that increase ion availability in soils is now advancing rapidly and is commonly seen as a solution to the problem of low micronutrient availability to crop plants (Grotz and Guerinot, 2002). Many of the micronutrients at which such research is aimed have similar $K_d$ values to radionuclides (Table 1). The families of proteins involved in nutrient uptake are rapidly being identified (e.g. Maser et al., 2001) and the genetic manipulation of plant uptake mechanisms to increase uptake of nutrient ions is increasingly possible (e.g. Guerinot,

2001). Increased As uptake rates by plants have already been reported (Dhankher et al., 2002). Much work has focused on the selectivity of plant uptake mechanisms and shown either that transporters have broad substrate ranges or that site directed mutagenesis to alter selectivity might be possible. Such research indicates that it certainly might be possible to increase plant uptake of $^{137}Cs$, $^{90}Sr$, $^{99}Tc$, $^{36}Cl$, $^{35}S$ and $^{60}Co$ but also that increased uptake of radionuclides that are less similar to nutrients, such as those of U and Pu, might also be possible.

It is tempting to be carried away by the possibilities of new technologies, but it is equally easy to spurn the opportunities they can provide. The manipulation of rhizosphere availability and plant uptake rates of a variety of plant nutrients is already well under way. It is clearly possible that techniques similar to those employed in nutrient uptake research might prove useful in manipulating rhizosphere availability and plant uptake of radionuclides. It has even been suggested that genetic modification of plants for phytoremediation has the potential to produce public faith in the potential of genetic modification to do good (Watanabe, 2001).

## 6. Conclusions and prospects

### 6.1. *Research priorities for phytoremediation of radionuclides*

There are numerous reasons why the decontamination of radionuclide polluted soils will be important in the 21$^{st}$ century, but the methods most widely used thus far are expensive, disruptive and probably cannot deal with the necessary volume of contaminated soil. The experience of radioecologists and the molecular transformation of the life sciences might help establish phytoremediation as a viable method of decontamination. This would be a significant development in radioecology. For radionuclides such as $^{90}Sr$, $^{99}Tc$, $^{36}Cl$ and $^{35}S$, soil to plant transfer can probably already be managed to achieve significant decontamination. For other radionuclides, such as $^{137}Cs$, it is primarily sorption to the soil that is limiting, but for other radionuclides such as $^{239}Pu$ it is also plant uptake. If industrial and environmental imperatives in the 21$^{st}$ century provide resources for phytoremediation they would profitably be spent investigating soil amendments for increasing the availability of specifically targeted radionuclides such as $^{137}Cs$ and $^{239}Pu$. Given the foundation that radioecologists have built, which includes knowledge of the effects of $NH_4^+$ salts, Ag-thiourea, zeolites, Na-tetraphenylborate and norabidine A, on Cs availability for example, it seems possible that investment of sufficient resources might identify mobilising agents for other radioecologically significant ions. The nuclear industry has certainly overcome greater scientific challenges. In conjunction with research on soil availability, research might be directed towards exploiting biodiversity and genetic manipulation to increase plant uptake rates of radionuclides. Radioecological knowledge suggests that there is biodiversity to exploit and plant scientists are already manipulating the uptake of a variety of elements. The simple modelling described in section 4 suggests that the manipulations in soil to plant transfer which are necessary to make phytoremediation viable are within the bounds of possibility.

Any research into the development of phytoremediation will not just have to overcome obstacles limiting soil to plant transfer. The greatest additional problem will be disposal of radioactively contaminated plant material. In fact, given that a significant impetus for developing decontamination regimes is that there is limited capacity in waste repositories, phytoremediation can be viewed as a waste reduction exercise. Safe ashing (Hall and Watt, 2002) or composting (Watt, 2004) procedures will probably have to be developed for contaminated biomass to achieve significant volume reduction. This might be a significant challenge at some contaminated industrial sites and might determine whether or not phytoremediation is actually cheaper than other methods of decontamination. The legislation relating to radioactively contaminated land is frequently driven by public pressure. As environmental legislation becomes trans-national and even global, public perceptions are likely to increase in importance even in those countries where it is currently of less account. There has been little sociological research into public perceptions of phytoremediating radioactively contaminated land. Some evidence from other phytoremediation trials suggests that local communities have responded very positively (Tucker and Shaw, 2000). Public perceptions can create or destroy entire markets for technologies and either fate might be possible for phytoremediation of radioactively contaminated land. Research into public perceptions might help assess the level of resources it is worth governments and nuclear industries investing in phytoremediation research. It is also possible that phytoremediation will give lower doses to operators than other decontamination options and this health physics criterion might prove to be decisive.

### 6.2. *Phytoremediation within the future of radioecology*

Radioecologists have a unique knowledge of the behaviour of contaminants in the environment (Shaw, 2005). This knowledge is unique not just because it relates to radionuclides that can be very toxic and expensive to extract from soils, but also because the monitoring and modelling of radionuclides have been carried out over a longer timescale than for any other group of environmental contaminants. Certain aspects of radionuclide behaviour in the environment have also been the focus of much attention, and resources and models for them are as complex as for almost any contaminant. Radionuclides of Tc, U and Pu also have quite complex environmental chemistries that have necessitated careful study. This radioecological knowledge has until now been primarily focused on monitoring and predicting the behaviour of radionuclides in the environment for dose assessments to humans. There is currently much debate in radioecology about turning attention towards dose assessments for organisms other than humans (see Chapter 11). We suggest that another outlet for radioecological knowledge might be the design of phytoremediation systems and that developing this strand in radioecology might also provide much relevant information for assessing doses to plants.

It is quite possible that developing soil decontamination techniques might be an important role of radioecology in the 21$^{st}$ century. Designing chemicals that can displace radionuclides from exchange sites and even clay interlayers would bring fundamental advances in our understanding of radionuclide interaction with soils. In some soils this is already good but in other soil types it is sparse. When the availability of radionuclides can be manipulated in a wide variety of soil types to optimise phytoremediation rates then radioecologists will not just have described the behaviour of radionuclides in soils, they will have an understanding of the interaction between soils and radionuclides that is sufficient to manipulate and control the process.

The uptake of radionuclides by plants is a crucial entry point for radionuclides into food chains. As the focus of dose assessments widens to include non-human species the variety of plant species in which radionuclide uptake is important will increase. Understanding the biodiversity landscape for radionuclide uptake by plants might, therefore, be important not just for selecting taxa for phytoremediation but also for dose assessments to organisms in the environment. Manipulation of the uptake systems that control plant concentrations and affect dose assessments might be important for radioecologists to test the controls on doses and understand them fully. Assessments of the biodiversity in radionuclide uptake by plants and how it might be manipulated might therefore have a variety of important roles in radioecology in the 21$^{st}$ century.

### 6.3. *Conclusions*

Phytoremediation is a nascent technology that might eventually be useful for decontaminating soils of radionuclides. Initially, it is likely to be most useful when contaminants are dispersed in the upper layers of soil at sites where plants capable of absorbing them can grow freely, a scenario which already occurs at many locations around the earth. Further research into the bioavailability of radionuclides in soils and the selection and manipulation of plant taxa will doubtless increase the number and type of scenarios in which phytoremediation is potentially feasible. There are challenges to be overcome but doing so is probably limited as much by resources for research as scientific possibilities. Although it has sometimes taken very significant resources, the nuclear industry has certainly overcome much greater challenges. Resources available for research are probably dependent on the future status of both the nuclear industry and environmental legislation. These are changeable, but the current absence of alternative decontamination methods and the presence of tightening legislation indicate that it is likely to be worth investing further resources into phytoremediation to build on the foundations of the first phase of field trials. Some of the challenges facing phytoremediation also coincide with those facing 21st century radioecology. The expertise of radioecologists will be indispensable in developing the technology of phytoremediation and this challenge should provide a significant new outlet for their energies.

## References


Abbazov, M.A., Dergunov, I.D., Milkin, R.G. (1978). Effect of soil properties on the accumulation of strontium-90 and cesium-137 in crops. *Soviet Soil Science* **10**, 52–56.

Ackerley, D.D. (2001). Comparative plant ecology and the role of phylogenetic information. In: Press, M.C., Scholes, J.D., Barker, M.G. (Eds.), *Physiological Plant Ecology*. Backwell Science, Oxford, pp. 391–414.

Ae, N., Arihara, J., Okada, K., Yoshihara, T., Johansen, C. (1990). Phosphorus uptake by pigeon pea and its role in cropping systems of the Indian subcontinent. *Science* **248**, 477–480.

Andersen, A.J. (1967). Investigations on the plant uptake of fission products from contaminated soils. 1. Influence of plant species and soil types on the uptake of radioactive strontium and caesium. Riso Report No. 170. Riso, Denmark.

Arabidopsis Genome Initiative (2000). Analysis of the genome sequence of the flowering plant Arabidopsis thaliana. *Nature* **408**, 796–813.

Askbrant, Sandalls, F.J. (1998). Root uptake of $^{137}Cs$ and $^{90}Sr$ by rye-grass on various soils in the CIS. *Journal of Environmental Radioactivity* **38**, 85–95.

Avery, S. (1996). Fate of caesium in the environment: Distribution between the abiotic and biotic components of aquatic and terrestrial ecosystems. *Journal of Environmental Radioactivity* **30**, 139–171.

Baes, C.F., Sharp, R.D. (1983). A proposal for estimation of soil leaching and leaching constants for use in assessment models. *Journal of Environmental Quality* **12**, 17–27.

Bennett, R., Willey, N. (2002). Soil availability, plant uptake mechanisms and soil to plant transfer of 99Tc—A review. *Journal of Environmental Radioactivity* **65**, 215–231.

Bell, N., Minski, C., Grogan, H.A. (1988). Plant uptake of radionuclides. *Soil Use and Management* **4**, 76–84.

Broadley, M.R., Willey, N.J. (1997). Differences in root uptake of radiocaesium by 30 plant taxa. *Environmental Pollution* **97**, 11–15.

Broadley, M.R., Willey, N.J., Meade, A. (1999). A method to assess taxonomic variation in Cs concentrations among flowering plants. *Enviromental Pollution* **106**, 341–349.

Brown, J.E., Kolstad, A.K., Brungot, A.L., Lind, B., Rudjord, A.L., Strand, P., Føyn, L. (1999). Levels of 99Tc in seawater and biota samples from Norwegian coastal waters and adjacent seas. *Marine Pollution Bulletin* **38**, 560–571.

Clemens, S., Palmgren, M.G., Kramer, U. (2002). A long way ahead: Engineering plant metal accumulation. *Trends in Plant Science* **7**, 309–314.

Coughtrey, P.J., Kirton, J.A., Mitchell, N.G. (1990). Transfer of radioactive cesium from soil to vegetation and comparison with potassium in upland grasslands. *Environmental Pollution* **62**, 281–315.

Cox, A.E., Joern, B.C. (1997). Release kinetics of non-exhcnagenable potassium in soils using sodium tetraphenylboron. *Soil Science* **162**, 588–598.

Cremers, A., Elsen, A., de Peter, P., Maes, A. (1988). Quantitative analysis of radiocaesium retention in soils. *Nature* **335**, 247–249.

Cummings, S.L., Bankert, L., Garret, A.R., Regnier, J.F. (1969). Cs-137 uptake by oat plants as related to the soil fixing capacity. *Health Physics* **17**, 145–148.

D'Souza, T.J., Mistry, K.B. (1980). Adsorption of gamma-emitting fission products and activation products by rice under flooded and unflooded conditions from two tropical soils. *Plant and Soil* **55**, 189–198.

Delmas, J., Grauby, A., Paulin, R., Rinaldi, C., Ndoye, R., Seckgassama, S. (1987). Etude expérimentale du transfert du 90Sr et du 137Cs d'un sol de la region du Cap Vert à l'arachide. *Radioprotection* **22**, 333–340.

Delvaux, B., Kruts, N., Cremers, A. (2000). Rhizospheric mobilisation of radiocaesium in soils. *Environmental Science and Technology* **34**, 1489–1493.

Delvaux, B., Kruyts, N., Maes, E., Smolders, E. (2001). Fate of radiocaesium in soil and rhizosphere. In: Gobran, G.R., Wenzel, W.W., Lombi, E. (Eds.), *Trace Elements in the Rhizosphere*. CRC Press, Baton Rouge, pp. 61–91.

Dhankher, O.P., Li, Y., Rosen, B.P., Shi, J., Salt, D., Senecoff, J.F., Sashti, N.A., Meagher, R.B. (2002). Engineering tolerance and hyperaccumulation of arsenic in plants by combining arsenate reductase and $\gamma$-glutamylcysteine synthetase expression. *Nature Biotechnology* **20**, 1140–1145.

Drissner, J., Burmann, W., Enslin, F., Heider, R., Klemt, E., Miller, R., Schick, G., Zibold, G. (1998). Availability of caesium radionuclides to plants—Classification of soils and role of mycorhizza. *Journal of Environmental Radioactivity* **41**, 19–32.

Durante, M., Pugliese, M. (2002). Estimates of radiological risk from depleted uranium weapons in war scenarios. *Health Physics* **82**, 14–20.

Dushenkov, S.D., Vasudev, D., Kapulnik, Y., Gleba, D., Fleischer, D., Ting, K.C., Ensley, B. (1997). Removal of uranium from water using terrestrial plants. *Environmental Science and Technology* **31**, 3468–3474.

Dushenkov, S., Mikheev, A., Prokhnevsky, A., Ruchko, M., Sorochinsky, B. (1999). Phytoremediation of radiocaesium contaminated soil in the vicinity of Chernobyl, Ukraine. *Environmental Science and Technology* **33**, 469–475.

Ebbs, S., Brady, D.J., Kochian, L.V. (1998). Role of uranium speciation in the uptake and translocation of uranium by plants. *Journal of Experimental Botany* **49**, 1183–1190.

Echevarria, G., Sheppard, M.I., Morel, J. (2001). Effect of pH on the sorption of uranium in soils. *Journal of Environmental Radioactivity* **53**, 257–264.

Entry, J.A., Watrud, L. (1998). Potential remediation of 137Cs and 90Sr contaminated soil by accumulation in Alamo switchgrass. *Water, Air and Soil Pollution* **104**, 339–352.

Entry, J.A., Watrud, L.S., Reeves, M. (1999). Accumulation of cesium-137 and strontium-90 from contaminated soil by three grass species innoculated with mycorrhizal fungi. *Environmental Pollution* **104**, 449–457.

Entry, J.A., Watrud, L.S., Reeves, M. (2001). Influence of organic amendments on the accumulation of 137Cs and 90Sr from contaminated soil by three grass species. *Water, Air and Soil Pollution* **126**, 385–398.

Fink, A. (1982). *Fertilisers and Fertilisation. Introduction and Practical Guide to Crop Fertilisation*. Verlag Chemie, Veinheim, Germany.

Firsakova, S.K., Zhuchenko, Y.M., Voigt, G. (2000). An example of rehabilitation strategies for radioactive contaminated areas in Belarus. *Journal of Environmental Radioactivity* **48**, 23–33.

Fredrikkson, L. (1970). Plant uptake of fission products IV. Uptake of 90Sr and 137Cs from some tropical and subtropical soils. *Lantbrukshogskolans Annaler* **36**, 61–89.

French, C.E., Rossner, S.J., Davies, G.J., Nicklin, S., Bruce, N.C. (1999). Biodegredation of explosives by transgenic plants expressing pentaeryhtitol tetranitrate reductase. *Nature Biotechnology* **17**, 491–494.

Frissel, M.J. (1992). An update of the recommended soil-to-plant transfer factors of 90Sr, 137Cs and transuranics. 8th report of the IUR working group on soil plant transfer. IUR Banlan, Belgium.

Frissel, M.J., van Bergeijk, K.E. (1989). Mean transfer values derived by simple statistical analysis. In: VIth annual report of the working group on soil-to-plant transfer, pp. 10–21. Grimselpass, Switzerland, 1989. Published by RIVM, the Netherlands.

Fuhrman, M., Lasat, M.M., Ebbs, S.D., Kochian, L.V., Cornish, J. (2002). Uptake of cesium-137 and strontium-90 from contaminated soil by three plant species: Application to phytoremediation. *Journal of Environmental Quality* **31**, 904–909.

Garaudée, S., Elhabiri, M., Kaly, D., Robiolle, C., Trendel, J.-M., Hueber, R., van Dorsselaer, A., Albrecht, P., Albrecht-Gary, A.-M. (2002). Allosteric effects in norbadione A. A clue for the accumulation process of $^{137}$Cs in mushrooms? *Chemical Communications* **9**, 944–945.

Gerzabek, M.H., Strebl, F., Temmel, B. (1998). Plant uptake of radionuclides in lysimeter experiments. *Environmental Pollution* **99**, 93–103.

Glass, D. (2000). Economic Potential of Phytoremediation. In: Raskin, I., Ensley, B.D. (Eds.), *Phytoremediation of Toxic Metals—Using Plants to Clean Up the Environment*. John Wiley and Sons, NY, USA, pp. 15–32.

Grotz, N., Guerinot, M.L. (2002). Limiting nutrients: An old problem with a new solution? *Current Opinion in Plant Biology* **5**, 158–163.

Guerinot, M.L. (2001). Improving rice yields—ironing out the details. *Nature Biotechnology* **19**, 466–469.

Hall, S., Watt, N. (2002). The Potential of Phytoextraction to remediate caesium-137 contaminated ground on nuclear licensed sites. *Nuclear Engineer* **43**, 27–31.

Harvey, P.H., Leigh-Brown, A.J., Maynard-Smith, J., Nee, S. (Eds.) (1996). *New Uses for New Phylogenies*. Oxford University Press, Oxford, UK.

Haywood, S.M., Smith, J.G. (1992). Assessment of potential doses at the Maralinga and Emu test sites. *Health Physics* **63**, 624–630.

Hinchee, R.E., Means, J.L., Burris, D.R. (1995). *Bioremediation of Inorganics*. Batelle Press, Columbus, OH, USA.

Hopkins, W. (1999). *Introduction to Plant Physiology*. John Wiley.

Howe, S.E., Davidson, C.M., McCartney, M. (2002). Determination of uranium concentration and isotopic composition by means of ICP-MS in sequential extracts of sediment from the vicinity of a uranium enrichment plant. *Journal of Analytical and Atomic Spectrometry* **17**, 497–501.

Huang, J.W., Blaylock, M.J., Kapulnik, Y., Ensley, B.D. (1998). Phytoremediation of uranium contaminated soils: Role of organic acids in triggering uranium hyperaccumulation in plants. *Environmental Science and Technology* **32**, 2004–2008.

Lasat, M.M., Norvell, W.A., Kochian, L.V. (1997). Potential for phytoextraction of 137Cs from a contaminated soil. *Plant Soil* **195**, 99–106.

Lasat, M.M., Fuhrmann, M., Ebbs, S.D., Cornish, J.E., Kochian, L.V. (1998). Phytoremediation of a radiocaesium contaminated soil: evaluation of cesium-137 bioaccumulation in the shoots of three plant species. *Journal of Environmental Quality* **27**, 165–169.

Lee, J.H., Hossner, L.R., Attrep, M., Kung, K.S. (2002). Uptake and translocation of plutonium in two plant species using hydroponics. *Environmental Pollution* **117**, 61–68.

López-Bucio, J., Nieto-Jacobo, M.F., Ramírez-Rodríguez, V., Herrera-Estrella, L. (2000). Organic acid metabolism in plants: from adaptive physiology to transgenic varieties for cultivation in extreme soils. *Plant Science* **160**, 1–13.

Ma, L.Q., Komar, K.M., Tu, C., Zhang, W.H., Cai, Y., Kennelley, E.D. (2001). A fern that hyperaccumulates arsenic—a hardy, versatile, fast-growing plant helps to remove arsenic from contaminated soils. *Nature* **409**, 579.

Marschner, H. (1995). *Mineral Nutrition of Higher Plants*, 2nd edn. Academic Press, London.

Maser, P., Thomine, S., Schroeder, J.I., et al. (2001). Phylogenetic relationship within cation transporter families of Arabidopsis. *Plant Physiology* **126**, 1646–1667.

Mason, C.F.W., Turney, W.R.J.R., Thomson, B.M., Lu, N., Longmire, P.A., Chisholm-Brause, C.I. (1997). Carbonate leaching of uranium from contaminated soils. *Environmental Science and Technology* **31**, 2707–2711.

Negri, M.C., Hinchman, R.R. (2000). The use of plants for the treatment of radionuclides. In: Raskin, I., Ensley, B.D. (Eds.), *Phytoremediation of Toxic Metals—Using Plants to Clean Up the Environment*. John Wiley and Sons, NY, USA, pp. 107–132.

Neunhauserer, C., Berreck, M., Insam, H. (2001). Remediation of soils contaminated with molybdenum using soil amendments and phytoremediation. *Water, Air and Soil Pollution* **128**, 85–96.

Nie, L., Shah, S., Rashid, A., Burd, G.I., Dixon, D.G., Glick, B.R. (2002). Phytoremediation of arsenate contaminated soil by transgenic canola and the plant growth-promoting bacterium Enterobacter cloacae CAL2. *Plant Physiology and Biochemistry* **40**, 355–361.

Nisbet, A.F., Woodman, R.F.M. (2000). Soil-to-plant transfer factors for radiocaesium and radiotrontium in agricultural systems. *Health Physics* **78**, 279–288.

Nishita, H., Steen, A.J., Larson, K.H. (1958). The release of Sr-90 and Cs-137 from Vina loam upon prologued cropping. *Soil Science* **86**, 195–201.

Nishita, H., Romney, E.M., Alexander, G.V., Larson, K.H. (1960). Influence of K and Cs on release of Cs-137 from three soils. *Soil Science* **89**, 167–176.

Nishita, H., Haug, R.M. (1972). Influence of clinoptilolite in Sr-90 and Cs-137 uptake by plants. *Soil Science* **114**, 149–157.

OSPAR (2003). Implementation of PARCOM Recommendation 91/4 on Radioactive Discharges. OSPAR Commission for the Protection of the Marine Environment of the North-East Atlantic, London. ISBN 1-904426-21-2.

Pachepsky, E., Crawford, D.J.W., Brown, J.L., Squire, G. (2001). Towards a general theory of biodiversity. *Nature* **410**, 923–930.

Pilon-Smits, E.A.H., de Souza, M.P., Lytle, C.M., Shang, C., Lugo, T., Terry, N. (1998). Selenium volatilisation and assimilation by hybrid poplar (Populus tremulaxalba). *Journal of Experimental Botany* **49**, 1889–1892.

Raskin, I. (1996). Plant genetic engineering may help with environmental clean up. *Proceedings of the National Academy of Sciences, USA* **93**, 3164–3166.

Reeves, R.D., Baker, A.J.M. (2000). Metal-accumulating plants. In: Raskin, I., Ensley, B.D. (Eds.), *Phytoremediation of Toxic Metals—Using Plants to Clean Up the Environment*, pp. 193–230.

Roed, J., Anderseen, K.G., Prip, H. (1996). The skim and burial plough: A new implement for reclamation of radioactively contaminated land. *Journal of Environmental Radioactivity* **33**, 117–128.

Romney, E.M., Neel, J.W., Nishita, J., Olafson, J.H., Larson, K.H. (1957). Plant uptake of Sr-90, Y-91, Ru-106, Cs-137 and Ce-144 from soils. *Soil Science* **83**, 369–376.

Romney, E.M., Wallace, A., Schultz, R.A., Kinear, J., Wood, R.A. (1981). Plant uptake of $^{237}$Np, $^{239,240}$Pu, $^{241}$Am and $^{244}$Cm from soils representing major food production areas of the United States. *Soil Science* **132**, 40–59.

Rugh, C.L., Senecoff, J.F., Meagher, R.B., Merkle, S.A. (1998). Development of transgenic poplar for mercury phytoremediation. *Nature Biotechnology* **16**, 925–928.

Shahandeh, H., Lee, J.H., Hossner, L.R., Loeppert, R.H. (2001). Bioavailability of uranium and plutonium to plants in soil-water systems and the potential of phytoremediation. In: Gobran, G.R., Wenzel, W.W., Lombi, E. (Eds.), *Trace Elements in the Rhizosphere*. CRC Press, pp. 93–126.

Shahandeh, H., Hossner, L.R. (2002). Enhancement of uranium phytoaccumulation from contaminated soils. *Soil Science* **167**, 269–280.

Shaw, G. (2005). Applying radioecology in a world of multiple contaminants. *Journal of Environmental Radioactivity* **81**, 117–130.

Shaw, G., Wadey, P., Bell, J.N.B. (2004). Radionuclide Transport Above a Near-Surface Water Table: IV. Soil migration and crop uptake of $^{36}$Cl and $^{99}$Tc, 1990 to 1993. *Journal of Environmental Quality* **33**, 2272–2280.

Sheppard, M.I., Thibault, D.H. (1990). Default soil solid/liquid partition co-efficients, Kds, for four major soil types: A compendium. *Health Physics* **59**, 471–482.

Simon, S., Bouville, A. (2002). Radiation doses to local populations near nuclear weapons test sites worldwide. *Health Physics* **82**, 706–725.

Smith, J.T., Comans, R.N.J., Beresford, N.A., Wright, S.M., Howard, B.J., Camplin, W.C. (2000). Chernobyl's legacy in food and water. *Nature* **405**, 141.

Starr, C. (2000). Powerful reactions. *Millennium Essay Nature* **406**, 679.

Takahashi, M.T., Nakanishi, H., Kawasaki, S., Nishizawa, N.K., Mori, S. (2001). Enhanced tolerance of rice to low iron availability in alkaline soils using barley nicotianamine aminotransferase genes. *Nature Biotechnology* **19**, 466–469.

Tilman, D., Cassman, K.G., Matson, P.A., Naylor, R., Polasky, S. (2002). Agricultural sustainability and intensive production practices. *Nature* **418**, 671–677.

Tucker, R.K., Shaw, J.A. (2000). Phytoremediation and public acceptance. In: Raskin, I., Ensley, B.D. (Eds.), *Phytoremediation of Toxic Metals—Using Plants to Clean Up the Environment*. John Wiley, NY, USA, pp. 33–42.

USDA (1999). Soil Survey: A Basic System of Soil Classification for Making and Interpreting Soil Surveys, 2nd edition, United States Department of Agriculture, Agriculture Handbook No. 436. USDA Natural Resources Conservation Service.

US Department of Energy, Office of Environmental Management (1995). Estimating the Cold War Mortgage: The 1995 Baseline Environmental Management Report DOE/EM-0232.

US Environmental Protection Agency (1997). Recent Developments for In Situ Treatment of Metal Contaminated Soils.

US Environmental Protection Agency (1999). Understanding variation in partition coefficient, Kd, values, Vol. II, Review of Geochemistry and available Kd values for cadmium, cesium, chromium, lead, plutonium, radon, strontium, thorium, tirtium and uranium. EPA 402-R-99-004B.

Van Bergeijk, K.E., Noordijk, H., Lembrechts, J., Frissel, M.J. (1992). Influence of soil pH, soil type and soil organic matter content on soil-to-plant transfer of radiocaesium and strontium as analysed by a non-parametric method. *Journal of Environmental Radioactivity* **15**, 265–276.

Van Rees, K.C.J., Comerford, N.B. (1986). Vertical root distrubution and stontium uptake of a slash pine stand on a Florida spodosol. *Soil Science Society of America Journal* **50**, 1042–1046.

Vandenhove, H., Van Hees, M., De Brouwer, S., Van Den Casteele, C.M. (1996). Transfer of radiocaesium from podsol to ryegrass as affected by AFCF concentration. *Science of the Total Environment* **187**, 237–245.

Victorova, N., Voitesekhovitch, O., Sorochinsky, B., Vandenhove, H., Konoplev, A., Konopleva, I. (2000). Phytoremediation of Chernobyl contaminated land. *Radiation Protection Dosimetry* **92**, 59–64.

Wallace, E.A., Romney, E.M., Wood, R.A. (1982). The role of stable Cs in plant uptake of Cs-137. *Soil Science* **134**, 71–75.

Watt, N.R. (2004). Assessing the potential of phytorextraction to remediate land contaminated with $^{137}$Cs at nuclear power station sites. PhD Thesis, Applied Sciences, University of the West of England, Bristol, UK.

Watanabe, M.E. (2001). Can bioremediation bounce back? *Nature Biotechnology* **19**, 1111–1115.

White, P.J., Broadley, M.B. (2000). Tansley Review: Mechanisms of caesium uptake by plants. *New Phytologist* **147**, 241–256.

White, P.J., Broadley, M.B. (2001). Chloride in soils and it uptake and movement within the plant: A review. *Annals of Botany* **88**, 967–988.

Willey, N.J. (2001). Phytoremediation—Solar Powered Soil Decontamination. Biochemist Aug 2001, 18–20.

Willey, N.J., Martin, M.H. (1997). A comparison of stable caesium uptake by six grass species of contrasting growth strategy. *Environmental Pollution* **95**, 311–317.

Willey, N.J., Hall, S.C., Mudiganti, A. (2001). Assessing the potential of phytoextraction at a site in the UK contaminated with 137Cs. *International Journal of Phytoremediation* **3**, 321–333.

Willey, N.J., Fawcett, K., Wilkins, J. (2002). The use of phylogeny to simplify predictions of radionuclide concentrations in crop plants. Report to the UK Food Standards Agency.

Willey, N.J., Tang, S., Watt, N.R. (2005). Predicting inter-taxa differences in plant uptake of cesium-134/137. *J. Env. Qual.* **34**, 1478–1489.

Willey, N.J., Fawcett, K. (2006a). A phylogenetic effect on strontium concentrations in angiosperms. *Environmental and Experimental Botany* **57**, 258–269.

Willey, N.J., Fawcett, K. (2006b). Inter-taxa differences in root uptake of $^{103/106}$Ru by plants. *Journal of Environmental Radioactivity* **86**, 227–240.

Willey, N.J., Wilkins, J. (2006). An analysis of intertaxa differences in sulphur concentration in angiosperms. *Journal of Plant Nutrition and Soil Science* **169**, 717–727.

Yasuda, H., Uchida, S., Muramatsu, Y., Yoshida, S. (1995). Sorption of manganese, cobalt, zinc, strontium and cesium onto agricultural soils: statistical analysis on effects of soil properties. *Water, Air and Soil Pollution* **83**, 85–96.

Yu, J., et al. (2002). A draft sequence of the rice genome (Oryza sativa L. ssp. Indica). *Science* **296**, 79–91.

Zerulla, W., Barth, T., Dressel, J., Erhadt, K., Horchler von Locquenghein, K., Pasda, G., Raedle, M., Wissemeier, A.H. (2001). 3,4-dimethylpyrazole phosphate (DMPP): A new nitrification inhibitor for agriculture and horticulture. *Biology and Fertility of Soils* **34**, 79–84.

# Transfer of radionuclides in animal production systems

Gabriele Voigt[a], Brenda J. Howard[b], Nicholas A. Beresford[b]

[a]*International Atomic Energy Agency, Director Agency's Laboratories Seibersdorf, Wagramer Strasse 5, PO Box 200, A-1400 Vienna, Austria*
[b]*Natural Environment Research Council, Centre for Ecology and Hydrology, Lancaster Environment Centre, Library Avenue, Bailrigg, Lancaster, LA1 4AP*

## 1. Introduction

The definition of an ecosystem in modern times was introduced by the British plant ecologist Tansley (1935) and includes not only the community of organisms in an environment but also the whole complex of physical and chemical factors around them. Ecology therefore involves the study of organisms and their abiotic surroundings and concerns the movement of energy and materials through communities (see section 3 of Chapter 5). This concept is applicable to any scale and can be adapted to a drop of water containing protozoa, a lake or a clearly defined terrestrial ecosystem. As an extreme, the term can be even used for the whole world representing one ecosystem (Lovelock, 1979). Most ecosystems cannot be regarded as having definite borders but a minimum of three distinct components may be distinguished: biomass, energy flow and availability of chemicals or minerals. Understanding the flows of materials and energy within an ecosystem can help in quantifying the fluxes of radioactive substances, which lead to irradiation of humans and other species, within that ecosystem if it becomes contaminated due to an accidental or other release to the environment.

Irradiation of humans can occur via external and internal exposure to radionuclides. External exposure is due to irradiation by the passage of a cloud during an accident, irradiation from contaminated surfaces and, in medical investigations, due to X-ray or radionuclide therapies. Internal exposures occur from natural radionuclides and artificial radionuclides released into the environment either during normal operation of nuclear installations, accidental releases or as a consequence of nuclear weapons detonations. Radionuclides can enter the body either via wounds, through the skin, inhalation or ingestion of contaminated food and liquids. Due to the relatively high contribution of contaminated foodstuffs to the total ingestion dose in many circumstances, understanding, description and modelling of radionuclide transfer in foodchains is one of the key topics in radiation protection and radioecology.

RADIOACTIVITY IN THE ENVIRONMENT
VOLUME 10 ISSN 1569-4860/DOI: 10.1016/S1569-4860(06)10004-2

Animal products, especially milk and meat, are important components of the human diet and therefore play an important role in determining internal radiation doses. In the past, many experiments concerning the transfer of radionuclides to farm animals have been performed. Most of these have considered the radioisotopes of Cs, Sr, I and, to a lesser extent, H. Information published from these studies is utilised in radioecological models such as PATHWAY (Whicker and Kirchner, 1987), FARMLAND (Brown and Simmonds, 1995) and ECOSYS-87 (Müller and Pröhl, 1993), and assessed and summarised within key data review documents such as IAEA (1994).

As a consequence of the Chernobyl accident in 1986, experimental research on the transfer to farm animals increased, addressing mainly the transfer of radiocaesium and radioiodine under *realistic* conditions (Voigt, 1993; Voigt et al., 1993, 1996; Howard and Beresford, 2001; Howard et al., 2001). Considerable efforts were made to understand the factors affecting the transfer to ruminants (Howard and Beresford, 2001), countermeasure development (Voigt et al., 1993; Howard et al., 2001) and also spatial and temporal variations in transfer (Voigt and Kiefer, 2000; Gillett et al., 2001). Soon after the Chernobyl accident it became obvious that the impact of products contaminated with radiocaesium from animals in natural and semi-natural environments on human radiation exposure is of as much, if not more, importance as that from intensively reared animals, especially in the long-term after an accident (Howard et al., 1996).

In response, countermeasures to reduce especially radiocaesium, radioiodine and radiostrontium in animal products have been studied, tested and implemented, including those for animals living in semi-natural environments (Hove et al., 1990; Howard and Desmet, 1993; Howard et al., 2001). However, to apply countermeasures most effectively, the behaviour and transfer of radionuclides in animals need to be understood and properly modelled. This chapter therefore gives an overview on (i) the most important contamination pathways, (ii) fluxes and sinks of radionuclides, (iii) how transfer can be described and quantified and (iv) some potential countermeasures.

## 2. Animal production systems

Husbandry of animals is highly location-specific and dependent on the particular conditions of climate, tradition and developmental stage of the country being considered. In general, two production systems can be distinguished: private and commercial production. In private production, the yield is driven by personal (family) consumption and to a much lower extent by commercial considerations (selling of products in local markets) whereas in commercial systems yield is an important factor. The type of ecosystem will influence management practices: for natural and semi-natural environments, free range grazing by animals often occurs and there is minimal control by humans whereas in intensive agricultural systems feeding, housing and production are highly controlled and influenced by man.

Production systems within Europe vary widely from location to location and it is difficult to provide generic information since climatic conditions, cultural maintenance practices and management intensity of agricultural production differ spatially over the European continent. Generally, the following agricultural systems occur.

- Intensive agriculture (improved pasture).

Table 1
Typical diets for dairy cattle (650 kg live weight yielding 30 kg $d^{-1}$ milk) in southern and northern England (Nisbet et al., 1999) and areas of Germany (Kirchgeßner, 1987)

| Feed | Intake kg $d^{-1}$ (dm) | | | | | | | |
|---|---|---|---|---|---|---|---|---|
| | Summer 15.3–31.10 | | | | Winter 1.12–14.3 | | | |
| | n. E. | s. E. | G. a. | G. p. | n. E. | s. E. | G. a. | G. p. |
| Grass | 16 | 16 | 7 | 13 | – | – | – | – |
| Grass silage | – | | – | – | 13 | 5 | 7 | 13 |
| Maize silage | – | – | 6 | – | – | 10 | 6 | – |
| Fodder beet/kale | – | – | – | – | – | – | – | – |
| Compound feed | 8 | 8 | 9 | 9 | 11 | 5 | 9 | 9 |
| Straights | | – | – | – | 1 | 4 | | |
| Total intake | 24 | 24 | 22 | 22 | 24 | 25 | 22 | 22 |

n. E.: in northern England. s. E.: in southern England. G. a.: in Germany areas with arable land. G. p.: in Germany areas without arable land (mainly pasture).

- Extensive agriculture (sheep, goats, reindeer or cattle grazing upland and other unimproved areas; private production especially within the former Soviet Union).
- Forest or natural ecosystem (predominantly game species, including large mammals and birds).

The composition of animal feed depends on the season and varies spatially. As an example, Table 1 shows seasonal changes of the diet for dairy cattle in southern and northern England in comparison with Germany. In southern England, maize silage is only fed in autumn and winter combined with a reduction of compound feed. In northern England, grass silage predominates as the winter forage (climatic conditions not being suitable for maize production). For Germany, land cover reflects differences in the feed composition. In areas with arable land, maize silage is included in the ration for dairy cows during the whole year; in areas without arable land fresh grass is the dominant feed in the summer and grass silage in winter.

After large-scale contamination of the environment, the provenance of feedstuffs for different animal species is important. Information about import and export of feed is needed when considering potential sources of uncontaminated or contaminated feed, and is relevant for the application of countermeasures. Imported cereals can be used to provide uncontaminated feed. Figure 1 shows quantities of imported and exported wheat for different European countries (Fefac—European Federation of Feed Industries, 1996). Table 2 and Figure 2 show the utilisation of such feed by animal species in each country. These data provide examples though it should be noted that values may change significantly between years. Import/export data and the use of feeding compounds for different animal species in European Union (EU) countries allow the identification of countries with a high export rate for potentially contaminated or uncontaminated feedstuffs. Countries with a high import rate will be affected by the availability of uncontaminated feed in the event of a major incident such as a nuclear accident. The self-sufficiency of individual countries with respect to feedstuff production is shown in Table 3. Countries with a high export rate for cereals include France, whereas countries with high import rates include The Netherlands, Portugal and Belgium.

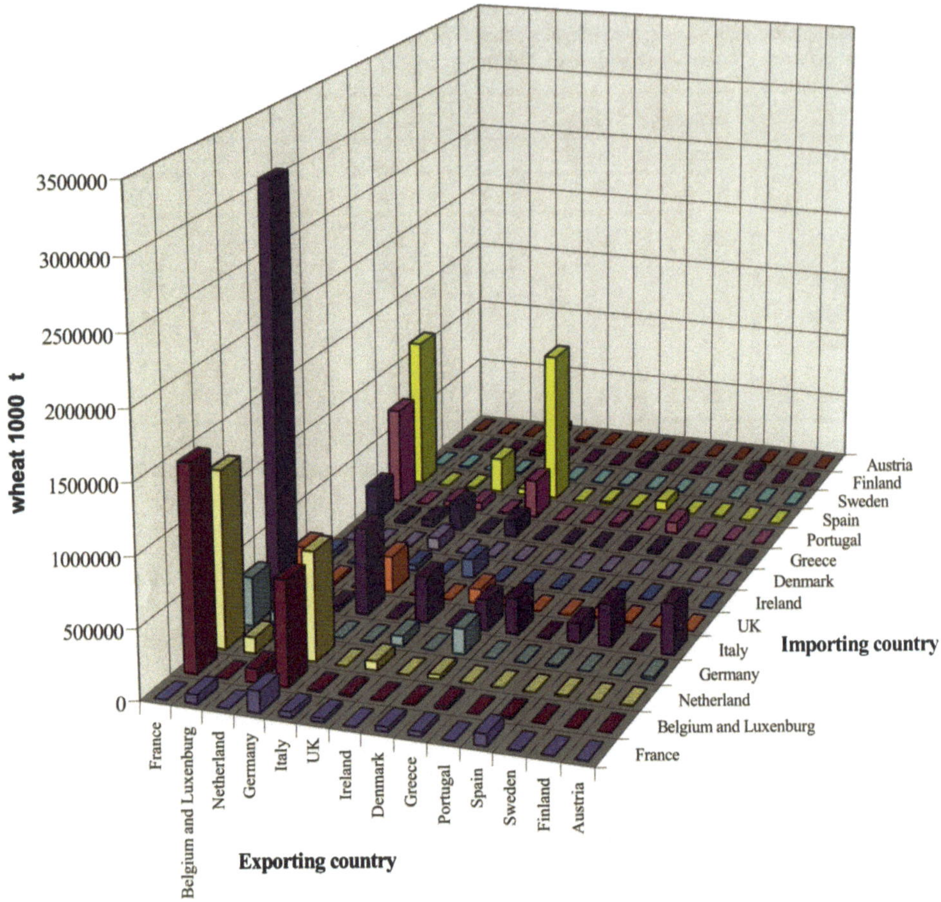


Fig. 1. Import and export data between different EU countries for wheat in 1000 t (Fefac—European Federation of Feed Industries, 1996).

Production systems (including animal production) in former Soviet Union (FSU) countries differ considerably from those in Western Europe and can be subdivided into three major groups, as follows.

(i) Intensive agricultural systems (state owned collective farms, or 'kolkhoz', of the communist period or their successors), which mainly consist of large-scale farming based on collective farms routinely using land rotation combined with ploughing and fertilisation to improve productivity.

(ii) Private farming, which is normally associated with family units. These small-scale farms mostly use animal manure to improve yields, particularly in the local vegetable (kitchen) gardens. In addition natural ecosystems, such as clearings in forests and unimproved pasture, are used to provide some of the winter fodder and as grazing land for animals.

Table 2
Industrial utilisation of compound feed (1000 t) for animal species in the European Union (EU) in 1996

| Country | Cattle | Pigs | Poultry | Others | Total |
|---|---|---|---|---|---|
| Belgium | 1197 | 3454 | 1062 | 149 | 5862 |
| Denmark | 1522 | 3429 | 651 | 144 | 5746 |
| Germany | 8020 | 6244 | 4136 | 926 | 19326 |
| Spain | 3210 | 6530 | 4100 | 1375 | 15215 |
| France | 4709 | 6671 | 9144 | 1540 | 22064 |
| Ireland | 1720 | 685 | 468 | 299 | 3172 |
| Italy | 3900 | 2300 | 4400 | 1100 | 11700 |
| The Netherlands | 4634 | 7230 | 3417 | 800 | 16081 |
| Austria | 183 | 93 | 460 | 156 | 892 |
| Portugal | 1090 | 1300 | 1330 | 240 | 3960 |
| Finland | 472 | 322 | 222 | 158 | 1174 |
| Sweden | 1210 | 700 | 500 | 150 | 2560 |
| UK | 4585 | 2534 | 4089 | 1205 | 12413 |
| Total | 36452 | 41492 | 33979 | 8242 | 120165 |

Greece and Luxembourg not included.

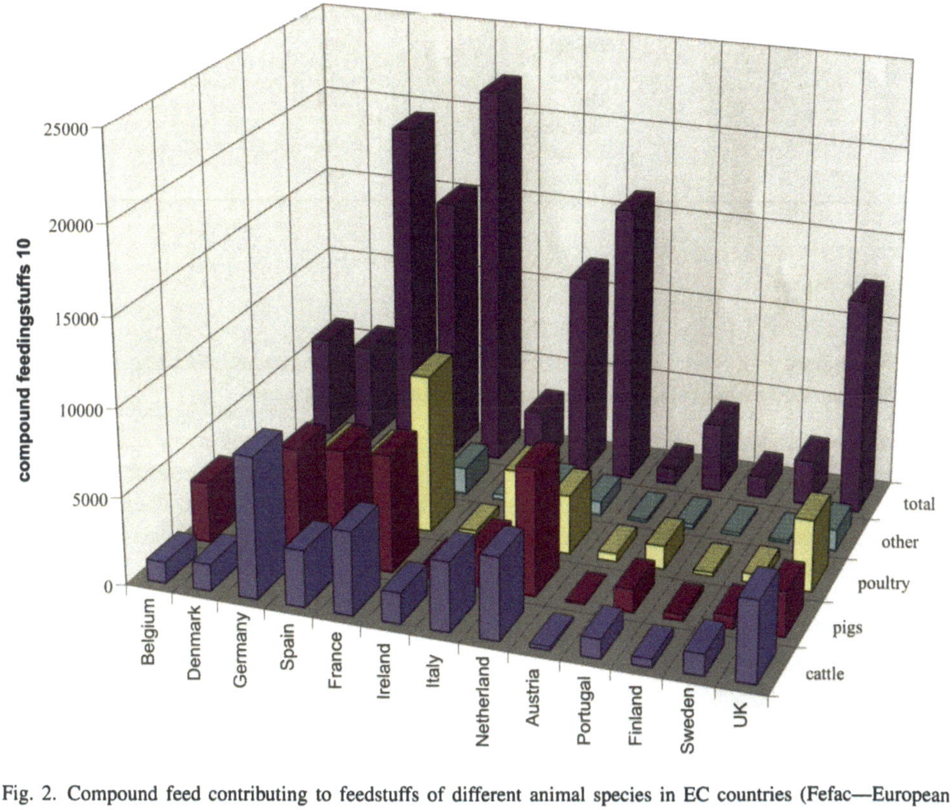


Fig. 2. Compound feed contributing to feedstuffs of different animal species in EC countries (Fefac—European Federation of Feed Industries, 1996).

Table 3
Percentage self-sufficiency in total cereals in different European Union countries (source: Eurostat http://epp.eurostat.cec.eu.int/)

| Country | Belgium | Denmark | Germany | Greece | Spain | France | Ireland |
|---|---|---|---|---|---|---|---|
| Self sufficiency % | 45 | 127 | 110 | 87 | 59 | 180 | 79 |
| Country | Italy | Netherlands | Austria | Portugal | Finland | Sweden | UK |
| Self sufficiency % | 78 | 25 | 105 | 35 | 109 | 112 | 111 |

Table 4
Example of production data from two production systems in 1994/95 in Russia and Ukraine (after Mehli and Strand, 1988)

| Food producing system | Food product | Quantity produced ($t\ y^{-1}$) | |
|---|---|---|---|
| | | Russian site | Ukrainian site |
| Collective[a] | Milk | 1241 | 1300 |
| | Meat | 155 | 350 |
| | Potatoes | | 813 |
| | Grain | | 1673 |
| | Sub-total | 1396 | 4136 |
| Private[b] | Milk | 22 | 543 |
| | Meat | 34 | 98 |
| | Vegetables | 72 | 113 |
| | Potatoes | 235 | 528 |
| | Sub-total | 363 | 1282 |

[a]Large scale farming.
[b]Small scale farming.

(iii) Food gathering, which includes collection of natural food products such as mushrooms, berries, freshwater fish and game from semi-natural ecosystems.

Collective farms produce food through intensive management of their major soil and animal resources. Typically, 2–5 villages are located within the area of a collective farm. Often a plot of land is allocated to village families to grow foodstuffs for their own use. Therefore, within the village, a parallel subsistence farming economy operates; income is partly based on participation in collective farm/large-scale farming and partly on exchange and sale of homegrown vegetables and animal products. Traditionally, private farms have one or two cows, and milk is used for self-consumption as well as food for animals or sold to (or bartered with) neighbours. Often the collective farm allocates private cattle owners some winter forage (as part payment for their labour) which is supplemented by harvesting from forests or scrubland. The grazing regime of privately owned cattle varies but, to some extent, relies on the utilisation of marginal land that is not used by the collective. This includes the use of riverbanks, natural pastures and clearings in the forest. Sometimes privately owned cows graze on fields, which are allocated to the farmers by the collective farm. The manure produced by the cattle is normally the only additional source of plant nutrients used for the private plots, although artificial fertilisers are occasionally available depending on the local economic situation. Table 4 provides comparative production data for collective and private farming systems in Russia and Ukraine.

In heavily contaminated areas, the extensive use of countermeasures after the Chernobyl accident (especially ploughing and fertilising) made the ecological distinction between private and collective systems less well defined. Additionally, the breakdown of the FSU resulted in a lot of changes in land management and economic systems which make differences in the farming systems less clear.

## 3. Transfer of radionuclides to animals

Contamination of food products from both intensively farmed, free ranging and wild animals can represent a major route of internal radiation exposure for humans from both routine and accidental releases of radioactivity into the environment. Predictive models to estimate the transfer of a given radionuclide to animal food products make use of published data on radionuclide behaviour in the animal, such as distribution and retention in different organs or tissues and subsequent excretion routes. Depending on the availability of experimental data, compartment models comparable to the models used in ICRP[1] documents for dose estimates to humans (e.g. ICRP, 1979) have been developed to describe empirical transfer rates between different key compartments once the radionuclide has entered the body. In simpler (animal) models, activity concentrations in an animal product of concern, and consequently the activity intake by humans, can be calculated based on the known activity intake with time and the activity concentration within the product considered.

Animals can be contaminated by three different routes: through the skin or by wound, by inhalation, and, most importantly, by ingestion of contaminated feed. Uptake through the skin or wounds is usually not an important route of contamination, although skin lesions can provide a direct route of entry for radionuclides into an animal's circulation system, especially for those radionuclides for which gastrointestinal absorption is low. Because of the importance of radioiodine, radiostrontium and radiocaesium for the internal human dose and the fact that most of the available information is for these radionuclides, this chapter mainly (but not exclusively) focuses on these radionuclides.

### 3.1. *Inhalation*

Theoretically, inhalation of radionuclides can be a potential contamination route since lung surfaces act as a site of gaseous exchange and are permeable to a wide range of elements. Radionuclides may be inhaled in different forms, including gaseous compounds, aerosols and particles, which will all have different physico-chemical characteristics and therefore different transfer, absorption or residence times in the lungs. The size of particulate-associated radionuclides is particularly important: large particles (diameter 5 to 30 $\mu$m) are deposited in the upper parts of the respiratory tract, whereas small particles (diameter 1 $\mu$m) can reach the lower parts of the pulmonary system, depositing in the alveoli and providing a concentrated local dose to the surrounding cells and tissue.

[1] International Commission on Radiological Protection.

During respiration, a fraction of the aerosol particles present deposit on the epithelium of the respiratory tract. Particles are then subject to interactions with fluids, cells and tissues of the respiratory tract the retention of which can be described by a time dependent function:

$$R(t) + f(t, R(0)), m(t), N(t) \ldots), \qquad (1)$$

where: $R(t)$ = fraction of particles retained at time $t$, $m$ = mass of particles, $n$ = number of particles, $t$ = time.

Particle clearance is usually defined as the fraction of particles eliminated in a given time. In general, clearance is defined for different anatomical sites such as the alveolar region (peripheral lung), conducting airways of lungs and extra thoracic airways. Clearance involves displacing particles from their initial deposition site to protect the lung's primary function of gas exchange of oxygen and carbon dioxide. Different clearance mechanisms operate at various locations in the respiratory tract and depend on the physical, chemical and biological properties of the deposited foreign particulate matter. These can be summarised as follows.

*Intra-thoracic airways*

- Particle clearance by transport and subsequent transfer to the gastro-intestinal tract. Deposited particles are returned to the throat by aereal transport and may then pass along the digestive tract, where they are subject to gastro-intestinal uptake processes.
- Mucociliary clearance by ciliary and mucus transport in the airway channels from the terminal bronchioli towards the trachea and vice versa. The mean linear mucous velocities after inhalation are about $5.3 \pm 1.3$ mm min$^{-1}$. Scheuch et al. (1996) observed complete clearance 6 h after inhalation of monodisperse gamma emitting particles with a geometric diameter larger than 6.5 $\mu$m. For aerosols of 3 $\mu$m geometric diameter only 5% were cleared rapidly (i.e. within 10 h after application), followed by half-lives of 5 and 30 days or even 100 days (alveolar region) depending on where particles were finally deposited. A clear correlation between slowly cleared fractions and the geometric particle diameter occurred: the greater the diameter the lower the slow fraction which was cleared (Scheuch et al., 1996).
- Cough clearance by mucus propelled towards the larynx by fast exhalation.

*Clearance from peripheral lungs*

- Long term particle clearance by transport towards the ciliated airways, to the hilar lymph nodes, to interstitial sites, to subpleural spaces, retention and relocation on the alveolar epithelium, dissolution into the epithelium lining fluid, intracellular dissolution by alveolar macrophages and other phagocytic cells.

The ability of radionuclides to pass through the pulmonary membranes varies considerably; despite the low transfer rates for actinides (e.g. plutonium) they are often more readily absorbed via the lungs than via the gastrointestinal (GI) tract. Various experimental data, in particular on the transfer mechanisms in the alveoli of dogs, have indicated how the transfer and clearance of selected radionuclides are controlled (Kreyling et al. 1986, 1988, 1991, 1998).

Due to their limited solubility, noble gases may generally be neglected as a source of contamination of animals. However, gaseous iodine is readily absorbed. Observations soon after the

Chernobyl accident have shown that cows housed indoors during deposition and fed on uncontaminated feed exhibited significant radioiodine inhalation (G. Voigt, unpublished data). Also, inhalation of resuspended material with dust can be a route of radioactivity intake to animals in areas with strong and frequent winds combined with high soil erosion. Although inhalation is often not a major contamination route of agricultural animals it should not be ignored.

### 3.2. *Ingestion and gastro-intestinal absorption*

The most important transfer pathway to animals is the ingestion of contaminated feed, soil and drinking water. Intake via drinking water generally contributes little to total radionuclide intake, except soon after an accident for volatile radionuclides or radionuclides attached to respirable particles. Radionuclide intake via soil can be significant, but the availability for gut absorption of soil-associated radionuclides (especially radiocaesium) is generally low, depending on soil properties. Hence, it is mainly the ingestion of contaminated feed, and the processes influencing daily intake and absorption, which determine the radionuclide content in animals.

The extent of internal absorption from the GI tract is generally the most important factor for many radionuclides. Different values of absorption in the GI tract are recommended for monogastric (e.g. pigs) and ruminant species (e.g. cattle or sheep) due to the different nature and function of their stomachs. Ruminants have four stomachs: the first two stomachs (rumen and reticulum) are where fermentation of chewed vegetation occurs as the gut microflora digest the vegetation and some absorption of breakdown products takes place. The anaerobic, reducing environment of the rumen can lead to changes in radionuclide speciation and therefore may change bioavailability. In the case of radiocobalt, rumen microflora synthesise vitamin $B_{12}$ from inorganic cobalt and hence radiocobalt ingested in an inorganic form is absorbed as vitamin $B_{12}$ (which subsequently affects the distribution and residence times in tissues) (Voigt et al., 1988). Subsequently, the fermented rumen fluid passes into the omasum and abomasum and is subjected to enzymatic digestion, which is similar to that in monogastric animals. Fermentation in non-ruminants occurs in the caecum in mammals and the crop of some birds, but its importance is lower than rumen fermentation.

Absorption through the GI tract is very dependent on the chemical form of the radionuclide and the biological matrix with which it is associated. Additionally, the age of the animal and its nutritional and physiological status influence the extent of absorption. In general, neonate animals absorb radionuclides more rapidly and more efficiently due to the immature composition of the gut wall (Kistner et al., 1987); in ruminants absorption of radiocaesium (from suckled milk) is approximately 100% prior to the opening of the oesophageal grove, thereafter it is approximately 80% (Mayes et al., 1992). The nutritional status (fasting or non-fasting conditions) and especially the supply of homologous or stable elements are important factors influencing the uptake of homeostatically controlled elements such as radioiodine or radiostrontium. In general radioecological models use average or most likely values for absorption as default values. However, most of these values have been obtained experimentally by feeding ionic radioisotopes or vegetation contaminated by fallout as a consequence of accidental releases (especially since the Chernobyl accident) or on the basis of concentrations of stable elements. Such data may not adequately take into account differences in bioavailability.

Table 5
True absorption coefficients for ruminants for selected radionuclides and sources (summarised from Beresford et al. (2000b), with additions for Ru, Ag and Ce)

| Radio-nuclide | Source | True absorption | Species | Reference |
|---|---|---|---|---|
| Ag | $AgNO_3$ | $5.6 \times 10^{-2}$ | Sheep | Beresford et al. (1998a) |
| Ce | $CeCl_3$ | $1.9 \times 10^{-4}$ | Sheep | Beresford et al. (1998a) |
| Cs | Vegetation | 0.8–0.9 | Sheep | Beresford et al. (1995) |
| | Chernobyl fallout | 0.7 | Sheep | Mayes et al. (1996) |
| | Chernobyl exclusion zone pasture | 0.2 | Cows | Beresford et al. (2000a, 2000b) |
| | Marine discharges associated with silt | 0.1 | Sheep | Beresford et al. (1992) |
| | Milk | 1.0 | Lambs | Mayes et al. (1992) |
| I[a] | Several sources | 1.0 | Cows | Vandecasteele et al. (2000) |
| | NaI | 1.0 | Goat/sheep | Beresford et al. (1995) |
| Sr[b] | Silage, hay and $SrCl_2$ | 0.1 | Sheep | Vandecasteele and van Hees (1995) |
| | Associated with clinoptilolite | 0.2 | Goat | Vandecasteele and van Hees (1995) |
| Ru | $RuCl_3$ | $3.4 \times 10^{-3}$ | Sheep | Beresford et al. (1998a) |
| Pu | Soil associated | $<7 \times 10^{-5}$ | Sheep | Beresford et al. (2000a, 2000b) |
| | Vegetation | $1.2 \times 10^{-5}$ | Sheep | Ham et al. (1989) |

[a]For different stable iodine intakes.

[b]For animals receiving an adequate calcium diet.

Absorption within radioecological studies was traditionally determined as the difference in dietary intake and faecal excretion (referred to as apparent absorption). For some radionuclides this approach might provide a reliable estimate of absorption, but it is too insensitive to measure absorption from sources with small differences in bioavailability. This is because the calculation does not take into account recycling (excretion of absorbed radioactivity from the blood into faeces and re-absorption along the GI tract) of some radionuclides. For radionuclides such as $^{99}$Tc, $^{131}$I or $^{129}$I, endogenous secretion into the GI tract from the circulatory system (e.g. via saliva or bile) is a significant route of excretion. Furthermore, endogenous excretion of radiocaesium in ruminants is typically 20–25% of that absorbed (Mayes et al., 1996).

Mayes et al. (1996) introduced the *true absorption coefficient* ($A_t$), which is defined as the fraction of ingested radionuclide transferred across the gut wall. $A_t$ is best determined by using a dual isotope technique (one isotope ingested with diet, one directly injected into the jugular vein) and thus is especially suitable for studies of radionuclides which have more than one radioactive isotope (e.g. $^{134}$Cs and $^{137}$Cs). The technique allows the direct estimation of endogenous faecal excretion or the rate of uptake from the diet into the plasma by comparing isotopic ratios. The true absorption coefficient has been measured for a number of different radionuclides and sources in a range of ruminant farm animals. An overview of these values for the most important radionuclides and their different sources is given by Beresford et al. (2000b), and some examples are provided in Table 5.

As already mentioned, an important factor for absorption is *bioavailability*. In the past this has been determined by sequential extraction where easily (water or weak acid) dissolved radionuclides have been considered to be bioavailable. However, a number of authors have reported *in-vitro* extraction procedures, which are suggested as methods of estimating the potential bioavailability of radionuclides for absorption (Barth, 1977; Salbu et al., 1992; Singleton et al., 1992; Beresford et al., 1995; Cooke et al., 1997). In their review of these methods, Beresford et al. (2000b) concluded the following.

- Reliable techniques have been developed for radiocaesium which have been calibrated against measurements in animals.
- *In-vitro* extraction procedures for radiostrontium imply differences in bioavailability which are not observed in animals.
- Limited available data for Pu suggest that *in-vitro* incubations provide a measure of relative bioavailabilities.

### 3.3. *Distribution and retention of radionuclides in animal tissues*

Once a radionuclide has entered the blood it will be distributed within organs and the body tissues and excreted with a biological half-life which is characteristic for each compartment. The biological half-life depends on the physico-chemical nature of the radionuclide and the metabolic features of the radionuclide in question. To allow for radionuclide distribution between tissues, compartment models representing the different organs or tissues have been developed to describe the pathways and excretion routes of radionuclides.

Generalised radionuclide compartment models have been described by Coughtrey and Thorne (1983) and also in the ICRP reports used to derive ALI (Annual Limits of Intake) and DAC (Derived Air Concentrations) values for humans. For radioiodine, the model described by Crout and Voigt (1996) is used to predict the transfer of radioiodine to cow's milk, and a similar iodine model exists for goat (Crout et al., 2000). Models have also been published for $^{14}C$, $^{3}H$ and $^{35}S$ in sheep tissues (Crout et al., 1998).

## 4. Quantification of the transfer of radionuclides in animal products

### 4.1. *Generic approaches to quantifying transfer*

The advantages of common models to describe the behaviour of radionuclides in different animals was recognised by Stara et al. (1971) who demonstrated relationships between the long component of biological half-life of radiocaesium and live-weight. Unfortunately, such approaches did not progress beyond this analysis for many years. More recently, Nalezinski et al. (1996) derived expressions between live-weight and radiocaesium transfer coefficients from a (limited) compilation of literature data. Separate expressions were derived for ruminants and monogastrics (comprising hens and pigs). Similarly, models for radioisotopes of a number of heavy metals, parameterised against data from bespoke studies using sheep, have been capable of predicting activity concentrations in dairy cattle by adjusting the model rate coefficients by the ratio of the metabolic live-weights of sheep and cattle (Beresford et al., 1997, 1998d). However, Beresford et al. (1998a) found that variation in radiocaesium transfer to muscle of individual sheep (including pregnant and lactating adults, and growing lambs) could not be explained by differences in live-weight. This variation was well explained by relationships derived on the basis of dry-matter intake.

There have been developments of more generic models for a number of other radionuclides. The derivation of a relationship to predict the transfer of radiostrontium to the milk of dairy animals based upon dietary calcium intake and the concentration of calcium in milk will be discussed later (Howard et al., 1997; Beresford et al., 1998b). Crout et al. (1998) suggested a model describing the transfer of $^{3}$H, $^{14}$C and $^{35}$S to sheep tissues based upon a simplified representation of processes such as protein synthesis, degradation and respiration. As a consequence of the approach, the models for the individual radionuclides shared many common parameter values. Whilst the models described well the experimental data on which they were parameterised, their applicability is currently limited to animals under similar management and physiological conditions. However, models for these radionuclides based upon a metabolic understanding have great potential as they could predict the influence of dietary intake, physiological status and form of radionuclide within the diet. The concept has recently been applied to the prediction of transfer coefficients for $^{3}$H in food producing animals by Galeriu et al. (2001).

These developments in modelling represent potentially useful methods of predicting the transfer of radionuclides to animal derived food products. However, the work is now being conducted largely as desk studies with little experimental validation. For instance, there are no experimental data with which to compare the predictions of the model of Galeriu et al. (2001) for $^{3}$H concentrations within poultry. Similarly, the relationship presented by Beresford et al. (1998b) to predict the transfer of radiostrontium to the milk of dairy animals could be adapted to predict Sr transfer to other animal derived food products—but there are currently limited data with which to compare/develop this approach.

### 4.2. *The concept of aggregated transfer factors*

One of the simpler methods to describe the transfer of radionuclides to animal products is the aggregated transfer factor $T_{ag}$. This factor describes the activity concentration in the product related to the deposited activity ($\mathrm{Bq\,kg^{-1}/Bq\,m^{-2}}$). This measure was introduced to allow a rapid prediction of the expected activity concentration in a given animal (and plant) product used for human consumption. It is quite obvious that this factor changes with time according to the effective ecological half-lives of radionuclides in different ecosystems. For radiocaesium the following steady state model based on experimental data in the Chernobyl affected areas has been proposed by Voigt et al. (1996).

$$T_{ag}(t) = (A\mathrm{e}^{-k_1 t} + B\mathrm{e}^{-k_2 t}) \times T_{ag}(0), \tag{2}$$

where: $T_{ag}(t)$ = aggregated transfer factor at time = $t$ ($\mathrm{m^2\,kg^{-1}}$), $T_{ag}(0)$ = aggregated transfer factor at time = 0 (1986) ($\mathrm{m^2\,kg^{-1}}$), $A$ = empirical parameter (fitted to 0.814), $B$ = empirical parameter ($1-A = 0.186$), $k_1$ = empirical first order decay constant fast pool ($0.6935\,\mathrm{y^{-1}}$), $k_2$ = empirical first order decay constant slow pool ($0.06935\,\mathrm{y^{-1}}$).

The $T_{ag}$ concept is not extensively used in agricultural production systems (although it is in common usage in the FSU) and should not be used in the first few weeks after a deposition event when interception and surface contamination are important. However, it is valuable

Table 6
Examples of $T_{ag}$ values for $^{137}$Cs in animal products ($m^2 kg^{-1}$) used in the RESTORE-EDSS (Voigt et al., 1996)

| Animal product | Peat | Sand | Loam | Clay |
|---|---|---|---|---|
| Milk | 0.6 | 0.2 | 0.1 | 0.05 |
| Meat | 0.444 | 0.15 | 0.075 | 0.035 |
| Wild boar meat | 6 | 2 | 1 | 0.5 |
| Roe deer meat | 18 | 6 | 3 | 1.5 |

in semi-natural environments where ecological half-lives can be very long and it is difficult to conduct detailed transfer studies to parameterise appropriate models. In Table 6 some $T_{ag}$ values for radiocaesium in different animal products are given.

## 4.3. *The concept of transfer coefficients*

Ward et al. (1965) initially defined the transfer coefficient $F_m$ or $F_f$, expressed in $d\,kg^{-1}$, as the activity concentration in the product related to the daily activity intake under equilibrium conditions or at the time of slaughter. The transfer coefficient was rapidly adopted as the basis for quantifying transfer to both milk ($F_m$) and meat ($F_f$) for all radionuclides and single values were recommended for many radionuclide/animal product combinations (e.g. Ng et al., 1977, 1982; Coughtrey, 1990; IAEA, 1994).

The suitability and validity of this parameter, so often used in radioecological models, were discussed later by the same authors (Ward and Johnson, 1986, 1989). One of the major problems is the need for equilibrium conditions, which are rarely achieved due to the restricted lifetime of animals. Therefore, for a variety of radionuclides this might lead to an underestimate, especially of $F_f$ values. A second problem is that the physical and chemical form, source and stable isotope or analogue intake of the radionuclide will influence the transfer coefficient. Therefore, those values determined in laboratory studies with ionic radionuclides often do not reflect reality. There are a variety of other animal specific influences on transfer such as the physiological status of the animal (lactating, non-lactating), age, species and breed. Variations of the transfer coefficients therefore can vary by 2 to 3 orders of magnitude.

When the time sequence of radionuclide activity concentrations in animal tissues is obtained by continuous feeding experiments, the resulting curves can be described best by exponential equations. For example, the time dependent activity concentration of Cs in cow's milk can be described by a two compartment model with a short and long biological half-life (Voigt et al., 1989).

$$c_m(T) = I f_m \int_0^T c_f(t) x \left\{ a\lambda_1 e^{-(\lambda_{1\lambda}+\lambda_r)(T-t)} + (1-a)\lambda_2 e^{-(\lambda_2+\lambda_r)(T-t)} \right\} dt, \tag{3}$$

where: $c_m(T)$ = activity concentration in milk ($Bq\,kg^{-1}$) at time $T$, $c_f(t)$ = activity concentration in feed ($Bq\,kg^{-1}$) at time $t$, $I$ = daily intake of feed ($kg\,d^{-1}$), $f_m$ = equilibrium transfer coefficient ($d\,kg^{-1}$), $\lambda_1$, $\lambda_2$ = biological transfer rates ($d^{-1}$), $\lambda_r$ = radioactive decay rate ($d^{-1}$), $a$ = fraction of fast excretion.

The recommendation of single transfer coefficient values for radionuclides which are either themselves, or through a stable analogue, subject to homeostatic control has been shown to be inappropriate for radiostrontium (Howard et al., 1997), radioiodine (Crout et al., 2000; Vandecasteele et al., 2000) and tritium (Galeriu et al., 2001). However, relationships can be derived to enable transfer coefficients to be estimated for such radionuclides which allow for the effect of metabolic control or the effect of stable elements or analogues as countermeasures. The following description of such relationships is based on a summary of the overview on animal transfer by Howard et al. (2001).

Since the 1950s it has been known that the dietary intake rate of calcium influences the rate of radiostrontium transfer to milk (Comar et al., 1956; Comar, 1966). Currently, calcium is one of the few practical dietary supplements which could be used as a countermeasure to reduce radiostrontium transfer to milk. However, published reduction factors are of little value as the reduction achieved depends upon the basal dietary calcium intake of the animal. Howard et al. (1997) derived a relationship between the radiostrontium transfer coefficient to milk ($F_m$Sr;) and the dietary calcium intake of dairy cattle.

$$F_m\mathrm{Sr} = \frac{\mathrm{OR}_{\text{milk-diet}} \cdot [\mathrm{Ca}]_{\text{milk}}}{I_{\mathrm{Ca}}}, \tag{4}$$

where: $\mathrm{OR}_{\text{milk-diet}}$ = the observed ratio of the comparative transfers of strontium and calcium from the diet to milk; $[\mathrm{Ca}]_{\text{milk}}$ = the concentration of calcium in milk ($\mathrm{g\,kg^{-1}}$); $I_{\mathrm{Ca}}$ = the daily intake of Ca ($\mathrm{g\,d^{-1}}$).

On the basis of published evidence $\mathrm{OR}_{\text{milk-diet}}$ was assumed to be a constant of 0.11 for conditions of adequate calcium intake. Initially, this relationship could only be validated for cattle since there was insufficient data for other ruminant species. However, it has since been confirmed by data for dairy goats and sheep (Beresford et al., 1998b). The generic nature of the equation was validated using these data, as shown in Figure 3 in which the predicted lines represent typical values of Ca in cattle and sheep milk.

The relationship indicates that previously recommended $F_m$ values for dairy ruminants (e.g. IAEA, 1994) are inappropriate for agricultural feeding regimes in most countries since they correspond to low levels of calcium intake. The relationship suggested by Howard et al. (1997) for milk could be substituted with suitable values for meat or other animal derived food products.

Iodine is an important trace element required by the thyroid for hormone synthesis, and the metabolism and excretion of radioiodine is controlled by an individual's stable iodine status. Transfer of iodide to both the thyroid and mammary glands is an active process (Falconer, 1963), which is saturated at high iodide plasma concentrations. This results in a change in $F_m$ for radioiodine with varying stable iodine intake (Figure 4). The effect can be explained by considering the relative transfer of iodide to the thyroid and mammary glands. As the stable dietary iodine increases, the proportion of the daily radioiodine intake which is transferred to the thyroid declines because of the constant rate of uptake of iodide by the thyroid. Thus, a smaller fraction of the daily iodide intake (both stable iodine and radioiodine isotopes) is transferred to the thyroid and the proportion available for transfer to the mammary gland increases. At high stable iodine daily intakes, this effect is offset by saturation of the transfer from plasma to milk.

In the case of technetium, for which no stable isotopes are known, the specific radioisotope of the element affects rates of transfer to animal tissues. Ennis et al. (1988) showed that the

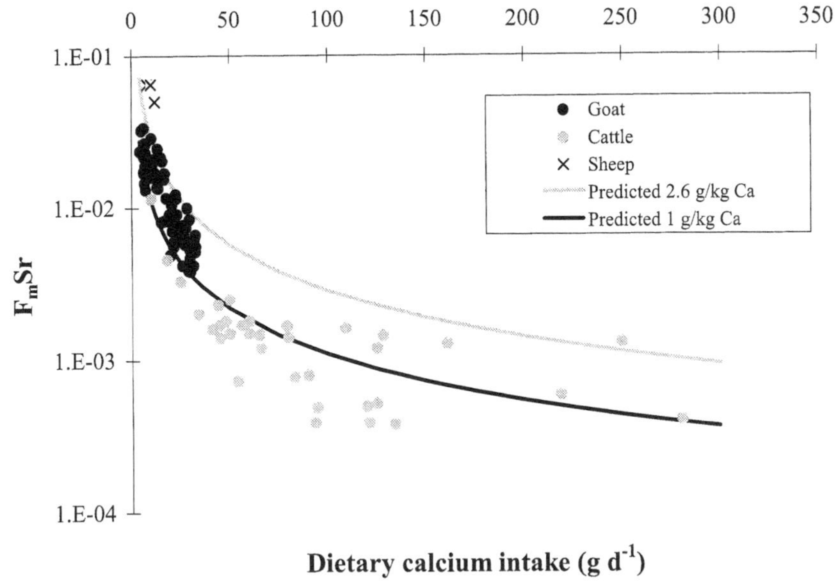


Fig. 3. Comparison between stable calcium intake and $F_m$ for strontium. The lines represent predicted values from equation (1) based upon calcium contents in milk of 1 g kg$^{-1}$ (typical for cattle) and 2.6 g kg$^{-1}$ (typical for sheep), respectively (Beresford et al., 1998b, with additional data from Beresford et al., 2000a).

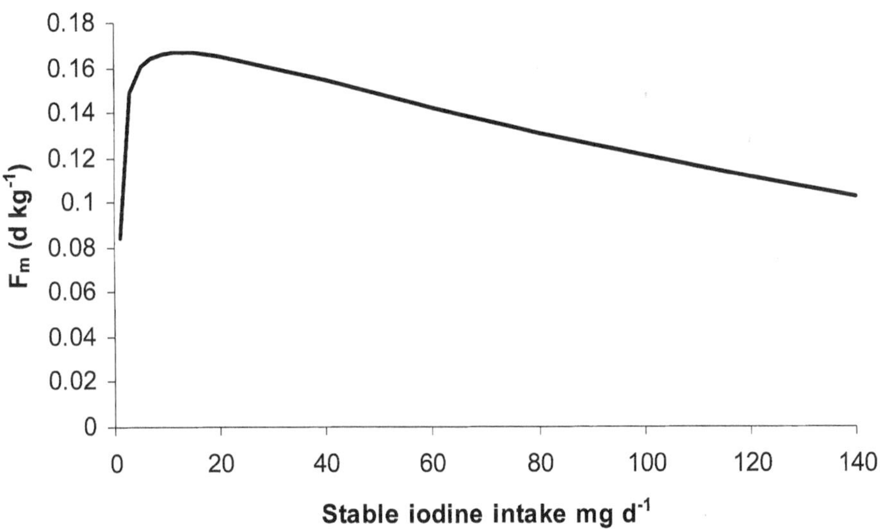


Fig. 4. Predicted variation in radioiodine transfer coefficient with varying stable iodine intakes in dairy goats; stable iodine intake rates include those representative of those prevailing under normal practice (1–20 mg d$^{-1}$) to the maximum tolerable advised intake for ruminants (50 mg kg$^{-1}$ dry matter), after Howard et al. (2001).

Table 7
Measured transfer coefficients for different isotopes of Tc to ruminant products

| Animal | Tc isotope | $F_m$ | References |
|---|---|---|---|
| Goat milk | $^{95m}Tc$ | $8.5 \times 10^{-4}$–$>1.3 \times 10^{-3}$ | Jones (1983, 1989)[a]; Ennis et al. (1988, 1989) |
| | $^{99m}Tc$ | $1.5 \times 10^{-4}$ | Ennis et al. (1988, 1989) |
| | $^{99}Tc$ | $1.1 \times 10^{-2}$ | Ennis et al. (1988, 1989) |
| Cow milk | $^{95m}Tc$ | $1.4 \times 10^{-4}$–$<1.7 \times 10^{-4}$ | Voigt et al. (1988); Johnson et al. (1988) |
| | $^{99m}Tc$ | $4.2 \times 10^{-6}$–$2.3 \times 10^{-5}$ | Johnson et al. (1988); Wiechen et al. (1983) |
| Beef | $^{99m}Tc$ | $7.3 \times 10^{-7}$ | Johnson et al. (1988) |
| Goat meat | $^{99m}Tc$ | $2.2 \times 10^{-4}$ | Johnson et al. (1988) |

[a]Estimated from data given in the paper.

transfer coefficient for Tc to goat milk is isotope dependent due to the greatly different specific activities of the Tc isotopes; $F_m$ increased with decreasing specific activity. The difference was thought to be due to the slower rate of reduction to an "unavailable" oxide form of Tc in the gut for $^{95m}Tc$ ($8.4 \times 10^{14}$ Bq g$^{-1}$ Tc) and more so for $^{99}Tc$ ($6.3 \times 10^{8}$ Bq g$^{-1}$ Tc) compared with $^{99m}Tc$ ($1.5 \times 10^{16}$ Bq g$^{-1}$ Tc) (Ennis et al., 1988). Jones (1989) showed that the reduction of bioavailability occurred in the forestomachs of ruminants and observed a higher absorption in monogastric pigs. In Table 7, reported transfer coefficient values for ruminants are shown for different Tc isotopes; only one measurement is for $^{99}Tc$, a beta emitter, compared with nine measurements for the more easily measured gamma-emitting isotopes ($^{95m}Tc$ and $^{99m}Tc$). The use of transfer coefficients determined for $^{95m}Tc$ and $^{99m}Tc$ is invalid for predicting the behaviour of $^{99}Tc$ as it would underestimate transfer. This is a notable deficiency since $^{99}Tc$ is the major long-lived environmental contaminant and is discharged by reprocessing plants such as Sellafield. The methodology used by Van Bruwaene et al. (1986) of labelling $^{99}Tc$ sources with $^{95m}Tc$ presents a mechanism whereby transfer values can be derived for a realistic isotopic form ($^{99}Tc$) using a gamma emitter.

Galeriu et al. (2001) have developed a model for predicting transfer coefficients of organically bound tritium and tritiated water to animal products based upon metabolic parameters. The authors suggest that concentration ratios are more robust than transfer coefficients for estimating transfer of $^{3}H$ to animal products because they are less influenced by metabolically varying parameters such as dry matter intake and production rate.

Many other radionuclides have to be considered in radiological assessments and there are relatively few data from which to derive suitable models which quantify their transfer to grazing animals. The few recent studies of these lesser-studied radioisotopes have generally obtained lower values for gut absorption and transfer coefficients than previously recommended values commonly used to predict transfer to animal derived food products (Beresford et al., 1998c). As an illustration, Table 8 presents recently determined transfer coefficient values for Ag, Ce, and Ru to lamb meat and compares them with previously recommended values. Confirmation of the validity of these new estimates and studies of other poorly quantified radionuclides may lead to reduced estimations of doses to populations as a consequence of nuclear power generation. Obviously, transfer rates may depend on the chemical form of radionu-

Table 8
Comparison of recently derived and previously recommended $F_f$ values for Ag, Ce and Ru to lamb meat

| Radionuclide | Recently derived $F_f$ (d kg$^{-1}$)[a] | Previously recommended $F_f$ (d kg$^{-1}$)[b] |
|---|---|---|
| Ag | $4.8 \times 10^{-4}$ | $2.7 \times 10^{-4}$–$2 \times 10^{-2}$ |
| Ce | $2.5 \times 10^{-4}$ | $2.0 \times 10^{-4}$–$1.0 \times 10^{-2}$ |
| Ru | $6.3 \times 10^{-4}$ | $1.0 \times 10^{-2}$–1.5 |

[a]From Beresford et al. (1998c).

[b]Collated from the recommendations for sheep meat of Coughtrey and Thorne (1983), Coughtrey (1990), IAEA (1994), Brown and Simmonds (1995).

clides and it is important to use parameter values which are derived using environmentally appropriate forms of each radionuclide.

### 4.4. *The concept of physiological animal models*

For the description of radionuclide transfer to any product the use of compartment models will also give information on the time-dependent transfer and excretion rates of radionuclides to the target tissues, especially those which are consumed. Compartment models can be adapted to take account of the physiological condition of the animal, explicitly representing factors, for instance, which affect milk yield (Figure 5). Such models should therefore provide better predictions of radionuclide transfer than equilibrium transfer coefficients. However, detailed knowledge of the metabolism and storage of individual radionuclides in different species is necessary for modelling and these data are not available for the majority of radionuclides.

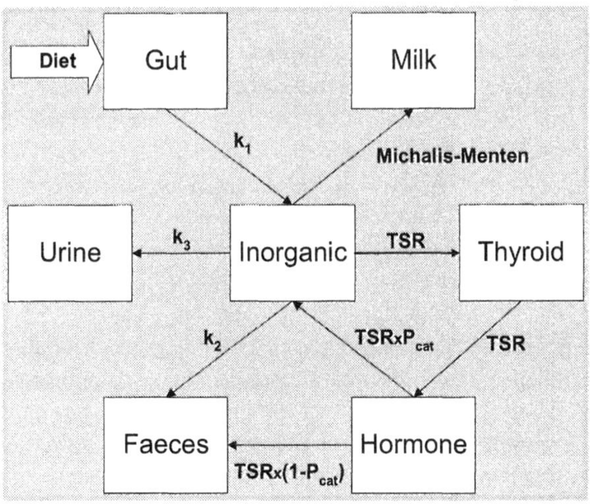


Fig. 5. Example of a physiological model.

## 5. Countermeasures

Generally, animal countermeasures are implemented to reduce the transfer of radionuclides to animal food products and thus reduce human ingestion doses. For radiocaesium, radioiodine and radiostrontium, animal products such as meat and milk are generally considered to be the main contributors to the dose. In general, the most relevant countermeasures for animal products are:

- providing uncontaminated food
- administration of binding agents to animals
- feeding stable analogues, and
- changes in behaviour habits (change of diet, food processing, hunting restrictions).

For semi-natural ecosystems, human based countermeasures become more important, because animal-based measures are difficult to manage.

Many individual countermeasures have been developed, especially over the decade following the Chernobyl accident. A comprehensive compilation of countermeasures for animal products is given by Howard et al. (2001). In practice, many of these have mainly been implemented within the former Soviet Union (FSU), but also in Norway, Sweden and the UK. However, the applicability of most of these measures to other countries such as the European Union Member States has not been fully and critically assessed. Furthermore, countermeasure research has largely focused on the effectiveness of individual methods, although some attention has also been given to cost benefit analyses (i.e. the monetary cost per unit of dose averted—see, for example Chapter 6). There has been little consideration of how to combine these individual countermeasures within a sustainable restoration strategy for a contaminated area comprising urban, rural and industrial systems. In designing restoration strategies to ensure the long-term sustainability of large and varied contaminated areas, there is a requirement to adopt a more holistic approach rather than simply selecting cost-effective countermeasures.

A detailed description and evaluation of the usefulness of countermeasures in different ecosystems, taking into account other aspects than dose reduction effectiveness, was initiated by the International Union of Radioecologists (Voigt et al., 2000a, 2000b) and has been extended by Howard et al. (2005). A novel and integral part of this approach is to include social and ethical issues, including public perceptions and communication, explicitly within the evaluation of restoration strategies (Oughton et al., 2004). This should contribute to a higher degree of public understanding of remedial measures and improve confidence in information given by authorities and in the event of a nuclear accident.

The acceptability of animal based countermeasures from the viewpoint of the producer is considered in Table 9. However, for the application of countermeasures the acceptability to the consumer of different countermeasures also has to taken into account. Changing consumption behaviour can reduce the transfer of radionuclides to human populations. Such changes in behaviour depend first on normal, pre-accident consumption behaviour, which varies tremendously over an area such as Europe and, secondly, on the perception of risks due to radioactively contaminated foodstuffs. This perception will be closely related to the social, psychological and economic conditions of the persons affected and will change with time before and after an accident. A clear demonstration is the behaviour of people in the FSU countries affected by Chernobyl fallout, when restrictions on the use of both ecosystems

Table 9
List of animal based countermeasures (Voigt and Kiefer, 2000)

| Animal based countermeasures | Effectiveness (Reduction)[a] | Accept ability | Cost | Radio nuclide | Variation | Comments |
|---|---|---|---|---|---|---|
| Limiting intake of radionuclides | | | | | | |
| • house animals to prevent the adsorption through skin or lung from a passing cloud | high | variable[1] | variable[1] | I | spatial | [1]depends on the possibility to house the animals |
| • provide un- or less contaminated feed | high (up to 100%) | variable[2] | variable[2] | all | spatial/ seasonal | [2]depends on availability of 'clean' food or the possibility to move animals to 'clean' pastures. |
| – to animals prevented from grazing by housing or fenced in enclosed areas | high (up to 100%) | variable[3] | variable[3] | all | spatial/ seasonal | [3]depends on management; (outdoor grazing; permanent or temporary grazing system) |
| – to dairy animals and for meat animals in their final stage of fattening. Use contaminated land for non-dairy animals[5] | variable[6] (up to 100%) | high | variable | all | spatial | [5]young growing animals and breeding stock or animals not intended for immediate slaughter<br>[6]depends on biological half-lives and time period before slaughter |
| • change the feeding composition to selected crops which accumulate low levels of radionuclides | variable[7] (up to 90%) | high | low | all | spatial | [7]depends on the availability of selected crops and the traditional feeding practice;<br>perennial grasses and leguminous tend to be more contaminated than cereals, tubers and root crops; may adversely affect animal productivity. |
| • select lower contaminated pasture for grazing | variable | high | low | all | spatial | relevant for spatially different deposition patterns |
| • groundwater instead of surface water for drinking | high | high[8] | variable[8] | all | spatial | [8]depends on the availability of groundwater |
| • increase stubble height | medium low | medium | low | all | spatial | avoids harvesting plant parts with resuspended soil adhering to it. |
| • increase grazing intensity | | | | | | |

Table 9
(*Continued* )

| Animal based countermeasures | Effectiveness (Reduction)[a] | Accept ability | Cost | Radio nuclide | Variation | Comments |
|---|---|---|---|---|---|---|
| Reducing the adsorption of ingested radionuclides | | | | | | |
| • administer chemical binders/analogues; | | | | | | toxicity of high doses; binding of other elements can cause health problems in long term application |
| – i.e. Prussian Blue[b] | (up to 90%)[9] | | low | | | [9]maximum effect when applied daily |
| – added to concentrate | high | high | | Cs | | |
| – as slow release boli | med (25–80%) | high[10] | | Cs | | [10]most appropriate for grazing animals |
| – provided in salt-licks | variable[11] (0–80%) | high | | Cs(Sr) | | [11]most effective when the Na content of the vegetation is low |
| –supplement diet with clay minerals (like bentonite vermiculite mordenite) or zeolites (like clinoptilolite) | high (up to 80%) | high | low | Cs (Sr?) | | most suitable for animals fed daily |
| – provide alginates with diet | up to 75% for monogastrics; at least 50% for ruminants | | 12 | Sr, (Cs?) | | [12]Currently prohibited |
| –provide high calcium diets to reduce milk activity concentration | variable[13] | high | low | Sr | | feed calcium salts or natural feeds rich in Ca<br>[13]depends on the status of the Ca supply—health problems with high Ca/P ratio in fodder |

Table 9
(*Continued*)

| Animal based countermeasures | Effectiveness (Reduction)[a] | Acceptability | Cost | Radionuclide | Variation | Comments |
|---|---|---|---|---|---|---|
| Blocking the uptake into and the transport through organs—increasing the excretion of a previously accumulated body burden | | | | | | |
| – administer chemical analogues/binder or stable forms of elements | Medium low | | | I | spatial | toxicity of high doses food product may be also influenced. |
| Controlling | | | | | | |
| – use live monitoring to avoid slaughter of contaminated animals or monitor carcasses | high | (high) | variable | Cs | | requires semi skilled labour |
| Change slaughter time/hunting time | variable (up to 80%) | medium | variable | (all) Cs | spatial season | most appropriate for game with seasonal feeding habits |
| Alter animal species | variable (up to 80%) | low | high | all | spatial | replace sheep or goat with cattle—environmental, social and cultural difficulties |
| Change production system milk to meat | high | ? | high | Sr | | environmental, social and cultural difficulties |
| Destruction | high (up to 100%) | low[14] | very high | All | spatial | [14]depends on the economical situation as well as ethical and moral attitude |
| Divert animal products from human to animal consumption | up to 100% | low | high | | | social obstacles |

[a]Reduction: $R = (C_N - C_R)/C_N$; $C_R$: reduced concentration, $C_N$: normal concentration without applying cms.

[b]The term 'Prussian Blue' refers to a number of ferric hexacyanoferrates; mainly ammonium ferric cyanoferrate (AFCF).

Table 10
Considerations for consumers of potentially contaminated animal products in the early and later phases following an accidental release of radioactivity to the environment

| Products | Potential countermeasures for the first[a] and second phases after deposition of radionuclides for the consumers. Accept or not? |
|---|---|
| Milk and milk products (cow, sheep, goat) | • Stop consumption<br>• Use products after binders (AFCF or clay minerals) have been applied<br>• Use processed milk products instead of fresh milk<br>• Use conserved milk instead of fresh milk[a] |
| Meat (cow, sheep, goat, pork) | • Stop consumption<br>• Sell products from animals after binders (AFCF or clay minerals) have been given to the animals<br>• Discard cooking fluids after boiling<br>• Apply food processing (salt/pickle)<br>• Consume products from animals which are fed with clean food before slaughter |
| Wild game | • Stop consumption<br>• Change the hunting season<br>• Discard water after boiling<br>• Sell products from animals after binders (AFCF or clay minerals) have been applied |
| Fish | • Stop consumption<br>• Species selection<br>• Sell products from fish from fish farms instead of freshwater fish<br>• Discard water after boiling<br>• Apply food processing (salt, pickle, boil etc.) |

[a]These actions are only useful in the first few days after an accident.

and food products had to be made. These imposed restrictions were obeyed in the short term after the accident but were often ignored after a few years. Therefore, the following questions concerning the acceptability of countermeasures should be considered for the producer and the consumer:

- Which products are important for consumption behaviour?
- Which of the recommended countermeasures are acceptable?

Some questions to be considered by consumers of animal food products are given in Table 10.

The processing of food can greatly reduce the intake of radioactivity by humans and reduction factors for animal food products have been compiled and taken into account. However, the question remains as to how many people are prepared to change their consumption habits, whether they will accept and consume treated products or if they will be willing to pay more for uncontaminated food.

Recently a network has been established which addresses stakeholders' and consumers' opinions with respect to agricultural countermeasures (www.ec-farming.net/). The practicability of, and attitudes towards, countermeasures are being evaluated based on the stakeholders' experiences and technical as well as social constraints.

## 6. Concluding remarks

During recent years novel data from animal experiments have become a rarity and, with time, the number of research groups carrying out such research has declined so that few groups in Europe or elsewhere currently conduct such experiments. It is evident from this chapter that most of the data available are related to well studied radionuclides of Cs, Sr and I, and many have been obtained as a consequence of the Chernobyl accident. Fewer studies have dealt with radionuclides which are important from other perspectives such as Tc, S, Cl, U and $^{3}H$ for which data are sparse. In addition to the need for data on transfer to animal products, which constitute major human dietary components, we also need data for unusual animal-derived products which can be important for individual consumers whose dietary habits may expose them to radionuclide ingestion to a greater degree than the 'average' population.

## References

Barth, J. (1977). Application of the artificial rumen and simulated bovine gastrointestinal fluids procedure in the study of the bioavailability of transuranics. In: White, M.G., Dunaway, P.B. (Eds.), *Proceedings of A Symposium on Transuranics in Terrestrial and Aquatic Environments*. NTIS, Springfield, USA, pp. 419–433.

Beresford, N.A., Mayes, R.W., Howard, B.J., Eayres, H.F., Lamb, C.S., Barnett, C.L., Segal, M.G. (1992). The bioavailability of different forms of radiocaesium for transfer across the gut of ruminants. *Radiat. Prot. Dosim.* **41**, 87–91.

Beresford, N.A., Barnett, C.L., Mayes, R.W., Pollaris, K., Vandecasteele, C.M., Howard, B.J. (1995). The use of an in-vitro technique to predict the absorption of dietary radiocaesium by sheep. *Radiat. Envir. Biophys.* **34**, 191–194.

Beresford, N.A., Mayes, R.W., Crout, N.M.J., Dawson, J.M., Maceachern, P.J., Dodd, B.A., Soar, J.B., Barnett, C.L., Lamb, C.S. (1997). Transfer of heavy metals from feeding stuffs and environmental sources into meat, milk and other foods of animal origin. Part 1. Arsenic, Cadmium and mercury. Final report. 68 pp. Ministry of Agriculture, Fisheries and Food.

Beresford, N.A., Mayes, R.W., Barnett, C.L., MacEachern, P.J., Crout, N.M.J. (1998a). Variation in the metabolism of radiocaesium between individual sheep. *Radiat. Environ. Biophys.* **37**, 277–281.

Beresford, N.A., Mayes, R.W., Hansen, H.S., Crout, N.M.J., Hove, K., Howard, B.J. (1998b). Generic relationship between calcium intake and radiostrontium transfer to milk of dairy ruminants. *Radiat. Environ. Biophys.* **37**, 129–131.

Beresford, N.A., Crout, N.M.J., Mayes, R.W., Howard, B.J., Lamb, C.S. (1998c). Dynamic distribution of radioisotopes of cerium, ruthenium and silver in sheep tissues. *J. Environ. Radioact.* **38**, 317–338.

Beresford, N.A., Mayes, R.W., Crout, N.M.J., Dawson, J.M., MacEachern, P.J., Dodd, B.A., Soar, J.B., Barnett, C.L., Lamb, C.S. (1998d). Transfer of heavy metals from feeding stuffs and environmental sources into meat, milk and other foods of animal origin. Part II. Lead. Final Report. 41 pp. Ministry of Agriculture, Fisheries and Food.

Beresford, N.A., Mayes, R.W., Colgrove, P.M., Barnett, C.L., Bryce, L., Dodd, B.A., Lamb, C.S. (2000a). A comparative assessment of the potential use of alginates and dietary calcium manipulation as countermeasures to reduce the transfer of radiostrontium to the milk of dairy animals. *J. Environ. Radioact.* **51**, 321–342.

Beresford, N.A., Mayes, R.W., Cooke, A.I., Barnett, C.L., Howard, B.J., Lamb, C.S., Naylor, G.P.L. (2000b). The importance of source-dependent bioavailability in determining the transfer of ingested radionuclides to ruminant derived food products. *Environ. Sci. Technol.* **34**, 4455–4462.

Brown, J., Simmonds, J.R. (1995). *FARMLAND A Dynamic Model for The Transfer of Radionuclides Through Terrestrial Foodchains*. National Radiological Protection Board, Chilton, UK. NRPB-R273.

Comar, C.L. (1966). Transfer of strontium-90 into animal produce. In: Russell, R.S. (Ed.), *Radioactivity and Human Diet*. Pergamon Press, Oxford, pp. 247–275.

Comar, C.L., Wasserman, R.H., Nold, M.M. (1956). Strontium–calcium discrimination factors in the rat. *Proc. Soc. Exptl. Biol. Med.* **92**, 859–863.

Cooke, A.I., Rimmer, D.L., Weekes, T.E.C., Green, N., Wilkins, B.T. (1997). Transfer of radiocaesium, plutonium and americium to sheep after ingestion of contaminated soil. In: Prost, R. (Ed.), *Contaminated Soils*. INRA, Paris, CD:\Data\comminuic\082.PDF; CD-Rom.

Coughtrey, P.J. (1990). *Radioactivity Transfer to Animal Products*. Commission of The European Communities, Luxembourg.

Coughtrey, P.J., Thorne, M.C. (1983). *Radionuclide Distribution and Transport in Terrestrial and Aquatic Ecosystems: A Critical Review of Data*, vol. 1. Balkema, AA, Rotterdam.

Crout, N.M.J., Voigt, G. (1996). Modelling the dynamics of radioiodine in dairy cows. *J. Dairy Science* **79**, 254–259.

Crout, N.M.J., Mayes, R.W., Beresford, N.A., Lamb, C.S., Howard, B.J. (1998). A metabolic approach to simulating the dynamics of C-14, H-3 and S-35 in sheep Tissues. *Radiat. Environ. Biophys.* **36**, 243–250.

Crout, N.M.J., Beresford, N.A., Mayes, R.W., MacEachern, P.J., Barnett, C.L., Lamb, C.S., Howard, B.J. (2000). A model of radioiodine transfer to goat milk incorporating the influence of stable iodine. *Radiat. Envir. Biophys.* **39**, 59–65.

Ennis, M.E., Johnson, J.E., Ward, G.M., Boamah, K.N. (1989). Technetium metabolism by lactating goats. *Health Phys.* **57**, 321–330.

Ennis, M.E. Jr, Johnson, J.E., Ward, G.M., Voigt, G.M. (1988). A specific activity effect in the metabolism of technetium. *Health Phys.* **54**, 157–160.

Falconer, I.R. (1963). Iodide metabolism of the thyroid and mammary glands during lactation in sheep. *J. Endocr.* **25**, 533–539.

Fefac—European Federation of Feed Industries (1996). Statistical information from the EC-DG VI (http://europa.eu.int/comm/dg06/agrista).

Galeriu, D., Crout, N.M.J., Melintescu, A., Beresford, N.A., Peterson, S.R., van Hees, M. (2001). A metabolic derivation of tritium transfer factors in animal products. *Radiat. Envir. Biophys.* **40**, 325–334.

Gillett, A.G., Crout, N.M.J., Absalom, J.P., Wright, S.M., Young, S.D., Howard, B.J., Barnett, C.L., McGrath, S.P., Beresford, N.A., Voigt, G. (2001). Temporal and spatial prediction of radiocaesium transfer to food products. *Radiat. Environ. Biophys.* **40**, 227–235.

Ham, G.J., Harrison, J.D., Popplewell, D.S., Curtis, E.J.C. (1989). The distribution of $^{137}$Cs, plutonium and americium in sheep. *The Science of The Total Environment* **85**, 235–244.

Hove, K., Pedersen, Ø., Garmo, T.H., Hansen, H.S., Staaland, H. (1990). Fungi: A major source of radiocesium contamination of grazing ruminants in Norway. *Helth Phys.* **59**, 189–192.

Howard, B.J., Desmet, G.M. (Eds.) (1993). *Relative Effectiveness of Agricultural Countermeasure Techniques (REACT), Proceedings of a Workshop Organiz-ed By the Commission of the European Communities, Brussels, 1–4 October 1991. Sci. Total Environ.*, vol. 137.

Howard, B.J., Johanson, K.J., Linsley, G.S., Hove, K., Pröhl, G., Horyna, J. (1996). Transfer of radionuclides by terrestrial food products from semi-natural ecosystems. In: *Second Report of the VAMP Terrestrial Working Group*. IAEA, Vienna, pp. 49–79. IAEA-TECDOC-857.

Howard, B.J., Beresford, N.A., Mayes, R.W., Hansen, H.S., Crout, N.M.J., Hove, K. (1997). The use of dietary calcium intake of dairy ruminants to predict the transfer coefficient of radiostrontium to milk. *Radiat. Environ. Biophys.* **36**, 39–43.

Howard, B.J., Beresford, N.A. (2001). Advances in animal radioecology. In: *Radioactive Pollutants: Impact on The Environment and Humans, Proc. of Int. Symp. in Aix-en-Provence, France, 3–7 Sept 2001*. EDP Science, Les Ulis, pp. 187–207.

Howard, B.J., Beresford, N.A., Voigt, G. (2001). Countermeasures for animal products: a review of effectiveness and potential usefulness after an accident. *J. Environ. Radioactivity* **56**, 115–137.

Howard, B.J., Beresford, N.A., Nisbet, A., Cox, G., Oughton, D.H., Hunt, J., Alvarez, B., Andersson, K.G., Liland, A., Voigt, G. (2005). The STRATEGY project: decision tools to aid sustainable restoration and long-term management of contaminated agricultural ecosystems. *Journal of Environmental Radioactivity* **83**, 275–295.

IAEA (1994). Handbook of Transfer Parameter Values for The Prediction of Radionuclide Transfer in Temperate Environments, International Atomic Energy Agency, *Technical Report Series No. 364*. IAEA, Vienna.

ICRP (1979). Publication 30: Limits for intakes of radionuclides by workers. *Annals of the ICRP* **2** (3/4).

Johnson, J.E., Ward, G.M., Ennis, M.E. Jr., Boamah, K.N. (1988). Transfer coefficients of selected radionuclides to animal products 1 Comparison of milk and meat from dairy cows and goats. *Health Phys.* **54**, 161–166.

Jones, B. (1983). Metabolism of technetium in goats. *International Journal of Radiation Isotopes—Applications and Instrumentation Part A* **34**, 837–840.

Jones, B.-E.V. (1989). Technetium metabolism in goats and swine. *Helth Phys.* **57**, 331–336.

Kirchgeßner, M. (1987). *Tierernährung*. DLG-Verlag, Frankfurt (Main), Germany.

Kistner, G., Wiggenhauser, W., Krestel, R., Buheitel, G., Erzberger, A., Roedler, H.D. (1987). Age-related biokinetics of cerium-141-chloride in rats. In: Gerber, G.B., Metivier, H., Smith, H. (Eds.), *Proceedings 'Age-related factors in radionuclide metabolism and dosimetry'*. Martinus Nijhoff Publishers, Dordrecht/Boston/Lancaster, pp. 221–228.

Kreyling, W.G., Ferron, G.A., Haider, B. (1986). Metabolic fate of inhaled Co aerosols in beagle dogs. *Health Phys.* **51**, 773–795.

Kreyling, W.G., Schuhmann, G., Ortmaier, A., Ferron, G.A., Karg, E. (1988). Particle transport from the lower respiratory tract. *J. Aerosol Med.* **1**, 351–370.

Kreyling, W.G., Andre, S., Collier, C.G., Ferron, G.A., Metevier, H., Schuhmann, G. (1991). Interspecies comparison of lung clearance after inhalation of monodisperse, solid cobalt oxide aerosol particles. *J. Aerosol Sci.* **22**, 509–535.

Kreyling, W.G., Hodgson, A., Guilmette, R.A., Scarlett, C., Damin, A., Stadling, G.N. (1998). Interspecies comparison of lung clearance using momodisperse terbium oxide particles. *Rad. Prot. Dos.* **79**, 1–4.

Lovelock, J.E. (1979). *Gaia: A New Look at Life on Earth*. Oxford University Press, Oxford.

Mayes, R.W., Eayres, H.F., Beresford, N.A., Lamb, C.S., Howard, B.J. (1992). Changes with age in the absorption of radiocaesium by sheep. *Radiat. Prot. Dosim.* **41**, 83–86.

Mayes, R.W., Beresford, N.A., Howard, B.J., Vandecasteele, C.M., Stakelum, G. (1996). The use of the true absorption coefficient as a measure of bioavailability of radiocaesium in ruminants. *Radiat. Environ. Biophys.* **35**, 101–109.

Mehli, H., Strand, P. (1988). Fluxes of $^{137}Cs$ in Chernobyl affected areas in Russia and Ukraine. Norwegian Radiation Protection Authority report, Oslo-As.

Müller, H., Pröhl, G. (1993). ECOSYS-87: A dynamic model for assessing radiological consequences of nuclear accidents. *Health Phys.* **64**, 232–252.

Nalezinski, S., Rühm, W., Wirth, E. (1996). Development of a general equation to determine the transfer factor feed-to-meat for radiocesium on the basis of the body mass of domestic animals. *Health Phys.* **70**, 717–721.

Ng, Y.C., Colsher, C.S., Quinn, D.J., Thompson, S.E. (1977). Transfer coefficients for the prediction of the dose to man via the forage–cow–milk pathways from radionuclides released to the biosphere, Lawrence Livermore National Laboratory, UCRL-51939.

Ng, Y.C., Colsher, C.S., Thompson, S.E. (1982). Transfer coefficients for assessing the dose from radionuclides in meat and eggs, Lawrence Livermore National Laboratory, NREG CR-2976, UCID-19464.

Nisbet, A.F., Woodman, R.F.M., Brown, J., Smith, J.G., Wilkins, B.T. (1999). Derivation of working levels for animals feedstuffs for use in the event of a future nuclear accident. NRPB report 299.

Oughton, D., Forsberg, E.-M., Bay, I., Kaiser, M., Howard, B. (2004). An ethical dimension to sustainable restoration and long-term management of contaminated areas. *Journal of Environmental Radioactivity* **74**, 171–183.

Salbu, B., Østby, G., Garmo, T.H., Hove, K. (1992). Availability of caesium isotopes in vegetation estimated from incubation and extraction experiments. *Analyst* **117**, 487–491.

Scheuch, G., Stahlhofen, W., Heyder, J. (1996). An approach to deposition and clearance measurements in human airways. *J. Aerosol Med.* **9**, 35–41.

Singleton, D.L., Livens, F.R., Beresford, N.A., Howard, B.J., Barnett, C.L., Mayes, R.W., Segal, M.G. (1992). Development of a laboratory method to predict rapidly the availability of radiocaesium. *Analyst* **117**, 505–509.

Stara, J.F., Nelson, N.S., Della Rosa, R.J., Bustad, L.K. (1971). Comparative metabolism of radionuclides in mammals: a review. *Helth Phys.* **20**, 113–137.

Tansley, A.G. (1935). The use and abuse of vegetational concepts and terms. *Ecology* **16**, 284–307.

Van Bruwaene, R., Gerber, G.B., Kirchman, R., Garten, C.T. Jnr., Vankerkom, J., Bonotto, S., Mathieu, T., Cogneau, M. (1986). Absorption and retention in sheep of technetium administered into the rumen as pertechnetate or bound to food. In: Desmet, G., Myttenaere, C. (Eds.), *Technetium in the Environment*. Elsevier, London, pp. 333–338.

Vandecasteele, C., van Hees, M. (1995). In: Howard, B.J., Mayes, R.W., Assimakopoulos, P., Crout, N.M.J., Voigt, G., Vandecasteele, C., Zelenka, J., Hove, K., Hinton, T. (Eds.), *Transfer of radionuclides in animal production systems*. Institute of Terrestrial Ecology, Grange-over-Sands. Final report to CEC. CEC contract F13P-CT92-0006.

Vandecasteele, C.M., Van Hees, M., Hardeman, F., Voigt, G.M., Howard, B.J. (2000). The true absorption of iodine and effect of increasing stable iodine in the diet. *J. Environ. Radioactivity* **47**, 301–317.

Voigt, G. (1993). Chemical methods to reduce the radioactive contamination of animals and their products in agricultural ecosystems. *Science of the Total Environment* **137**, 205–255.

Voigt, G., Henrichs, K., Proehl, G., Paretzke, H.G. (1988). Measurements of transfer coefficients from $^{137}$Csm, $^{60}$Co, $^{54}$Mn, $^{22}$Na, $^{131}$I and $^{95m}$Tc from feed into milk and beef. *Journal Rad. Environ. Biophys.* **27**, 143–152.

Voigt, G., Pröhl, G., Müller, H., Bauer, T., Lindner, J.P., Probstmeier, G., Röhrmoser, G. (1989). Determination of the transfer of cesium and iodine from feed into domestic animals. In: *The Science of The Total Environment*, vol. 85, pp. 329–338. September 1989.

Voigt, G., Pröhl, G., Müller, H., Bauer, T., Lindner, J.P., Probstmeier, G., Röhrmoser, G. (1993). $^{137}$Cs transfer after Chernobyl from fodder into chicken meat and eggs. *Helth Phys.* **65**, 141–146.

Voigt, G., Rauch, F., Paretzke, H.G. (1996). Long-term behavior of radiocesium in dairy herds in the years following the Chernobyl accident. *Helth Phys.* **71**, 370–373.

Voigt, G., Kiefer, P. (2000). Influence of spatial variation on countermeasures with special regards to animal food products. GSF-Report 10/2000, Munich.

Voigt, G., Eged, K., Hilton, J., Howard, B.J., Kis, Z., Nisbet, A.F., Oughton, D.H., Rafferty, B., Salt, C.A., Smith, J.T., Vandenhove, H. (2000a). In: *Countermeasures: Radioecological and Social Impacts*. NRPA, Osteras, Norway, p. 23. IUR working document.

Voigt, G., Eged, K., Hilton, J., Howard, B.J., Kis, Z., Nisbet, A.F., Oughton, D.H., Rafferty, B., Salt, C.A., Smith, J.T., Vandenhove, H. (2000b). Countermeasures: radioecological and social impacts. A wider perspective on the selection of countermeasures. *J. Rad. Prot. Dosim.* **92**, 45–48.

Ward, G.M., Johnson, J.E., Stewart, H.F. (1965). Deposition of fallout $^{137}$Cs on forage and its transfer to cows' milk. In: Kelement Jr., A. (Ed.), *Proc. 2nd AEC Symp. on Fallout*. National Technical Information Centre, Oak Ridge, Tennessee.

Ward, G.M., Johnson, J.E. (1986). Validity of the term transfer coefficient. *Health Phys.* **50**, 411–414.

Ward, G.M., Johnson, J.E. (1989). Assessment of milk transfer coefficients for use in prediction models of radioactivity transport. *Sci. Total Environ.* **85**, 287–294.

Whicker, F.W., Kirchner, T.B. (1987). PATHWAY: A dynamic food chain model to predict radionuclide ingestion after fallout deposition. *Health Phys.* **52**, 717–737.

Wiechen, A., Heine, K., Hagemeister, H. (1983). Ein Beitrag zur Frage des Transfers von Technetium in die Milch. *Atomenergie und Kerntechnik* **42**, 199–200.

# Radionuclides in grassland, heath, mire and mountain ecosystems

Carol A. Salt

*School of Biological and Environmental Sciences, University of Stirling, FK9 4LA, UK*

## 1. Introduction

On a global scale terrestrial ecosystems have received the largest inputs of anthropogenic radionuclides from nuclear weapons tests and from the accident at the Chernobyl nuclear power plant (UNSCEAR, 2000) (see Table 1). Weapons testing fallout was concentrated in the middle and northern latitudes of the Northern Hemisphere and its distribution was strongly correlated with rainfall. Chernobyl fallout by contrast was largely confined to Europe (De Cort et al., 1998) and weather patterns as well as rainfall intensity played an important role in its deposition. It can be generalised that within a geographical area, landscapes with higher annual rainfall will be more affected than those with less rainfall. High rainfall areas in middle and northern latitudes typically support less intensive land uses and are characterised by low soil nutrient reserves and high content of soil organic matter. Under these conditions radiocaesium mobility for example can be very high (Sanchez et al., 1999).

The Chernobyl accident has greatly increased our awareness of the importance of semi-natural ecosystems as sources of radioactively contaminated food for humans (Howard and Howard, 1997). Prior to the accident mainly the lichen–reindeer pathway had been well researched in relation to nuclear weapons testing (Liden and Gustafsson, 1967). Semi-natural ecosystems form the transition between natural ecosystems with no human intervention and intensively managed agricultural systems. Low plant productivity and limited scope for intensification restrict land uses to forestry, low-density livestock farming, hunting and collecting

Table 1
Major global radionuclide releases in PBq ($10^{15}$ Bq) (UNSCEAR, 2000)

| | $^{90}$Sr | $^{137}$Cs | $^{239/240}$Pu |
|---|---|---|---|
| Nuclear weapons tests | 622 PBq | 948 PBq | 11 PBq |
| Chernobyl accident | 10 PBq | 85 PBq | 0.07 PBq |

RADIOACTIVITY IN THE ENVIRONMENT
VOLUME 10 ISSN 1569-4860/DOI: 10.1016/S1569-4860(06)10005-4

of mushrooms, berries and other 'free foods'. Deposition of radiocaesium in these systems after the Chernobyl accident led to surprisingly high contamination of food products such as milk, meat, mushrooms, berries and honey (Desmet et al., 1990).

This chapter will focus on the behaviour of long-lived anthropogenic radionuclides in semi-natural ecosystems encompassing grasslands, heaths, mires and mountain areas. Intensively managed permanent grassland systems are included since these show many similarities with other grasslands. Although the majority of research to date has focused on the contamination of food products and its prediction, an attempt is made here to draw together the more ecological aspects such as the distribution and fluxes of radionuclides in the different compartments of an ecosystem.

## 2. Global distribution of fallout and affected ecosystems

The deposition of radionuclides from nuclear weapons tests carried out mainly between 1951 and 1962 was non-uniform, with fallout predominantly occurring in the northern hemisphere and peaking in the mid-latitudes between 40° and 50°. Figure 1 shows the world-wide spatial distribution of the cumulative deposition of $^{90}Sr$ predicted by UNSCEAR (1969). Other long-lived isotopes of human health concern, $^{137}Cs$ and $^{239+240}Pu$, are assumed to have a similar pattern. Wet deposition in the mid-latitudes accounted for about 80–90% of fallout but dry deposition was a more important mechanism in areas of low precipitation (Wright et al., 1999). The very high deposition predicted for fairly arid parts of central North-America (Figure 1) could also be linked to the direct effect of weapons testing sites. It is known that the map underestimates fallout in areas of Chile around 40° latitude receiving 1000–4000 mm of rain (Schuller et al., 2002).

When the pattern of $^{90}Sr$ weapons fallout reconstructed by UNSCEAR (Figure 1) is compared with the global vegetation zones identified by Walter (1985), the areas most affected

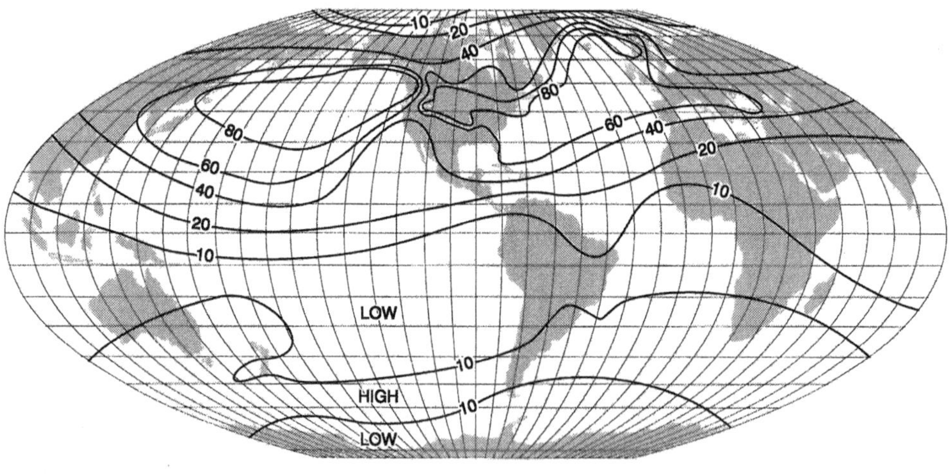


Fig. 1. Global Sr deposition in mCi km$^{-2}$ (10 mCi km$^{-2}$ = 37 MB km$^{-2}$) based on soils collected 1965–1967 (UNSCEAR, 1969); reproduced by permission of UNSCEAR.

coincide with the Temperate Nemoral, Arid-Temperate, Cold-Temperate Boreal and mountain zones of the northern mid-latitudes. Based on this the natural vegetation in the most affected areas would be predominantly forest with mountain communities above the treeline, mires in low-lying areas with water surplus and prairies in more arid areas. Due to human disturbance these systems have been altered or replaced especially in the temperate zone where due to agricultural activity 40% of land is now grassland (Briggs and Courtney, 1985). In Europe only 15.6% of land can be classed as undisturbed. Globally this rises to 27% (Hannah et al., 1994). At more northerly latitudes, despite lower deposition, natural ecosystems such as taiga and tundra communities were significantly affected by fallout due to their higher sensitivity (see Chapter 7).

The distribution of fallout after the accident at the Chernobyl Nuclear Power Plant (CNPP) was greatly influenced by the meteorological conditions prevailing during the release, which lasted for about 10 days (26 April–5 May 1986). The resulting deposition patterns for isotopes of Sr, Cs and Pu are very different from those of weapons test fallout. Sr-90 and $^{239/240}Pu$ were dispersed mostly locally, mainly around the CNPP. Cs-134 and $^{137}Cs$ were more widely dispersed with considerable fallout over the Former Soviet Union, and Northern and Central Europe (De Cort et al., 1998). Only approx. 20% of $^{137}Cs$ was dispersed beyond Europe. Figure 2 shows the spatial distribution of $^{137}Cs$ on 10 May 1986 across Europe and includes residual levels of global weapons fallout (De Cort et al., 1998).

Although radiocaesium deposition following the Chernobyl accident was well correlated with rainfall it does not mirror long-term rainfall patterns in the way that global weapons fallout does. This is due to fallout being concentrated over only 10 days and spreading from one point source. Consequently it is more difficult to generalise which ecosystems were most affected. Soil properties were crucial in determining the bioavailability and mobility of radiocaesium as explained later.

## 3. The material cycles in ecosystems

Ecosystems are characterised by two interacting subsystems, an open energy system and a more or less closed cyclical materials subsystem (Dickinson and Murphy, 1998). The materials, i.e. the nutrients are limited by external supply and have to be cycled within the biosphere to maintain a functioning ecosystem. Autotrophic plants have a central role in this cycling through the acquisition of nutrients from soil solution and atmosphere. Nutrients are then passed along food webs and ultimately detritivores break down organic matter, releasing the nutrients back into the soil. The rates at which nutrients and radionuclides move through food chains are determined by energy flows (Whicker and Schultz, 1982). Heal and Horrill (1983) stressed the importance of primary production and decomposition for the retention and circulation of radionuclides in ecosystems.

Anthropogenic radionuclides typically enter terrestrial ecosystems through wet and dry deposition onto plant and soil surfaces, input routes which are also important for many nutrients. However, the rapidity and extent to which a specific radionuclide becomes involved in nutrient cycles depends mainly on its chemical form and its similarity to nutrient elements. For example, radionuclides present within hot particles will be gradually released into the soil through leaching and weathering (Sandalls et al., 1993). Foliar interception can lead to direct uptake

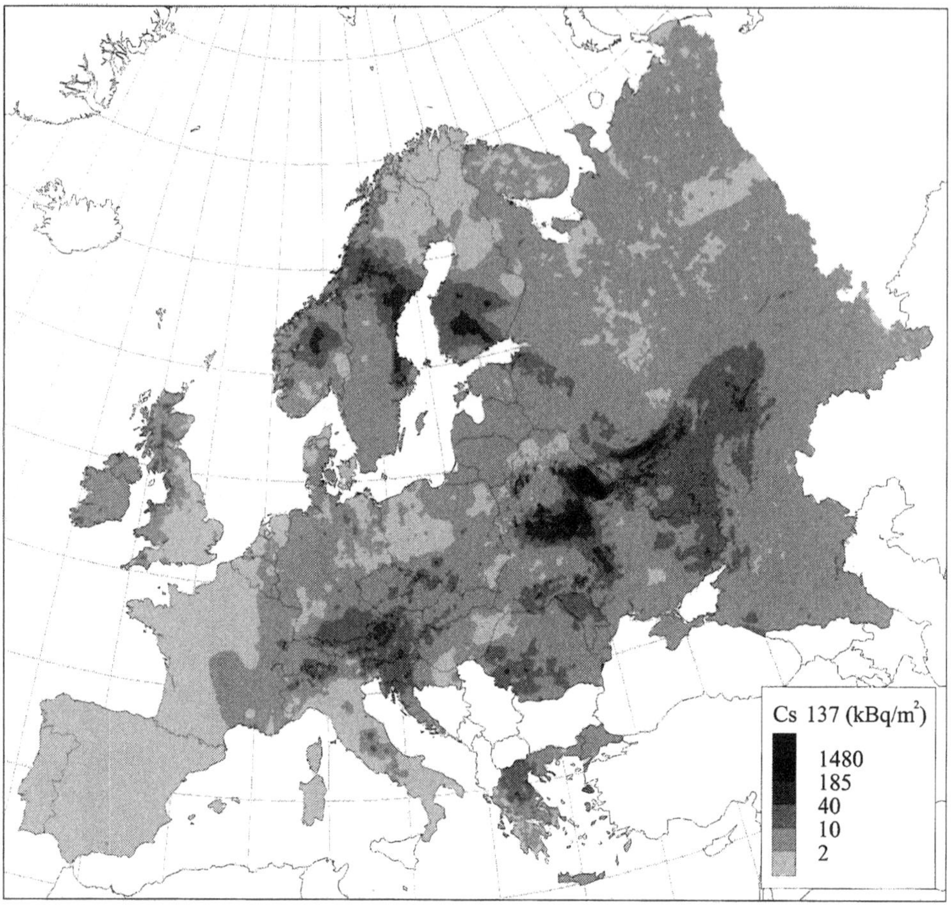


Fig. 2. European map of total $^{137}Cs$ deposition (global fallout and Chernobyl) in kBq $m^{-2}$ decay-corrected to 10 May 1986 (produced by S.M. Wright, CEH Lancaster, from De Cort et al., 1998).

of soluble radionuclides into leaves and distribution within the plant depending on the permeability of the cuticle (Moorby, 1964; Handley and Babcock, 1970). Less soluble deposits such as Pu-bearing particles can be retained until the foliage dies or is harvested (Hinton and Pinder, 2001). Intercepted particles can become dislodged or washed off by precipitation (Miller and Hoffman, 1983). The degree of foliar retention will depend on climate, plant biomass and type of plant surface, and the properties of the deposit.

Deciduous plants will release a large proportion of their foliar radionuclide content into the soil during senescence while evergreen species are likely to retain radionuclides for longer due to their longer leaf life-span (Grime, 1989). Ultimately, whether through removal from plant surfaces, leaching and decomposition of plant matter, or ingestion and excretion by herbivores, radionuclides intercepted by foliage will enter the soil. Figure 3 illustrates the losses of $^{137}Cs$ from cryptogams to soil compartments in an alpine plant community in Central Norway (Bretten et al., 1992). The subsequent fate of the radionuclide is largely determined by

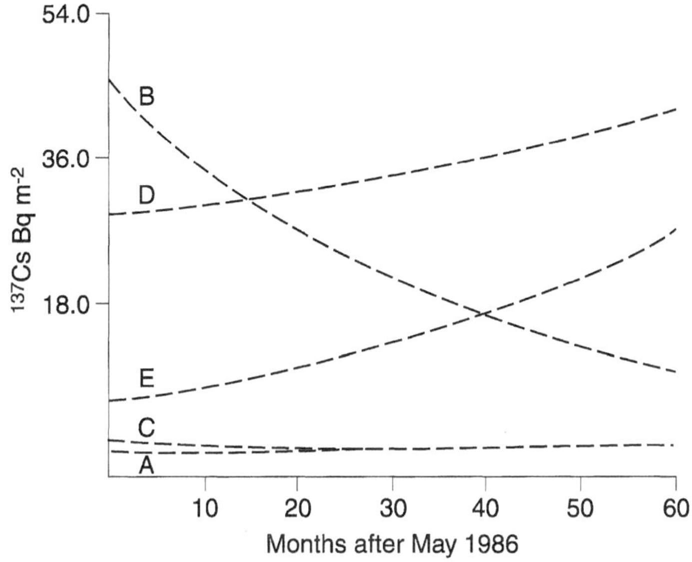


Fig. 3. Redistribution of Chernobyl $^{137}$Cs in an alpine plant community in Central Norway 1986–1990 (Bretten et al., 1992); reproduced by permission of the Royal Society of Chemistry.

interactions with soil colloids and the influence of soil pH and redox conditions. Following discrete deposition events such as the 10-day release after the Chernobyl accident (De Cort et al., 1998) radionuclides are likely to reach equilibrium with biogeochemical cycles more rapidly compared to fallout spread over many years during the weapons testing era. In the latter example surface contamination of vegetation remained an important long-term process. Once aerial inputs have ceased or become small relative to the accumulated ecosystem inventory, processes governing radionuclide availability for root uptake become dominant controls. The central role of the soil solution as an interface via which inputs, losses and cycling of radionuclides occur is illustrated in Figure 4.

The natural abundance of stable isotopes and nutrient analogues is very important in predicting the behaviour of long-lived radionuclides though this applies less to actinides and lanthanides (Whicker and Schultz, 1982). When radionuclides are chemically similar to essential nutrients, e.g. $^{137}$Cs to K, $^{90}$Sr to Ca, they become closely involved in nutrient cycles. However, they cannot physiologically substitute the nutrients. As discussed later, it is expected that radionuclide behaviour will gradually become similar to that of the natural stable isotopes, e.g. $^{133}$Cs and $^{88}$Sr.

## 4. The importance of soil properties for bioavailability

The soil has a central role as temporary storage medium for nutrients, and soil properties largely determine the availability of nutrients for plant root uptake. The same applies to anthropogenic radionuclides but their availability is often low due to binding to clay, humus and

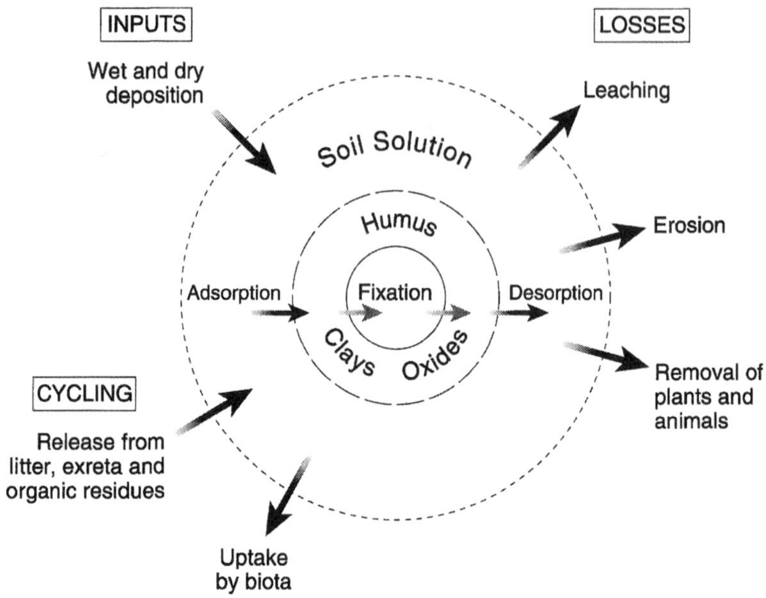


Fig. 4. Ecosystem inputs, outputs and cycling between biotic and soil compartments (adapted from Till, 1981).

oxide fractions (Livens and Rimmer, 1988). Technetium-99 in the pertechnetate form is a notable exception showing high mobility especially under aerobic conditions but its relevance in the terrestrial environment is largely restricted to waste repositories and crop systems where seaweed is used as a fertiliser (Webster et al., 2003). For the ecosystems discussed in this chapter the influence of soil chemistry on the mobility of Cs and Sr isotopes is of greatest biological relevance.

The behaviour of strontium is governed by adsorption onto exchange sites on clays and humus while caesium can be irreversibly fixed in clay minerals, especially illites (Alexakhin and Krouglov, 2001). This leads to greater bioavailability and migration of Sr than Cs in mineral soils (Squire and Middleton, 1966) and partly explains the greater focus of research on Sr behaviour during and after the main phase of weapons tests. On organic soils the bioavailability of Cs is greater than that of Sr (Askbrant and Sandalls, 1998).

Early work by Barber (1964) and Fredriksson et al. (1966) recognised the importance of organic matter for radiocaesium availability but the processes were poorly understood. Post-Chernobyl research has proven that even organic soils contain sufficient clay minerals to immobilise radiocaesium but interactions between clay minerals and organic matter inhibit fixation (Hird et al., 1995; Staunton et al., 2002). Numerous vegetation studies provide evidence of the high availability of radiocaesium to plants growing on organic soils which can continue for many years after deposition (Desmet et al., 1990; Sanchez et al., 1999). Figure 5 illustrates the close relationship between soil organic matter content and the uptake of $^{134}Cs$ by *Agrostis capillaris* (measured as the soil/plant concentration ratio) in an experiment with 23 artificially contaminated soils (adapted from Sanchez et al., 1999).

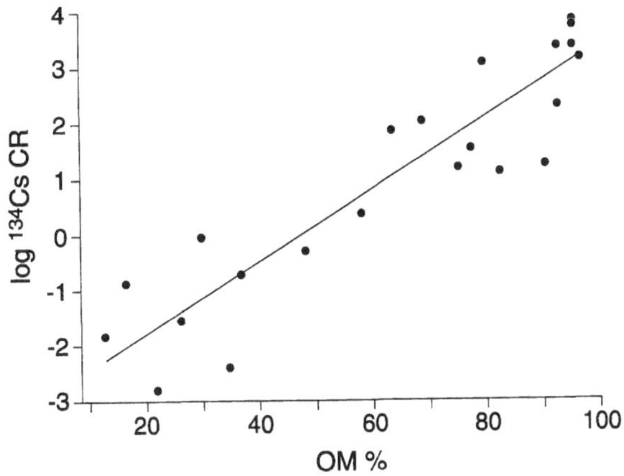


Fig. 5. Relationship between soil organic matter content and the log concentration ratio of $^{134}Cs$ in *Agrostis capillaris* (based on data from Sanchez et al., 1999).

The soil-to-plant transfer of radiocaesium can be predicted from the soil parameters % organic matter, % clay, exchangeable K and pH (Absalom et al., 2001). Studies using European soil data identified the most vulnerable ecosystems as those on soils rich in organic matter or with a coarse (sandy) texture (Howard, 2002). Figure 6 shows a map of the organic matter content of European soils. Following the Chernobyl accident, bioavailability of radiocaesium was particularly high where high deposition (see Figure 2) coincided with organic-rich soils (for example parts of Scandinavia). Despite more intermediate deposition levels, areas with highly organic soils (for example in parts of Ireland and the UK) also showed high bioavailability.

## 5. Time-dependent changes in bioavailability

The plant availability of soil-incorporated radiocaesium generally decreases over time as processes of adsorption, redistribution and fixation take place. This 'ageing' effect is much less pronounced in radiostrontium due to the dominance of exchange and complexation processes (Alexakhin and Krouglov, 2001). Fixation of radiocaesium is rapid in the presence of 2:1 type clay minerals (illite, vermiculite) (Sawhney, 1966).

Theoretically the behaviour of anthropogenically introduced radioisotopes should over time become similar to that of their stable analogues. However, as stable Cs and Sr are present inside the crystal structures of minerals, isotopic equilibrium with such radioisotopes is only likely at geological timescales. In humus-rich soils or soil layers a more rapid attainment of equilibrium than in mineral soils can be expected due to the greater relevance of exchange processes and for Sr also complexation (Juo and Barber, 1970) but there are few studies to draw on. Oughton (1989) found the same ratio of $^{137}Cs$ to $^{133}Cs$ in grass, in litter and in peaty topsoil (0–2 cm) one year after the Chernobyl accident and proposed that equilibrium will be reached fastest in organic soil fractions. This is supported by a study of *Vaccinium myrtillus* and several species of fungi in the humus-rich layers of a forest soil (Rühm et al., 1999) where $^{134+137}Cs$ had attained equilibrium with $^{133}Cs$ less than 10 years after deposition. In contrast

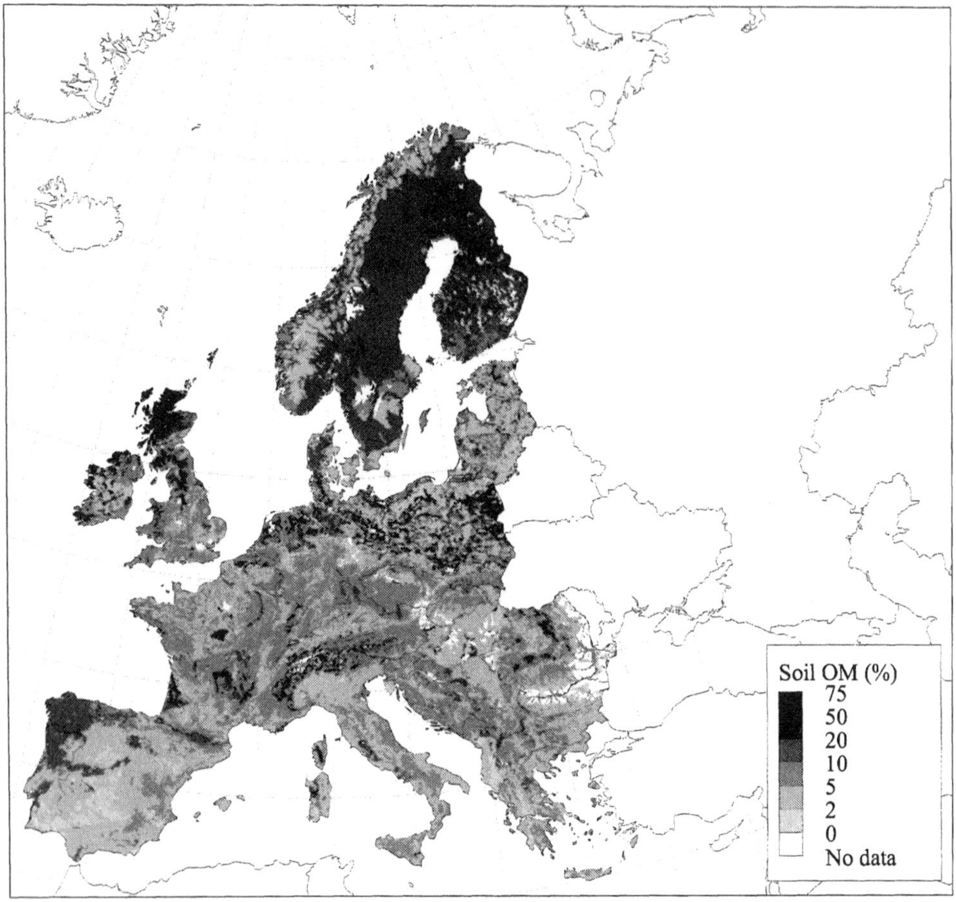


Fig. 6. Estimated soil organic matter (OM) derived from the European Soil Data Base (produced by S.M. Wright, CEH Lancaster, from Breuning-Madsen and Jones, 1995; Heineke et al., 1998).

uptake of $^{137}Cs$ by *Juncus trifidus* from the humus layer of subalpine soils was still 3–6 times higher than $^{133}Cs$ two years after the Chernobyl accident (Varskog et al., 1994). The rate at which radionuclides will equilibrate with their stable counterparts in organic soils could be significantly delayed by small amounts of clay minerals contained within a largely organic matrix.

The presence of $^{134}Cs$ (half-life 2.06 years) and $^{137}Cs$ (half-life 30.17 years) in the Chernobyl fallout provided a unique opportunity to distinguish between the plant availability of residual activity from weapons testing and that from the Chernobyl accident. Calculations are typically based on a $^{134}Cs$:$^{137}Cs$ ratio of 0.53 for Chernobyl fallout. As expected, some studies reported higher availability of Chernobyl-derived radiocaesium for plant uptake, for example in 1988 on heathland (McAulay et al., 1989) and in 1990–92 on grassland (Ehlken and Kirchner, 1996). But there are also studies in which no difference in the availability of the two types of deposit was found, for example on blanket mire in 1989 (Beresford et al., 1992) and in 1990 (McGee et al., 1992). Interpretation is hampered by the lack of soil information in some publications but it appears that on highly organic soils several years post-accident, plant availability of the two deposits is likely to be similar. Owing to the relatively short half-life of $^{134}Cs$, activity concentrations in most areas are now too low to distinguish between Chernobyl and weapons test deposits.

## 6. Partitioning and fluxes of radionuclides

The general characteristics of nutrient distribution and fluxes in terrestrial ecosystems can be applied to radionuclides provided they are bioavailable (Heal and Horrill, 1983). These are:

(a) The ecosystem capital is large relative to inputs and outputs.
(b) A high proportion of the capital is below ground.
(c) Internal cycling within plants and cycling via decomposers are important.
(d) Elements change between labile and stable soil pools.

In the case of anthropogenic radionuclides there is some difficulty in applying the concept of an 'ecosystem capital' since by definition they do not occur naturally. However, two approaches can be taken: (A) where a stable isotope exists the capital can be regarded as the sum of stable and radioactive isotopes or (B) the capital can be regarded as only the activity of the radioisotope present in the system at a given time. The latter approach has to be adopted where no stable forms exist and in such cases inputs can be large relative to the capital. Approach A is only useful when all isotopes are in equilibrium with each other. Since stable isotopes such as $^{133}Cs$ and $^{88}Sr$ are always present in quantities far exceeding radionuclides they will dominate ecosystem fluxes. For example the average content of $^{133}Cs$ in mineral soils is estimated at 3 $mg\,kg^{-1}$ (Ure and Berrow, 1982) while contamination with 1000 $Bq\,kg^{-1}$ of $^{137}Cs$ is only equivalent to 0.3 $ng\,kg^{-1}$.

### 6.1. *Inputs, outputs and redistribution*

Substantial inputs of anthropogenic radionuclides into terrestrial ecosystems occur most commonly via wet and dry deposition and depend on the nature of the contamination source. Smaller localised inputs occur via spillages and waste disposal. While authorised discharges can continually add to the existing radionuclide capital, fallout from weapons tests or accidental releases is time-limited and after cessation inputs will rapidly decline. Transfers between ecosystems may occur as a result of transport and redeposition, e.g. during flooding (Burrough et al., 1999). Sea-to-land transfer can redistribute a small proportion of marine discharges onto coastal land via sea spray (Nelis et al., 1994) and sediment transport (Tyler, 1999). Soil resuspension through wind and rain splash potentially causes secondary foliar contamination but should be regarded as a cyclical rather than an input process.

Outputs are defined as losses from the ecosystem rather than losses from a particular compartment that can be biologically retrieved. The key processes are vertical and horizontal leaching, erosion and removal via food products.

While it is relatively easy to identify the input and output pathways for radionuclides it is much more difficult to quantify them and detailed input/output studies are lacking. In the

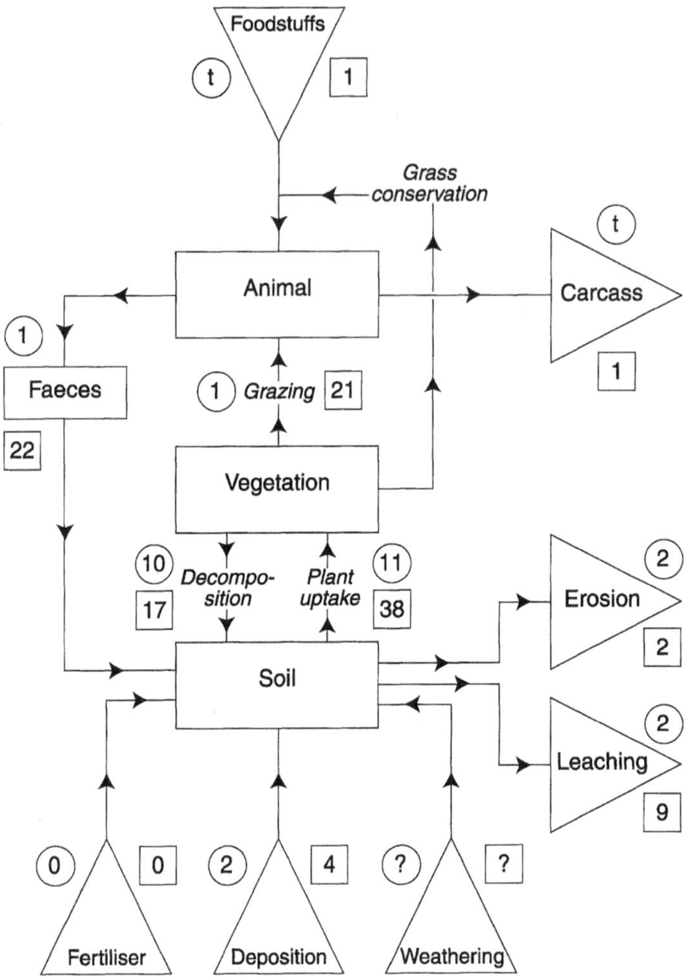


Fig. 7. Potassium budgets for two sheep grazing systems. Figures in circles are for a farm dominated by mire vegetation and figures in squares are for a farm with improved and unimproved grassland; t = trace (adapted from Briggs and Courtney, 1985; Newbould and Floate, 1978).

case of radiocaesium parallels can be drawn with ecosystem budgets for potassium but equivalent budgets for calcium are again lacking because it is less likely to be in short supply. Figure 7 shows potassium budgets for two grazing systems (adapted from (Newbould and Floate, 1978; Briggs and Courtney, 1985). Although the weathering of K-bearing minerals is the main source of K in non-fertilised systems (Mengel and Kirkby, 2001) nutrient budgets typically do not include an estimate of inputs via this route. This could explain the estimated net loss of K through erosion and leaching in grazed systems (Crisp, 1966; Briggs and Courtney, 1985) (see Table 2). On deep peat mineral weathering would only contribute potassium where weathering products are transferred downslope from higher lying mineral soils. Radio-

Table 2
Potassium inputs and outputs for an area of moorland in the Pennines, England. After (Crisp, 1966)

| | K (kg ha$^{-1}$ year$^{-1}$) |
|---|---|
| Precipitation input | 3.070 |
| Outputs | |
| Sale of sheep | 0.005 |
| Dissolved matter in stream | 8.970 |
| Drift of fauna in stream | 0.005 |
| Erosion of peat in stream | 2.060 |
| Total output | 11.040 |
| Net loss | 7.970 |

caesium, once fixed in the lattices of clay minerals, could conceivably be released through weathering but this slow process is not likely to yield a measurable input.

Vertical migration of radionuclides in soils can occur in the dissolved and the particulate phase, though much less is known about the latter. In soils with high adsorption capacity most radionuclides will migrate only very slowly, redistributing down the soil profile. Forsberg et al. (2000), experimenting with four European mineral soils, measured maximum migration rates for $^{137}$Cs and $^{90}$Sr of 11 mm year$^{-1}$ in a sandy loam. In a Swedish study maximum migration of 10 mm year$^{-1}$ was seen in an organic soil (Rosén et al., 1999). Estimates by Bunzl et al. (1994) for the migration of $^{137}$Cs, $^{241}$Am and $^{239+240}$Pu from weapons test fallout in a silty clay loam under permanent pasture ranged from 3 to 10 mm year$^{-1}$, increasing with soil depth. The distribution of Chernobyl-$^{137}$Cs indicated faster migration (7–50 mm year$^{-1}$).

Owing to the slow vertical migration of radiocaesium and radiostrontium, redistribution down the soil profile and beyond the rooting zone generally takes many decades or longer (Bunzl, 1987). However, there is evidence that radiocaesium migrates more rapidly in organic soils. For example, whilst a podzolic soil still retained the total inventory of $^{137}$Cs from weapons fallout 30 years after peak deposition, a neighbouring peat appeared to have lost 60% despite lying in a depression (Colgan et al., 1993). This loss estimate may appear high, but is supported by catchment studies showing that substantial redistribution and long-term losses to river systems can occur. Salo et al. (1984) estimated a cumulative loss of 25–52% of weapons fallout $^{90}$Sr from 4 Finnish catchments by 1981 and Kudelsky et al. (1996) calculated an annual loss of 0.3% of the $^{137}$Cs inventory from a peaty catchment in Belarus. Tyler and Heal (2000) successfully modelled the redistribution of $^{137}$Cs from weapons fallout in relation to topography and predicted 20% loss from a Scottish catchment over 30 years. The relative contributions of solute transport in throughflow and of particulate transport in overland flow and pipeflow are still uncertain. Annual radiocaesium losses through erosion of particulate matter are unlikely to exceed 1% of the inventory though losses of $^{90}$Sr could be as high as 5% (Bonnett, 1990).

Redistribution of surface-deposited radionuclides down a soil profile can be promoted by the action of soil fauna, so-called bioturbation. A study of upland soils linked the vertical displacement of $^{137}$Cs from weapons tests to the activity of earthworms and enchytraeids (Tyler et al., 2001). Also, on peat soils one might expect surface-deposits of radionuclides to become buried through peat formation which for example in Scotland adds about 1 mm of new material each year. However, the peat profile analysed by Tyler et al. (2001) suggests that radiocaesium is cycled back to the surface layer through plant uptake and decomposition.

The loss of radionuclides from grasslands and semi-natural ecosystems through removal of food products has been quantified in only few studies. For grassland and *Calluna* heath on organic soils it has been estimated that the $^{134}$Cs inventory would be reduced by about 0.1–0.2% per year through removal of sheep in autumn (Salt et al., 1992, 1994). This probably reflects the high bioavailability of the artificially applied $^{134}$CsCl which had been injected into the soil at the beginning of the growing season. Not unexpectedly much smaller amounts of total K are removed through sheep production, e.g. 0.0014% from unfertilised grassland (Perkins, 1978), 0.0032% from fertilised grassland (Till, 1981).

In a lysimeter study on mineral soil Strebl et al. (Chapter 2, this book) concluded that off-take through maize harvest (0.1%) was much more significant than the leaching loss of $^{137}$Cs (<0.0001%). However in semi-natural ecosystems on organic soils this could be reversed due to lower plant productivity coupled with greater leachability.

### 6.2. *Partitioning above and below ground*

The soil generally acts as a sink for radionuclides and typically at least 90% of the capital is below ground. Strong adsorption to soil colloids and complexation can lead to negligible plant root uptake from soils and hence very small above-ground inventories. In the case of plutonium contamination at Savannah River, South Carolina, >98% of $^{238}$Pu and $^{239,240}$Pu resided in the soil of abandoned fields that had been colonised by grasses, herbs and pine (Pinder et al., 1979). The plant inventory of Pu was split into 80% direct surface deposit and 20% resuspended deposit.

Partitioning into plant biomass is favoured when

(a) the radionuclide has a nutrient analogue,
(b) the nutrient analogue is in short supply,
(c) soil binding is low,
(d) plant biomass is high, and
(e) the plant species has a high uptake capacity.

In intensive agricultural systems soil fixation is the dominant factor for $^{137}$Cs leading to small plant inventories, e.g. 0.105% in maize biomass at harvest (see Section 5 of Chapter 2). The highest fluxes into above-ground biomass on mineral soils are reported for species with high uptake capability, for example 3% of total $^{137}$Cs in *Amaranthus retroflexus* (Lasat et al., 1998).

Table 3 summarises plant inventories measured for radiocaesium in herbaceous and dwarf shrub communities. Even on peat soils where radiocaesium bioavailability is generally at its highest, the above-ground inventory remains less than 20% of the total. This contrasts with the results of a pot experiment where as much as 72% of $^{137}$Cs and 89% of $^{90}$Sr have been found in grass biomass (Entry et al., 1999). The authors attributed this unusual result to extremely high root density and low available K.

The distribution of radionuclides inside plants often reflects similarities in behaviour with plant nutrients. For Cs, analogous to K, one can generalise that partitioning into younger

Table 3
Compilation of post-Chernobyl data on the percentage of total radiocaesium present in the above-ground plant biomass, with soil organic matter content where known

| Author | Sampling year | Vegetation | Soil type | % $^{137}Cs$ in above-ground biomass | % Soil OM |
|---|---|---|---|---|---|
| Harrison et al. (1990) | 1988 | Heather moor (Cumbria, England) | Stagnohumic gley | 10/8.4 | 94/69 |
| | 1988 | Heather moor (Cumbria, England) | Peat | 18 | 90 |
| | 1988 | Heather moor (Cumbria, England) | Humus-iron podzol | 7.7 | 54 |
| | 1988 | Heather moor (Cumbria, England) | Peaty podzol | 13 | 21 |
| | 1988 | Heather moor (Cumbria, England) | Brown podzolic soil | 8.2 | 21 |
| | 1988 | Heather moor (Cumbria, England) | Humic ranker | 14 | 50 |
| Livens et al. (1991) | 1989 | Alpine meadow (Italy) | Calcareous brown earth | 0.4 | 31 |
| | 1989 | Open hillside (Norway) | Ranker | 0.7 | 31 |
| | 1989 | Mire (Norway) | Peat | 2.5 | close to 100 |
| | 1989 | Heather moor (Scotland) | Podzol | 1.7 | 82 |
| | 1989 | Open hillside (Scotland) | Ranker | 2.5 | 40 |
| Gerzabek and Mohamad (1993) | 1990 | Alpine pasture (Nassfeld, Austria) | Alpine ranker | 6 | – |
| | 1990 | Alpine pasture (Nassfeld, Austria) | Alpine brown earth | 6.8 | – |
| | 1990 | Alpine pasture (Nassfeld, Austria) | Alpine brown earth | 7.1 | – |
| | 1990 | Alpine pasture (Nassfeld, Austria) | Alpine alluvium | 7.8 | – |
| Salt and Mayes (1991) | Exp. | Rye-grass clover pasture (Scotland) | Peaty podzol | 2 | 90 |
| Salt and Mayes (1993) | Exp. | Heather moor (Scotland) | Iron podzol | 9 | 52 |
| | Exp. | Agrostis-Festuca pasture (Scotland) | Iron-podzol | 0.5 | 43 |

tissues is favoured and active remobilisation from older tissues occurs via phloem transport. For example, in autumn the most actively growing parts of *Calluna vulgaris*, the long and the flowering shoots, contained about 45% of the above-ground pool of $^{134}Cs$ but contributed less than 20% to the above-ground biomass (Salt and Mayes, 1993). Sr like Ca is transported almost exclusively via the xylem and the tissues transpiring most actively, i.e. the leaves will accumulate Sr. Partitioning studies on crop plants, e.g. Coughtrey and Thorne (1983) and Nisbet and Shaw (1994), show that Sr and Ca distribution are frequently similar with foliage and stems representing the largest sink. Comparable partitioning studies of perennial plants are lacking.

There has been discussion in the literature concerning the potential role of fungal biomass as a below-ground storage reservoir for radiocaesium. Forest studies suggest that up to 50% of the $^{137}Cs$ inventory could be associated with fungal biomass (Olsen et al., 1990; Vinichuk and Johanson, 2003) and one grassland study stresses the potential for immobilisation by fungal hyphae on organic soils (Dighton et al., 1991). Due to methodological problems only indirect evidence has so far been presented and the estimates carry a large error. Fungi are only likely to form an important radiocaesium pool in humus-rich soil horizons or soils where fungal biomass tends to be higher than in predominantly mineral horizons (Ruzicka et al., 2000).

### 6.3. *Cycling and storage within plants*

Internal retention and cycling of nutrients in plants is assumed to be a widespread mechanism to minimise nutrient loss especially in plants growing in infertile habitats (Aerts and Chapin, 2000). In evergreen species, such as ericoid shrubs, retention of nitrogen and phosphorus is higher than in deciduous shrubs, forbs and graminoids but tissue concentrations tend to be lower. Seasonal reserve storage is believed to be more important in deciduous species. For example, in the graminoids *Eriophorum vaginatum* and *Molinia caerulea* the demand for nitrogen and phosphorus early in the growing season can be partially met by nutrients stored in overwintering organs (basal internodes, stems, rhizomes, roots) (Thornton and Millard, 1993). It has been postulated that mechanisms that contribute to nutrient use efficiency in nutrient-poor habitats will lead to greater persistence of radionuclides (Grime, 1989; Jones et al., 1998b). However, the focus of ecological research has been on the nutrients most likely to limit plant growth, namely N and P. Only rarely is K the limiting nutrient (Goodman and Perkins, 1968) and Ca supply is generally not regarded as limiting.

Plants remobilize nutrients from senescing tissues but K is generally less efficiently retrieved than N or P (Chapin, 1980; Marschner, 1995). Up to 70% of K can be translocated before leaf abscission but leaching is believed to be more important. The remobilisation of K and Cs from older leaves to young leaves, fruits and roots has been quantified in a range of crop plants (Wolterbeek and De Bruin, 1986a, 1986b; Mengel and Kirkby, 2001) with Cs being somewhat less mobile than K. Calcium and strontium cannot be remobilised since they lack mobility in the phloem; their transport inside the plant is via the xylem and is driven by the transpiration stream (Marschner, 1995).

While there is no doubt that remobilisation of K and Cs occurs in perennial plants, it is less clear whether they have the capacity to store K (and hence Cs) during winter. Underground storage of K and other nutrients is assumed to be important for rapid growth of tundra communities following snowmelt, but this is based on indirect proof from nutrient concentrations measured in plants (Chapin et al., 1975). While Jones et al. (1998b) provided good evidence of the retrieval of $^{137}Cs$ from old to new leaves in *E. vaginatum*, experimental problems prevented a clear distinction between remobilisation from storage organs, such as rhizomes, and root uptake. This problem is also common to field studies, (e.g. Bunzl and Kracke, 1989).

K is more readily leached from living and senescing leaves than N and P (Tukey, 1970) and radiocaesium is also lost via this route. In a 3-week leaching experiment deciduous graminoids (*Eriophorum vaginatum, Trichophorum cespitosum*) and an evergreen dwarf shrub (*Erica tetralix*) lost between 12 and 28% of $^{137}Cs$ in August (Jones et al., 1998a). In November when the graminoids were mostly senesced, foliar loss was only 1–5% but remained high at 14% in the evergreen shrub. A review of canopy throughfall data for evergreen and deciduous trees suggests that on an annual basis both groups lose similar amounts of nutrients but losses in evergreens are more spread out over the year (Aerts and Chapin, 2000).

Evergreen species do not appear able to retain radiocaesium more effectively from one year to another than deciduous species. A comparison of $^{134}Cs$ in evergreen *Calluna vulgaris* with a range of graminoids and forbs showed no obvious differences in the relative decline of activity concentrations between year 1 and year 2 following injection of $^{134}Cs$ into the soil (Salt and Mayes, 1993). Gradual fixation of $^{134}Cs$ in the soil seemed to be the overriding influence.

### 6.4. *Decomposition*

The rate at which nutrients and contaminants are released from litter determines how quickly they can be reabsorbed by biota. K is more rapidly released from *Sphagnum* litter than N, P and C, with only 2% remaining in the litter after 126 days, compared to >50% of P and N and 93% of C (Brock and Bregman, 1989). Decomposition of *Sphagnum* litter is slow compared to other mire plants, such as *Eriophorum vaginatum* and *Calluna vulgaris*, due to its high content of phenols (Coulson and Butterfield, 1978). Release of nutrients from litter of evergreen plants is thought to be lower than that of deciduous species due to a combination of lower nutrient concentrations and lower decomposition rate (Aerts and Chapin, 2000). The latter was confirmed by a litter leaching experiment in which K and $^{137}Cs$ were released more slowly from the evergreen *Calluna vulgaris* compared to *Agrostis capillaris* and *Trifolium repens* (Clint et al., 1992). In *Calluna* the total release of K and $^{137}Cs$ was similar, while in the two deciduous species the release of K was faster and greater than that of $^{137}Cs$, as also observed in litter from deciduous trees (Witkamp and Frank, 1970).

## 7. Seasonality

Flows of energy and materials in ecosystems are strongly influenced by seasonal changes in climate. The resulting seasonal patterns of plant productivity will influence the flux of radionuclides into the plant biomass and on to consumers. Assuming a constant uptake rate of a given radionuclide, increased plant productivity would increase the flux but not the activity concentration in the plant biomass. In reality uptake is not constant and many studies have observed temporal changes in activity concentrations. For radiocaesium and radiostrontium some studies have found significant correlations between these changes and seasonal

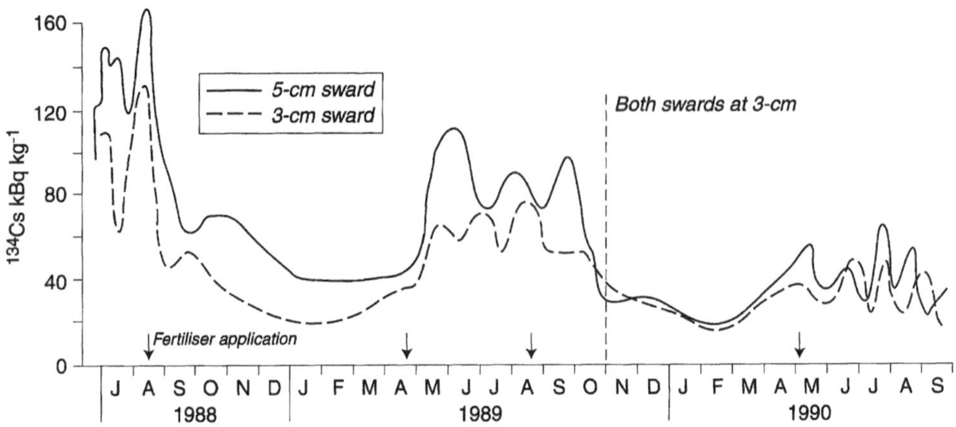


Fig. 8. $^{134}$Cs activity concentrations in 3-cm and 5-cm high swards of *Lolium perenne*, *Festuca rubra* and *Trifolium repens* grazed by sheep, north-east Scotland. Soil is a peaty podzol, artificially contaminated with 1500 kBq m$^{-2}$ in early June 1988 (Salt and Mayes, 1991; Salt, unpublished).

influences such as temperature, rainfall and soil moisture (Salt and Mayes, 1991; Ehlken and Kirchner, 1996).

Temporal patterns will change once an initial deposit onto vegetation becomes incorporated into the soil and root uptake begins to dominate vegetation contamination. Following the Chernobyl accident this led to higher plant $^{137}$Cs activity concentrations in the second year on soils with high bioavailability (Bunzl and Kracke, 1989).

It has to be stressed that infrequent sampling (i.e. less than monthly) or sampling for only part of a year make it difficult to detect seasonal patterns. Ideally sampling would be continued over 2 or more years. Also it can be difficult to recognise seasonal influences on the plant uptake of radiocaesium when these are accompanied by increasing fixation in soil minerals. Figure 8 provides an example of the combined influences of season and soil processes on $^{134}$Cs contamination in grass-dominated pasture following artificial contamination (Salt and Mayes, 1991; Salt, unpublished). While activity concentrations each year increased during the growing season and declined in autumn, the overall contamination level dropped year-on-year. The most likely cause is gradual immobilisation by the clay mineral illite which was identified in the organic surface horizon of the peaty podzol. Redistribution of $^{134}$Cs within the rooting zone could also be a contributory factor (Ehlken and Kirchner, 2002).

As discussed in Salt et al. (2004) there is no single generic seasonal pattern for radiocaesium emerging from the literature. Based on studies with sufficiently frequent sampling over one growing season two patterns may be recognised which also apply to potassium (Fleming and Murphy, 1968; Powell and Malcolm, 1974; Morton, 1977):

(a) spring rise, summer high, autumn decline, winter low,
(b) spring rise, decline through summer and autumn, winter low.

Pattern (a) is illustrated in Figure 9 with data for $^{137}$Cs from 1997/98 for unimproved grassland in Cumbria, England (Dale, 2000) and potassium concentrations (not shown) followed the same pattern. An earlier study in 1987–1989 on unimproved grassland in Cumbria had

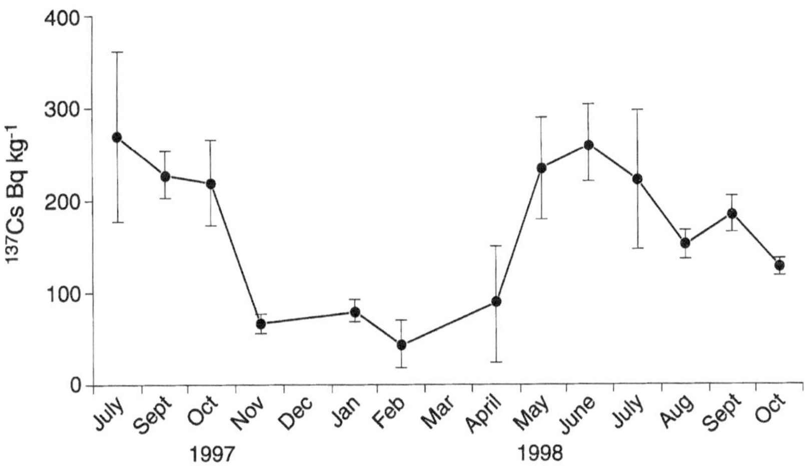


Fig. 9. $^{137}Cs$ activity concentration in sheep grazed vegetation dominated by *Carex nigra* and *Molinia caerulea* on a peaty podzol, Cumbria, England (Dale, 2000).

found similar seasonal patterns for $^{134}Cs$ and K (Sandalls and Bennett, 1992). Salt and Mayes (1993) suggest that Pattern (a) reflects plant growth and changes in the relative proportion of dead and green biomass. Pattern (b) is typically found when plant species, such as *Trichophorum cespitosum* and *Molinia caerulea*, are analysed individually (Bunzl and Kracke, 1989) (see Figure 10). The above-ground biomass of these species dies back completely in the autumn and new green tissue appears in spring. There is also evidence of a third seasonal pattern characterised by a winter peak in activity concentrations (Dale, 2000; Salt et al., 2004), possibly linked to a decrease in available K or an increase in exchangeable Ca in the soil.

Seasonal patterns of $^{133}Cs$ are poorly researched. A study of a range of higher plants failed to show any common trends between species though the authors speculate on the possible influence of soil moisture (Willey and Martin, 1995). Comparisons of temporal changes in $^{133}Cs$ and $^{137}Cs$ concentrations in *Trifolium repens* (Nishita et al., 1960) and *Agrostis capillaris* (Salt et al., 2004) show close correlation between the 2 isotopes. These two studies also illustrate that Cs and K can exhibit contrasting seasonal patterns but there are equally studies showing similar patterns for Cs and its nutrient analogue (Bunzl and Kracke, 1989; Dale, 2000). The factors responsible for these differences are most likely the relative availability of K and Cs in the soil, and plant-internal processes such as retranslocation and foliar leaching.

## 8. Plant species differences

There are many potential reasons why different plant species growing under the same environmental conditions and exposed to the same level of soil contamination will vary in radionuclide activity concentration. High initial interception of fallout and subsequent retention explain the high contamination levels found in lichens and mosses (Table 4). Lichens depend on nutrient inputs from dry and wet deposition and the lack of a waxy cuticle allows absorption over the entire lichen surface (Nash, 1996). Mosses also lack true roots and absorb nutrients

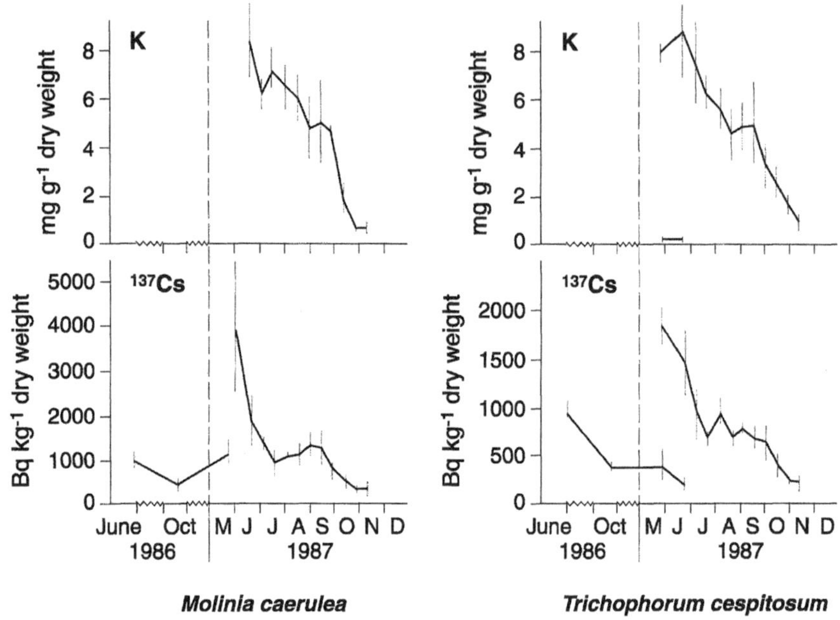


Fig. 10. $^{137}Cs$ and K activity concentrations in *Molinia caerulea* and *Trichophorum cespitosum* on a raised mire in Germany in 1986 and 1987 (Bunzl and Kracke, 1989); reproduced by permission of the Health Physics Society.

and pollutants from the atmosphere through their leaves. Metal pollutants are strongly held within the moss tissues (Onianwa, 2001).

It has been suggested that in higher plants growth strategy and nutrient use efficiency influence plant uptake of radiocaesium (Grime, 1989). Broadley and Willey (1997) observed higher Cs uptake in faster, compared to slower, growing grasses. Many post-Chernobyl studies have emphasised the high uptake of $^{137}Cs$ by the evergreen dwarf shrub *Calluna vulgaris* relative to other higher plant species such as graminoids as well as dwarf shrubs in the same family of the *Ericaceae* (Table 4). Problems arise with such comparisons because the species do not always grow in close proximity and differences in activity concentrations could be due to spatial differences in soil chemistry or deposition. For example, on dry soils *C. vulgaris* and *Erica cinerea* are typically found growing together but on wet soils *E. cinerea* is replaced by *E. tetralix*. Hence a direct comparison between these species is often difficult. When McGee et al. (1993) sampled all 3 species across 3 different farms, *C. vulgaris* consistently showed higher radiocaesium concentrations than the 2 *Erica* species. Data in Horrill et al. (1990) for five different sites confirm this for *C. vulgaris* when compared with *E. cinerea* as well as *Vaccinium myrtillus*. Taking all the evidence for species sampled in close proximity it can be generalised that grasses have lower radiocaesium concentrations than lichens, mosses and ericaceous shrubs. Within the family of the *Ericaceae*, *C. vulgaris* is typically the most contaminated species and this could be linked to the low potassium concentrations generally found in this species (Loach, 1968; Allen, 1989).

The fruiting bodies of ectomycorrhizal fungi found in forest habitats can accumulate even higher concentrations of radiocaesium than lichens and mosses (McGee et al., 2000). This

Table 4
Activity concentrations of $^{137}Cs$ in different plant species ($Bq\,kg^{-1}$ of d.w.) with the number of samples in brackets where known

| Study | | | Bunzl and Kracke (1986) | Bothmer et al. (1990) | Ekern et al. (1988) | Horrill et al. (1990) | McGee et al. (1992) |
|---|---|---|---|---|---|---|---|
| Geographical area | | | Raised mire Bavaria Germany | Forests and mires Harbo, Gävle Sweden | Mountain pasture Dovrefjell Norway | Mire, heath, grassland, Cumbria England | Blanket mire, heath Cuilcagh Mountains Ireland |
| Sampling year | | | 1984 | 1986–88 | 1987 | 1988 | 1990 |
| *Cladonia stellaris* | Lichen | | – | – | 55360 | – | – |
| *Spagnum spp.* | Moss | | – | – | – | 1589 (10) | – |
| *Polytrichum commune* | Moss | | – | – | – | 1992 (10) | – |
| *Calluna vulgaris* | Dwarf shrub | E | 1590 (5) | 13000 (15) | – | 915 (7) | 1690 (54) |
| *Erica tetralix* | Dwarf shrub | E | – | – | – | 1212 (3) | 324 (9) |
| *Erica cinerea* | Dwarf shrub | E | – | – | – | 52 (5) | – |
| *Vaccinium myrtillus* | Dwarf shrub | D | 740 (5) | 4100 (68) | 1565 | 141 (4) | – |
| *Vaccinium uliginosum* | Dwarf shrub | D | 510 (5) | 5900 (7) | – | – | – |
| *Vaccinium vitis-idaea* | Dwarf shrub | E | 310 (5) | 7500 (25) | – | – | – |
| *Vaccinium oxycoccus* | Dwarf shrub | E | 380 (5) | – | – | – | – |
| *Trichophorum cespitosum* | Sedge | D | 410 (5) | – | – | – | 1535 (6) |
| *Juncus effusus* | Rush | D | – | – | | 378 (14) | – |
| *Juncus squarrosus* | Rush | D | – | – | | – | 742 (9) |
| *Juncus trifidus* | Rush | D | – | – | 430 | – | – |
| *Molinia caerulea* | Grass | D | 550 (5) | – | – | – | – |
| *Nardus stricta* | Grass | D | – | – | – | 237 (13) | – |
| *Deschampsia flexuosa* | Grass | D | – | 1900 (30) | 2480 | – | – |

E = evergreen, D = deciduous.

Table 5
$^{134}Cs$ activity concentrations in two grasses (*L. perenne, D. flexuosa*) and one evergreen dwarf shrub (*C. vulgaris*) on 3 soils at Glensaugh experimental farm, north-east Scotland (data from Salt and Mayes, 1991, 1993; Salt, unpublished). Mean values for plant contamination in July/August 1–3 months following soil injection of $^{134}Cs$

| Soil type | Peaty podzol | | Iron podzol | | Brown forest soil | |
|---|---|---|---|---|---|---|
| $^{134}Cs$ added to soil ($kBq\,m^{-2}$) | 1500 | | 1500 | | 3000 | |
| pH | 4.3 | | 3.4 | | 5.8 | |
| Organic matter (%) | 90 | | 50 | | 8 | |
| Vegetation | Improved grassland | | Grassland/heath | | Improved grassland | |
| $^{134}Cs$ in plant ($kBq\,kg^{-1}$) | Mean | Stdev (*n*) | Mean | Stdev (*n*) | Mean | Stdev (*n*) |
| *Lolium perenne* | 80100 | 45400 (32) | – | – | 5469 | 2530 (15) |
| *Deschampsia flexuosa* | – | – | 34400 | 23200 (17) | – | – |
| *Calluna vulgaris* | 399000 | 165000 (6) | 218000 | 127000 (24) | – | – |

prompted research into the potential influence of mycorrhizal infection on radiocaesium uptake by higher plants, for example the ericoid mycorrhizae of *Calluna vulgaris* (Clint and Dighton, 1992) and vesicular-arbuscular mycorrhizae of grasses (Entry et al., 1999; Berreck and Haselwandter, 2001). The results have been contradictory though not unexpected, given that mycorrhizal infection is known to predominantly enhance the uptake of growth-limiting nutrients such as N and P but not K.

Experiments with 30 plant taxa and statistical analysis of a wide range of literature indicate that some plant taxa are characterised by high Cs uptake, e.g. *Chenopodium*, *Beta* and *Amaranthus*, while monocotyledons (e.g. grasses) generally show low uptake (Broadley and Willey, 1997; Broadley et al., 1999). Members of the *Ericaceae* and other families of woody species were not included in these studies. In the environment the major influence of soil type on radiocaesium bioavailability, in combination with taxonomic position and other factors such as root distribution, can lead to marked differences in the contamination of higher plant species within and between vegetation communities. Table 5 illustrates the effects of soil type and plant species.

## 9. The influence of management

The ecosystems under consideration in this chapter are most likely used for livestock farming, hunting and collection of berries. The main management tools are grazing control, mowing, fertilisation, reseeding, draining and burning though some systems will experience little or no management.

Over 40% of land in the temperate climate zone is grassland, largely replacing the natural vegetation, and livestock farming is a major land use (Briggs and Courtney, 1985). Under favourable climatic conditions pastures are either permanent, managed through fertilisation, reseeding and grazing control, or rotational with other crops. In some areas grazing by livestock has been replaced by cutting and indoor feeding all-year round. In more marginal areas

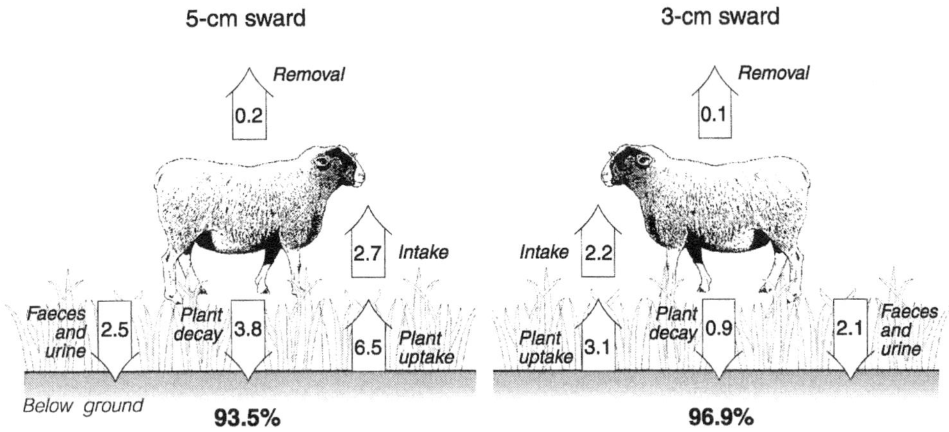


Fig. 11. Cycling of radiocaesium between soil, vegetation and sheep on improved grassland grazed to 3-cm and 5-cm height (based on Salt et al., 1992).

semi-natural pastures with low-productivity support lower stocking densities and receive less management.

Grazing and cutting regimes can affect radionuclide concentrations in pasture vegetation. Ehlken and Kirchner (1996) found higher $^{137}Cs$ concentrations in repeatedly clipped pasture compared to pasture harvested only once per year. They speculated that shallower rooting depth in response to defoliation would concentrate the roots more into the contaminated soil layer. However, it is equally possible that differences in growth and nutrient allocation play a role (Richards, 1993). Salt and Mayes (1991) measured higher $^{134}Cs$ concentrations in 5-cm compared to 3-cm tall grass swards grazed by sheep over two growing seasons (see Figure 8). In a pot experiment with defoliation by clipping this trend was confirmed for summer but not autumn measurements (Salt et al., 1997). On grazed swards the proportion of dead leaves increases with defoliation intensity (Hodgson, 1985). Since dead and senescing leaves have much lower radiocaesium concentrations than live tissues (Salt et al., 1997) this could lower the radiocaesium contamination of the whole sward. However, for radionuclide transfer to grazing animals the contamination in the ingested material is most relevant. On intensively managed grassland livestock will graze the surface layer of the sward fairly uniformly but on more heterogeneous vegetation diet selection can have a large impact on radionuclide intake (Salt et al., 1994).

As mentioned earlier only a very small proportion of the radionuclide inventory is removed in animal products. Figure 11 shows that most of the radiocaesium ingested by sheep is returned via faeces and urine to the pasture (Salt et al., 1992). In this study the proportion of $^{134}Cs$ removed via the sheep carcasses was estimated to be higher on the less severely grazed sward for two reasons, (a) higher dry matter intake by the sheep and (b) higher $^{134}Cs$ concentration in the vegetation.

Fertilisers can influence the soil chemical behaviour of radionuclides through ionic competition effects. The most well known examples are interactions of $^{137/134}Cs^{+}$ with $K^{+}$ and $NH_4^{+}$ released from mineral fertilisers, and of $^{90}Sr^{2+}$ with $Ca^{2+}$ released from liming agents (Nisbet, 1993; Shaw, 1993). In nutrient-poor soils with a low base saturation the fertiliser ions will most likely depress plant uptake of the radionuclide. Of the ecosystems considered here, only grasslands are likely to receive agrochemical inputs to boost plant productivity. Where the management regime maintains a high nutrient status and high base saturation, radionuclide availability will be low and fertiliser effects will be negligible. However, on marginal land, for example where organic soils have been improved to allow grass production, radionuclide mobility is more likely to be influenced by fertiliser inputs (Salt and Mayes, 1991).

The application of K fertiliser and lime to reduce radiocaesium and radiostrontium contamination of food products is a potentially useful countermeasure in a post-accident situation (IAEA, 1994). Effectiveness will be highest on unfertilised soils in low productivity systems but concern has been expressed about negative 'side-effects' especially in semi-natural ecosystems (Salt and Rafferty, 2001). Salt et al. (2002) reported lower concentrations of Mg, Ca and Mn in vegetation following K application especially on deep peat though the effects were not deemed severe enough to cause nutrient deficiencies in grazing sheep.

Controlled burning of heath dominated by the evergreen dwarf shrub *Calluna vulgaris* is widely practised in the UK. This encourages young growth, which is more nutritious for game birds such as red grouse and for domesticated sheep compared to mature shrubs (Gimingham, 1975). From experimental work it has been estimated that, depending on the temperature of the fire, 11–39% of the radiocaesium present in *Calluna* will be transferred to the smoke and deposited elsewhere (Horrill et al., 1995). Potentially the radiocaesium inventory could be reduced by 6% on a deep peat following a 'hot burn' and losses could be even greater where the fire progresses into the litter layer or soil. Between 17 and 24% of $^{137}Cs$ in the ash was leachable with water and could hence become mobilised when the ash is redeposited and exposed to rainfall.

Whilst the impact of heath burning will be geographically restricted, vegetation management aimed at intensifying livestock production will have more widespread relevance. Common practices such as liming, fertilising, draining, periodic ploughing and reseeding, especially in combination with each other, can result in substantial reductions in the soil-plant transfer of radiocaesium and radiostrontium. Such 'radical improvement' was a successful countermeasure in the FSU after the Chernobyl accident (Alexakhin, 1993).

## 10. Conclusions

Chemical properties, especially similarities with nutrient elements, determine the extent to which anthropogenic radionuclides will become involved in terrestrial nutrient cycling. Long-lived radioisotopes of caesium and strontium have received greatest attention due to their similarity with K and Ca and hence bioavailability, and their level of global deposition. Due to the focus of most studies on the human food chain, the best understood processes are those important for the soil-to-plant and plant-to-animal transfer. However, the vegetation of most grasslands and all heaths, mires and mountain ecosystems consists largely of perennial plant species and compared to arable crop systems storage and cycling are of greater importance. In the same way as it is for nutrients, the soil is typically the largest radionuclide reservoir and only a small proportion of the inventory is associated with the above-ground plant biomass. Interactions with soil colloids (clay and humus) greatly influence bioavailability as well as the extent to which radionuclides are lost from an ecosystem via leaching, erosion and removal of food products. Radiocaesium mobility is generally highest on organic soils with low K status whilst radiostrontium mobility is highest on sandy soils of low Ca status. Complex patterns of vegetation contamination arise through differences in soil properties in combination with plant-intrinsic factors and seasonal influences though evidence is largely restricted to radiocaesium. Catchment studies demonstrate how annual losses of $^{137}Cs$ and $^{90}Sr$ through solute and particulate transport can amount to significant reductions in the radionuclide inventory several decades after inputs have ceased.

Whilst our understanding of the behaviour of radiocaesium in the ecosystems reviewed here has greatly improved since the Chernobyl accident, internal cycling within plants and cycling via decomposers remain under-researched. More serious gaps exist for radiostrontium and other radionuclides.

## List of plant species mentioned

*Agrostis capillaris* L. (common bent)
*Amaranthus retroflexus* L. (common amaranth)
*Calluna vulgaris* (L.) Hull (heather)
*Cladonia stellaris* (Opiz) Pouzar and Vezda (star-tipped reindeer lichen)
*Deschampsia flexuosa* (L.) Trin. (wavy hair-grass)
*Erica cinerea* L. (bell heather)
*Erica tetralix* L. (cross-leaved heath)
*Eriophorum vaginatum* L. (hare's-tail cottongrass)
*Juncus effusus* L. (soft rush)
*Juncus squarrosus* L. (heath rush)
*Juncus trifidus* L. (three-leaved rush)
*Molinia caerulea* (L.) Moench (purple moor-grass)
*Nardus stricta* L. (matt-grass)
*Polytrichum commune* Hedw. (no common name, moss)
*Sphagnum* spp. (no common name, mosses)
*Trichophorum cespitosum* (L.) Hartm. (also *Scirpus cespitosus*) (deergrass)
*Trifolium repens* L. (white clover)
*Vaccinium myrtillus* L. (bilberry)
*Vaccinium oxycoccus* L. (cranberry)
*Vaccinium uliginosum* L. (bog bilberry)
*Vaccinium vitis-idaea* L. (cow berry)

Nomenclature follows:

Purvis, O.W., Coppins, B.J., Hawksworth, D.L., James, P.W., Moore, D.M. (1992). *The Lichen Flora of Great Britain and Ireland.* Natural History Museum, London.
Smith, A.J.E. (1993). *The Moss Flora of Britain and Ireland.* Cambridge University Press, Cambridge.
Stace, C. (1997). *New Flora of The British Isles.* Cambridge University Press, Cambridge.

## References

Absalom, J.P., Young, S.D., Crout, N.M.J., Sanchez, A., Wright, S.M., Smolders, E., Nisbet, A.F., Gillett, A.G. (2001). Predicting the transfer of radiocaesium from organic soils to plants using soil characteristics. *Journal of Environmental Radioactivity* **52**, 31–43.

Aerts, R., Chapin, F.S. (2000). The mineral nutrition of wild plants revisited: A re-evaluation of processes and patterns. *Advances in Ecological Research* **30**, 1–67.

Alexakhin, R.M. (1993). Countermeasures in agricultural production as an effective means of mitigating the radiological consequences of the Chernobyl accident. *Science of the Total Environment* **137**, 9–20.

Alexakhin, R.M., Krouglov, S.V. (2001). Soil as the main compartment for radioactive substances in terrestrial ecosystems. In: Brechignac, F., Howard, B.J. (Eds.), *Radioactive Pollutants. Impact on the Environment*. EDP Sciences, Les Ulis, pp. 149–174.

Allen, S.E. (Ed.) (1989). *Chemical Analysis of Ecological Materials*. Blackwell, Oxford.

Askbrant, S., Sandalls, J. (1998). Root uptake of $^{137}Cs$ and $^{90}Sr$ by rye-grass on various soils in the CIS. *Journal of Environmental Radioactivity* **38**, 85–95.

Barber, D.A. (1964). Influence of soil organic matter on the entry of caesium-137 into plants. *Nature* **204**, 1326–1327.

Beresford, N.A., Howard, B.J., Barnett, C.L., Crout, N.M.J. (1992). The uptake by vegetation of Chernobyl and aged radiocaesium in upland West Cumbria. *Journal of Environmental Radioactivity* **16**, 181–195.

Berreck, M., Haselwandter, K. (2001). Effect of the arbuscular mycorrhizal symbiosis upon uptake of cesium and other cations by plants. *Mycorrhiza* **10**, 275–280.

Bonnett, P.J.P. (1990). A review of the erosional behavior of radionuclides in selected drainage basins. *Journal of Environmental Radioactivity* **11**, 251–266.

Bothmer, S.V., Johanson, K.J., Bergstroem, R. (1990). Cesium-137 in moose diet considerations on intake and accumulation. *The Science of the Total Environment* **91**, 87–96.

Bretten, S., Gaare, E., Skogland, T., Steinnes, E. (1992). Investigations of radiocesium in the natural terrestrial environment in Norway following the Chernobyl accident. *Analyst* **117**, 501–503.

Breuning-Madsen, H., Jones, R.J.A. (1995). Soil profile analytical database for the European Union. *Danish Journal of Geography* **95**, 49–57.

Briggs, D.J., Courtney, F.M. (1985). *Agriculture and Environment*. Longman, London.

Broadley, M.R., Willey, N.J. (1997). Differences in root uptake of radiocaesium by 30 plant taxa. *Environmental Pollution* **97**, 11–15.

Broadley, M.R., Willey, N.J., Mead, A. (1999). A method to assess taxonomic variation in shoot caesium concentration among flowering plants. *Environmental Pollution* **106**, 341–349.

Brock, T.C.M., Bregman, R. (1989). Periodicity in growth, productivity, nutrient content and decomposition of *Sphagnum recurvum var. mucronatum* in a fen woodland. *Oecologia* **80**, 44–52.

Bunzl, K. (1987). Das Verhalten von Radionukliden im Boden. *Deutsche Tierärtzliche Wochenschrift* **94**, 325–380.

Bunzl, K., Kracke, W. (1986). Accumulation of fallout $^{137}Cs$ in some plants and berries of the family *Ericaceae*. *Health Physics* **50**, 540–542.

Bunzl, K., Kracke, W. (1989). Seasonal variation of soil-to-plant transfer of K and fallout $^{134,137}Cs$ in peatland vegetation. *Health Physics* **57**, 593–600.

Bunzl, K., Forster, H., Kracke, W., Schimmack, W. (1994). Residence times of fallout $^{239+240}Pu$, $^{238}Pu$, $^{241}Am$ and $^{137}Cs$ in the upper horizons of an undisturbed grassland soil. *Journal of Environmental Radioactivity* **22**, 11–27.

Burrough, P.A., Van der Perk, M., Howard, B.J., Prister, B.S., Sansone, U., Voitsekhovitch, O.V. (1999). Environmental mobility of radiocaesium in the Pripyat catchment, Ukraine/Belarus. *Water Air and Soil Pollution* **110**, 35–55.

Chapin, F.S. III (1980). The mineral nutrition of wild plants. *Annual Review of Ecological Systems* **11**, 233–260.

Chapin, F.S.I., VanCleve, K., Tieszen, L.L. (1975). Seasonal nutrient dynamics of tundra vegetation at Barrow, Alaska. *Arctic and Alpine Research* **7**, 209–226.

Clint, G.M., Dighton, J. (1992). Uptake and accumulation of radiocaesium by mycorrhizal and non-mycorrhizal heather plants. *New Phytologist* **121**, 555–561.

Clint, G.M., Harrison, A.F., Howard, D.M. (1992). Rates of leaching of $^{137}Cs$ and potassium from different plant litters. *Journal of Environmental Radioactivity* **16**, 65–76.

Colgan, P.A., McCann, P., McGee, E.J., McAulay, I.R. (1993). Short and long-term losses of $^{137}Cs$ from peatland soils. *Irish Journal of Agricultural and Food Research* **32**, 37–46.

Coughtrey, P.A., Thorne, M.C. (1983). *Radionuclide Distribution and Transport in Terrestrial and Aquatic Ecosystems*. Balkeema, Rotterdam.

Coulson, J.C., Butterfield, J. (1978). An investigation of the biotic factors determining the rates of plant decomposition on blanket bog. *Journal of Ecology* **66**, 631–650.

Crisp, D.T. (1966). Input and output of minerals for an area of Pennine moorland: The importance of precipitation, drainage, peat erosion and animals. *Journal of Applied Ecology* **3**, 327–348.

Dale, P.G. (2000). The Effect of Applying Potassium as a Countermeasure against Radiocaesium in Organic Soils. PhD Thesis, University of Stirling, Stirling.

De Cort, M., Dubois, G., Fridman, S.D., Germenchuk, M.G., Izrael, Y.A., Janssens, A., Jones, A.R., Kelly, G.N., Kvasnikova, E.V., Matveenko, I.I., Nazarov, I.M., Pokumeiko, Y.M., Sitak, V.A., Stukin, E.D., Tabachny, L.Y., Tsaturov, Y.S., Avdyushin, S.I. (1998). *Atlas of Caesium Deposition in Europe after the Chernobyl Accident*. Office for Official Publications of the European Communities, Luxembourg.

Desmet, G., Nassimbeni, P., Belli, M. (Eds.) (1990). *Transfer of Radionuclides in Natural and Semi-Natural Environments*. Elsevier, London.

Dickinson, G., Murphy, K. (1998). Ecosystems. A Functional Approach. Routledge, London.

Dighton, J., Clint, G.M., Poskitt, J. (1991). Uptake and accumulation of $^{137}$Cs by upland grassland soil fungi—a potential pool of Cs immobilization. *Mycological Research* **95**, 1052–1056.

Ehlken, S., Kirchner, G. (1996). Seasonal variations in soil-to-grass transfer of fallout strontium and cesium and of potassium in North German soils. *Journal of Environmental Radioactivity* **33**, 147–181.

Ehlken, S., Kirchner, G. (2002). Environmental processes affecting plant root uptake of radioactive trace elements and variability of transfer factor data: a review. *Journal of Environmental Radioactivity* **58**, 97–112.

Ekern, A., Hove, K., Garmo, T.H. (1988). Radiocaesium contamination of Norwegian mountain pastures and grazing animals after the Chernobyl accident. In: VI World Conference on Animal Production, Helsinki, p. 279.

Entry, J.A., Watrud, L.S., Reeves, M. (1999). Accumulation of $^{137}$Cs and $^{90}$Sr from contaminated soil by three grass species inoculated with mycorrhizal fungi. *Environmental Pollution* **104**, 449–457.

Fleming, G.A., Murphy, W.E. (1968). The uptake of some major and trace elements by grasses as affected by season and stage of maturity. *Journal of the British Grassland Society* **23**, 174–185.

Forsberg, S., Rosen, K., Fernandez, V., Juhan, H. (2000). Migration of $^{137}$Cs and $^{90}$Sr in undisturbed soil profiles under controlled and close-to-real conditions. *Journal of Environmental Radioactivity* **50**, 235–252.

Fredriksson, L., Garner, R.J., Scott Russell, R. (1966). Caesium-137, Radioactivity and Human Diet. In: Scott Russel, R. (Ed.). Pergamon Press, Oxford, pp. 319–354.

Gerzabek, M.H., Mohamad, S.A. (1993). Beurteilung des Radiocäsiumverhaltens auf Almböden oder diesen ähnlichen Böden, Rep. No. OEFZS-A-2680. Forschungzentrum Seibersdorf.

Gimingham, C.H. (1975). *An Introduction to Heathland Ecology*. Oliver and Boyd, Edinburgh.

Goodman, G.T., Perkins, D.F. (1968). The role of mineral nutrients in *Eriophorum* communities IV. Potassium supply as a limiting factor in an *E. vaginatum* community. *Journal of Ecology* **56**, 685–696.

Grime, J.P. (1989). The stress debate—symptom of impending synthesis? *Biological Journal of the Linnean Society* **37**, 3–17.

Handley, R., Babcock, K.L. (1970). Translocation of carrier-free $^{85}$Sr, $^{137}$Cs and $^{106}$Ru in woody plants. *Radiation Botany* **10**, 577–583.

Hannah, L., Lohse, D., Hutchinson, C., Carr, J.L., Lankerani, A. (1994). A preliminary inventory of human disturbance of world ecosystems. *Ambio* **23**, 246–250.

Harrison, A.F., Clint, G., Jones, H.E., Poskitt, J., Howard, B.J., Howard, D.M., Beresford, N.A., Dighton, J. (1990). Distribution and Recycling of Radiocaesium in Heather Dominated Ecosystems. Report to MAFF, Contract No N601, Rep. No. ITE T07050f1. Institute of Terrestrial Ecology, Merlewood Research Station, Grange-over-Sands.

Heal, O.W., Horrill, A.D. (1983). Terrestrial ecosystems: an ecological context for radionuclide research. In: Coughtrey, P.A. (Ed.), *Ecological Aspects of Radionuclide Release*. Blackwell, Oxford, pp. 31–46.

Heineke, H.J., Eckelmann, W., Thomasson, A.J., Jones, R.J.A., Montanarella, L., Buckley, B. (1998). Land Information Systems: Developments for Planning the Sustainable Use of Land Resources, Rep. No. EUR 17729 En. Office for Official Publications of the European Communities, Luxembourg.

Hinton, T.G., Pinder, J.E.I. (2001). A review of plutonium releases from the Savannah River Site, subsequent behaviour within terrestrial and aquatic environments and the resulting dose to humans. In: Kudo, A. (Ed.), *Plutonium in the Environment*, vol. 1. Elsevier, Amsterdam, pp. 413–435.

Hird, A.B., Rimmer, D.L., Livens, F.R. (1995). Total cesium-fixing potentials of acid organic soils. *Journal of Environmental Radioactivity* **26**, 103–118.

Hodgson, J. (1985). The control of herbage intake in the grazing ruminant. *Proceedings of the Nutrition Society* **44**, 339–346.

Horrill, A.D., Kennedy, V.H., Harwood, T.R. (1990). The concentrations of Chernobyl derived radionuclides in species characteristic of natural and semi-natural ecosystems. In: Desmet, G., Nassimbeni, P., Belli, M. (Eds.), *Transfer of Radionuclides in Natural and Semi-Natural Environments*. Elsevier, London, pp. 27–39.

Horrill, A.D., Kennedy, V.H., Paterson, I.S., McGowan, G.M. (1995). The effect of heather burning on the transfer of radiocesium to smoke and the solubility of radiocesium associated with different types of heather ash. *Journal of Environmental Radioactivity* **29**, 1–10.

Howard, B.J. (2002). Spatial analysis of vulnerable areas in central Europe. In: Schulte, E.-H., Kelly, G.N., Jackson, C.A. (Eds.), *Decision Support for Emergency Management and Environmental Restoration*. Office for Official Publications of the European Communities, Luxembourg, pp. 32–33.

Howard, B.J., Howard, D.C. (1997). The radioecological significance of semi-natural ecosystems. In: Lake, J.V. (Ed.), *Health Impacts of Large Releases of Radionuclides*, vol. 203. Wiley, Chichester, pp. 21–37.

IAEA (1994). Guidelines for Agricultural Countermeasures Following and Accidental Release of Radionuclides, Rep. No. 363. IAEA and FAO, Vienna.

Jones, D.R., Eason, W.R., Dighton, J. (1998a). Foliar leaching of $^{137}$Cs from *Eriophorum vaginatum* L., *Scirpus caespitosus* L. and *Erica tetralix* L. *Environmental Pollution* **99**, 247–254.

Jones, D.R., Eason, W.R., Dighton, J. (1998b). Investigation of spatial and temporal patterns of $^{137}$Cs partitioning in *Eriophorum vaginatum* L. In relation to its nutrient retrieval and storage strategy. *Journal of Environmental Radioactivity* **40**, 271–288.

Juo, A.S.R., Barber, S.A. (1970). The retention of strontium by soils as influenced by pH, organic matter and saturation cations. *Soil Science* **109**, 143–148.

Kudelsky, A.V., Smith, J.T., Ovsiannikova, S.V., Hilton, J. (1996). Mobility of Chernobyl-derived $^{137}$Cs in a peatbog system within the catchment of the Pripyat River, Belarus. *Science of the Total Environment* **188**, 101–113.

Lasat, M.M., Fuhrmann, M., Ebbs, S.D., Cornish, J.E., Kochian, L.V. (1998). Phytoremediation of a radiocesium-contaminated soil: Evaluation of cesium-137 bioaccumulation in the shoots of three plant species. *Journal of Environmental Quality* **27**, 165–169.

Liden, K., Gustafsson, M. (1967). Relationships and seasonal variation of $^{137}$Cs in lichen, reindeer and man in Northern Sweden 1961–1965. In: Aberg, B., Hungate, F.P. (Eds.), *Radioecological Concentration Processes*. Pergamon Press, Oxford, pp. 193–208.

Livens, F.R., Rimmer, D.L. (1988). Physico-chemical controls on artificial radionuclides in soil. *Soil Use and Management* **4**, 63–69.

Livens, F.R., Horrill, A.D., Singleton, D.L. (1991). Distribution of radiocaesium in the soil-plant systems of upland areas of Europe. *Health Physics* **60**, 539–545.

Loach, K. (1968). Relations between soil nutrients and vegetation in wet-heaths. II. Nutrient uptake by the major species in the field and in controlled conditions. *Journal of Ecology* **56**, 117–127.

Marschner, H. (1995). *Mineral Nutrition of Higher Plants*, 2nd edn. Academic Press, London.

McAulay, I.R., Colgan, P.A., Moran, D. (1989). Measurements on retention and transfer characteristics of radiocesium from poor quality upland soils to heather and from heather to sheep. *Science of the Total Environment* **85**, 159–167.

McGee, E.J., Synnott, H.J., Colgan, P.A. (1992). Availability of radiocaesium from Chernobyl and bomb fallout to peatland vegetation in Ireland. In: *VIII Report of The Working Group on Soil-Plant Transfer Factors*. IUR, Madrid, pp. 27–29.

McGee, E.J., Colgan, P.A., Synnott, H.J. (1993). A new method for prediction of radiocaesium levels in vegetation: evidence form Irish uplands. *Journal of Environmental Radioactivity* **18**, 53–70.

McGee, E.J., Synnott, H.J., Johanson, K.J., Fawaris, B.H., Nielsen, S.P., Horrill, A.D., Kennedy, V.H., Barbayiannis, N., Veresoglou, D.S., Dawson, D.E., Colgan, P.A., McGarry, A.T. (2000). Chernobyl fallout in a Swedish spruce forest ecosystem. *Journal of Environmental Radioactivity* **48**, 59–78.

Mengel, K., Kirkby, E.A. (2001). *Principles of Plant Nutrition*, 5 edn. Kluwer, Dordrecht.

Miller, C.W., Hoffman, F.O. (1983). An examination of the environmental half-time for radionuclides deposited on vegetation. *Health Physics* **45**, 731–744.

Moorby, J. (1964). The foliar uptake and translocation of caesium. *Journal of Experimental Botany* **15**, 457–469.

Morton, A.J. (1977). Mineral nutrient pathways in a *Molinietum* in autumn and winter. *Journal of Ecology* **65**, 993–999.

Nash, T.H. (1996). Nutrients, elemental accumulation and mineral cycling. In: *Lichen Biology*. Cambridge University Press, Cambridge, pp. 136–153.

Nelis, P.M., Branford, D., Unsworth, M.H. (1994). A model of the transfer of radioactivity from sea to land in sea spray. *Atmospheric Environment* **28**, 3213–3223.

Newbould, P., Floate, M.J.S. (1978). Nutrient cycling data. Agro-ecosystems in the UK. In: Frissel, M.J. (Ed.), *Cycling of Mineral Nutrients in Agricultural Ecosystems*, vol. 3. Elsevier, Amsterdam, pp. 33–48.

Nisbet, A.F. (1993). Effect of soil-based countermeasures on solid-liquid equilibria in agricultural soils contaminated with radiocesium and radiostrontium. *Science of the Total Environment* **137**, 99–118.

Nisbet, A.F., Shaw, S. (1994). Summary of a 5-year lysimeter study on the time-dependent transfer of $^{137}Cs$, $^{90}Sr$, $^{239}Pu$, $^{240}Pu$ and $^{241}Am$ to crops from 3 contrasting soil types. 2. Distribution between different plant-parts. *Journal of Environmental Radioactivity* **23**, 171–187.

Nishita, H., Romney, E.M., Alexander, G.V., Larson, K.H. (1960). Influence of K and Cs on release of Cs 137 from three soils. *Soil Science* **89**, 167–176.

Olsen, R.A., Joner, E., Bakken, L.R. (1990). Soil fungi and the fate of radiocaesium in the soil ecosystem. In: Desmet, G., Nassimbeni, P., Belli, M. (Eds.), *Transfer of Radionuclides in Natural and Semi-Natural Environments*. Elsevier, London, pp. 657–663.

Onianwa, P.C. (2001). Monitoring atmospheric metal pollution: A review of the use of mosses as indicators. *Environmental Monitoring and Assessment* **71**, 13–50.

Oughton, D.H. (1989). The Environmental Chemistry of Radiocaesium and other Radionuclides. PhD Thesis, University of Manchester, Manchester.

Perkins, D.F. (1978). The distribution and transfer of energy and nutrients in the *Agrostis-Festuca* grassland ecosystem. In: Heal, O.W., Perkins, D.F. (Eds.), *Production Ecology of British Moors and Montane Grasslands*, vol. 27. Springer Verlag, New York, pp. 375–395.

Pinder, J.E.I., Smith, M.H., Boni, A.L., Corey, J.C., Horton, J.H. (1979). Plutonium inventories in two old-field ecosystems in the vicinity of a nuclear-fuel reprocessing facility. *Ecology* **60**, 1141–1150.

Powell, C.F., Malcolm, D.C. (1974). Seasonal and spatial variation in the mineral content of some upland vegetation types in Scotland. *Journal of Applied Ecology* **11**, 1173–1184.

Richards, J.H. (1993). Physiology of plants recovering from defoliation. In: Baker, M.G., Crush, J.R., Humphries, L.R. (Eds.), *XVII International Grassland Congress*. Palmerston North, Hamilton, Lincoln and Rockhampton, pp. 85–94.

Rosén, K., Öborn, I., Lönsjö, H. (1999). Migration of radiocaesium in Swedish soil profiles after the Chernobyl accident, 1987–1995. *Journal of Environmental Radioactivity* **46**, 45–66.

Rühm, W., Yoshida, S., Muramatsu, Y., Steiner, M., Wirth, E. (1999). Distribution patterns for stable $^{133}Cs$ and their implications with respect to the long-term fate of radioactive $^{134}Cs$ and $^{137}Cs$ in a semi-natural ecosystem. *Journal of Environmental Radioactivity* **45**, 253–270.

Ruzicka, S., Edgerton, D., Norman, M., Hill, T. (2000). The utility of ergosterol as a bioindicator of fungi in temperate soils. *Soil Biology and Biochemistry* **32**, 989–1005.

Salo, A., Saxén, R., Puhakainen, M. (1984). Transport of airborne $^{90}Sr$ and $^{137}Cs$ deposited in the basins of the five largest rivers in Finland. *Aqua Fennica* **1**, 21–31.

Salt, C.A., Mayes, R.W. (1991). Seasonal variations in radiocaesium uptake by reseeded hill pasture grazed at different intensities by sheep. *Journal of Applied Ecology* **28**, 947–962.

Salt, C.A., Mayes, R.W., Elston, D.A. (1992). Effects of season, grazing intensity and diet composition on the radiocaesium intake by sheep on re-seeded hill pasture. *Journal of Applied Ecology* **29**, 378–387.

Salt, C.A., Mayes, R.W. (1993). Plant uptake of radiocaesium on heather moorland grazed by sheep. *Journal of Applied Ecology* **30**, 235–246.

Salt, C.A., Mayes, R.W., Colgrove, P.M., Lamb, C.S. (1994). The effects of season and diet composition on the radiocaesium intake by sheep grazing on heather moorland. *Journal of Applied Ecology* **31**, 125–136.

Salt, C.A., Kay, J.W., Donaldson, L., Woolsey, J.M. (1997). The influence of defoliation intensity, season and leaf age on radiocaesium concentrations in *Agrostis capillaris*. *Journal of Applied Ecology* **34**, 1177–1189.

Salt, C.A., Rafferty, B. (2001). Assessing potential secondary effects of countermeasures in agricultural systems: a review. *Journal of Environmental Radioactivity* **56**, 99–114.

Salt, C.A., Dale, P.G., Howard, B.J. (2002). Application of potassium on organic soils contaminated with radiocaesium—Is it effective in the field and is it likely to affect ecosystems negatively? In: Borretzen, P., Jolle, T., Strand, P. (Eds.), *Proceedings of the International Conference on Radioactivity in the Environment, Monaco, 1–5 Sept. 2002*. International Union of Radioecology, Monaco.

Salt, C.A., Kay, J.W., Jarvis, K.E. (2004). The influence of season and leaf age on concentrations of radiocaesium ($^{137}$Cs), stable caesium ($^{133}$Cs) and potassium in *Agrostis capillaris*. *Environmental Pollution* **130**, 359–369.

Sanchez, A.L., Wright, S.M., Smolders, E., Naylor, C., Stevens, P.A., Kennedy, V.H., Dodd, B.A., Singleton, D.L., Barnett, C.L. (1999). High plant uptake of radiocesium from organic soils due to Cs mobility and low soil K content. *Environmental Science and Technology* **33**, 2752–2757.

Sandalls, J., Bennett, L. (1992). Radiocaesium in upland herbage in Cumbria, UK: a three year field study. *Journal of Environmental Radioactivity* **16**, 147–165.

Sandalls, F.J., Segal, M.G., Victorova, N. (1993). Hot particles from Chernobyl: a review. *Journal of Environmental Radioactivity* **18**, 5–22.

Sawhney, B.L. (1966). Kinetics of cesium sorption by clay minerals. *Soil Science Society of America Proceedings* **30**, 565–569.

Schuller, P., Voigt, G., Handl, J., Ellies, A., Oliva, L. (2002). Global weapons' fallout $^{137}$Cs in soils and transfer to vegetation in south-central Chile. *Journal of Environmental Radioactivity* **62**, 181–193.

Shaw, G. (1993). Blockade by fertilizers of caesium and strontium uptake into crops—effects on the root uptake process. *The Science of the Total Environment* **137**, 119–133.

Squire, H.M., Middleton, L.J. (1966). Behaviour of Cs$^{137}$ in soils and pastures. A longterm experiment. *Radiation Botany* **6**, 413–423.

Staunton, S., Dumat, C., Zsolnay, A. (2002). Possible role of organic matter in radiocaesium adsorption in soils. *Journal of Environmental Radioactivity* **58**, 163–173.

Thornton, B., Millard, P. (1993). The effects of nitrogen supply and defoliation on the seasonal internal cycling of nitrogen in *Molinia caerulea*. *Journal of Experimental Botany* **44**, 531–536.

Till, A.R. (1981). Cycling of plant nutrients in pastures. In: Morley, A.R. (Ed.), *Grazing Animals, World Science, B1*. Elsevier, Amsterdam, pp. 33–53.

Tukey, H.B.J. (1970). The leaching of substances from plants. *Annual Review of Plant Physiology* **21**, 305–324.

Tyler, A.N. (1999). Monitoring anthropogenic radioactivity in salt marsh environments through in situ gamma-ray spectrometry. *Journal of Environmental Radioactivity* **45**, 235–252.

Tyler, A.N., Heal, K.V. (2000). Predicting areas of $^{137}$Cs loss and accumulation in upland catchments. *Water Air and Soil Pollution* **121**, 271–288.

Tyler, A.N., Carter, S., Davidson, D.A., Long, D.J., Tipping, R. (2001). The extent and significance of bioturbation on $^{137}$Cs distributions in upland soils. *Catena* **43**, 81–99.

UNSCEAR (1969). Report of the United Nations Scientific Committee on the Effects of Atomic Radiation. Official Records of the General Assembly, 24th Session, Suppl. 13/A7613. United Nations, New York.

UNSCEAR (2000). Sources and Effects of lonising Radiation. United Nations Scientific Committee on the Effects of Atomic Radiation Report 2000 to the General Assembly, Annex C and J. United Nations, New York.

Ure, A.M., Berrow, M.L. (1982). The elemental constituents of soil. In: Bowen, H.J.M. (Ed.), In: *Environmental Chemistry. A Review of the Literature published up to mid-1980*, vol. 2. The Royal Society of Chemistry, London, pp. 96–101.

Varskog, P., Næumann, R., Steinnes, E. (1994). Mobility and plant availability of radioactive Cs in natural soil in relation to stable Cs, other alkali elements and soil fertility. *Journal of Environmental Radioactivity* **22**, 43–53.

Vinichuk, M.M., Johanson, J.J. (2003). Accumulation of $^{137}$Cs by fungal mycelium in forest ecosystems of Ukraine. *Journal of Environmental Radioactivity* **64**, 27–43.

Walter, H. (1985). *Vegetation of the Earth*. Springer, Berlin.

Webster, S., Salt, C.A., Howard, B.J. (2003). Sea-to-land transfer of technetium-99 through the use of contaminated seaweed as an agricultural soil conditioner. *Journal of Environmental Radioactivity* **70**, 127–137.

Whicker, F.W., Schultz, V. (1982). *Radioecology: Nuclear Energy and the Environment*. CRC Press, Boca Raton.

Willey, N.J., Martin, M.H. (1995). Annual patterns of Cs-133 concentration in British upland vegetation. *Chemosphere* **30**, 717–724.

Witkamp, M., Frank, M.L. (1970). Effects of temperature, rainfall and fauna on transfer of $^{137}$Cs, K, Mg and mass in consumer-decomposer microcosms. *Ecology* **51**, 467–474.

Wolterbeek, H.T., De Bruin, M. (1986a). The import and redistribution of several cations and anions in tomato leaves. *Journal of Experimental Botany* **37**, 331–340.

Wolterbeek, H.T., De Bruin, M. (1986b). Xylem and phloem import of $Na^+$, $K^+$, $Rb^+$, $Cs^+$ and $Sb(SO_4)_2^-$ in tomato fruits: differential contributions from stem and leaf. *Journal of Experimental Botany* **37**, 928–939.

Wright, S.M., Howard, B.J., Strand, P., Nylen, T., Sickel, M.A.K. (1999). Prediction of $^{137}Cs$ deposition from atmospheric nuclear weapons tests within the Arctic. *Environmental Pollution* **104**, 131–143.

# Radionuclides in forest ecosystems

George Shaw

*Division of Agricultural & Environmental Sciences, University of Nottingham, Nottingham, NG7 2RD, UK*

## 1. Forests as receptors of atmospheric pollutants

Maintenance of semi-natural ecosystems in a normally functioning state is a key priority in the sustainable management of the Earth's resources, yet such ecosystems have proved to be particularly vulnerable to a multitude of anthropogenic disturbances. Among these disturbances the introduction of pollutants and contaminants following atmospheric deposition has resulted in several notorious ecological problems at both the regional and global scales. Forests, in particular, appear to have suffered disproportionately due to their tendency to scavenge atmospheric pollutants with greater efficiency than vegetation types such as grassland with lower rates of deposition (McCune and Boyce, 1992). Prolonged, chronic deposition of sulphur dioxide into forests of Europe, notably Scandinavia and the infamous 'Black Triangle'[1] of central Europe, resulted in sometimes spectacular forest decline following acidification of the forest soil and the associated syndrome of ecological effects (Pawlawski, 1997). By contrast, the acute deposition of radioactive contaminants into forests of Europe and the former Soviet countries after the Chernobyl accident has not, in all but the most extreme cases close to the reactor (Davydchuk, 1997), resulted in visible forest damage. However, a range of more insidious effects of this contamination has become apparent since 1986, including the unexpectedly high degree of transfer of radiocaesium to forest products such as game and edible vegetable products.

In Europe, forests cover approximately one third of the land area, amounting to 3.2 million km$^2$ (UNECE/FAO, 2000). This huge natural resource has recently been comprehensively mapped by Schuck et al. (2003) from Portugal in the west to the Urals in the east. Comparison of the geographical distribution of forests with the deposition pattern of radiocaesium in 1986 (Figure 2 of Chapter 5) indicates the degree to which forests across this large landmass have been affected, to a greater or lesser degree, by radioactive contamination from the Chernobyl accident. Prior to 1986, forests worldwide received chronic low level deposition of $^{137}$Cs, Pu isotopes and other radionuclides during the major period of atmospheric weapons testing in the early 1960s (Bunzl and Kracke, 1988). From these experiences, forests are known to

[1] Northern Czech Republic, south east Germany and southern Poland.

RADIOACTIVITY IN THE ENVIRONMENT
VOLUME 10 ISSN 1569-4860/DOI: 10.1016/S1569-4860(06)10006-6

act as efficient receptors of both chronic and acute releases of radioactive pollutants to the atmosphere. This raises several concerns, including the impact of radionuclides on human and animal populations which exploit the forest as a resource, and also the potential long term accumulation of long-lived radionuclides.

## 2. Relevant time scales

Historically, studies on the large scale contamination of forests with radionuclides can be traced to the southern Ural mountains of Russia where, at Kyshtym in 1957, one of the world's largest nuclear accidents occurred. This resulted in severe contamination of birch and pine forests with several radionuclides of which $^{90}$Sr was the most significant long-term contaminant (see section 5). Radioecological studies in the area affected were initially carried out in secret but have since been reported in detail (Sokolov et al., 1993). Decontamination of forests since this accident has effectively been controlled by the 28 year physical half life of $^{90}$Sr. Similarly, studies after Chernobyl have shown that exports of $^{137}$Cs (with a physical half life of 30 years) from forests are small. Measurements by Nylén and Grip (1997) have indicated that $^{137}$Cs losses by seepage through forest soils in northern Sweden were less than 0.02% in the 8 years following deposition, leading to an estimated ecological half life for caesium greater than 4200 years. These two examples both indicate the potentially long time scales over which radioactive contamination of forests could persist if contaminated with longer-lived radionuclides than $^{90}$Sr or $^{137}$Cs.

In terms of the recycling of radionuclides within forest ecosystems subject to contamination from the Chernobyl accident, Shcheglov et al. (2001) identified three key time periods: the first involving rapid mechanical weathering of radionuclides from the canopy to the underlying soil (1–3 months), the second involving 'biological decontamination' of the canopy (growth dilution, leaf fall) and the onset of significant root uptake of radionuclides (2–3 years) and the third involving root uptake and an approach to equilibrium over a period of 3–10 years and beyond. These three periods correspond to what Bergman (1994) has described as the 'early phase', the 'medium term' and the 'long term' (Figure 1). During the early phase, exposure of organisms to short-lived radionuclides may be high, but declines into the medium term. On a time scale of decades or longer it is to be expected that longer-lived radionuclides will

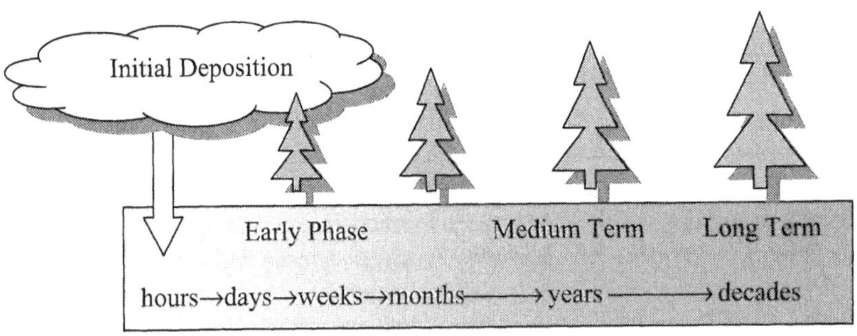


Fig. 1. Temporal phases to be considered following acute deposition of radionuclides to a forest ecosystem.

become incorporated into natural elemental cycles, especially in the case of elements such as strontium and caesium which each possess naturally occurring stable elements already present within ecosystems (Yoshida et al., 2004). Smith and Beresford (2005) have speculated on a time line for the contaminated ecosystems adjacent to the Chernobyl reactor. Over a period of 50 to 500 years radiation doses to organisms are likely to be dominated by $^{90}$Sr and $^{137}$Cs. On a time scale of 500 to 10,000 years $^{239}$Pu and $^{240}$Pu, with physical half lives of $2.4 \times 10^4$ and $6.6 \times 10^3$ years, respectively, are likely to give the most significant radiation doses, although these will be small compared with present day exposures.

These extremely long time scales are similar to those during which temperate and boreal forests developed following the retreat of glacial ice in the northern hemisphere. Furthermore, safety calculations for radioactive waste disposal facilities must be made for similar time periods extending into the future (Crossland et al., 2005). The behaviour of radionuclides with a variety of physical half lives in forests over a wide range of time scales is thus relevant to the assessment of radiation impacts under different scenarios.

## 3. Pathways and processes within forests

Figure 2 shows the generalised pathways and processes by which essential elements and pollutants alike recirculate within a typical forest ecosystem. Radionuclides also follow the same pathways and the basic methods used to quantify and understand the behaviour of radionuclides in forests have been borrowed and adapted from studies of nutrient and 'conventional' pollutant fluxes (UNECE, 1993). Most instances of radionuclide contamination of forests have resulted from deposition from the atmosphere. During deposition, radionuclides either in gaseous, dissolved or particulate forms are intercepted by the leaves/needles and branches of the canopy. The degree of interception is strongly controlled by the type of forest and the time of year. Coniferous forests, which account for a large percentage of commercial forestry in the northern hemisphere, have high projected leaf area indices[2] throughout the year, while deciduous forests have much reduced leaf area indices from late autumn through to early spring. Forests are known to be susceptible to enhanced inputs of pollutants from the atmosphere because of their efficient capture of particles under dry conditions (dry deposition) and of water droplets from clouds and mists (occult deposition). In the case of dry deposition, forest canopies have a high degree of aerodynamic roughness, leading to high friction velocities[3], compared with other vegetation types (Kinnersley et al., 1994). This results in efficient transfer of particulate material from the atmospheric boundary layer to tree canopies and the underlying ground surface. Ould-Dada et al. (2002) presented data on the deposition of $^{137}$Cs and $^{241}$Am to a dense Sitka spruce (*Picea sitchensis*) plantation adjacent to the Sellafield nuclear complex in the UK which indicate a dramatic enhancement in deposition within the plantation compared with the surrounding fields. These authors also showed that deposition was significantly enhanced at the forest edge, in both wind tunnel experiments and at the edge of the Sitka spruce plantation facing the Sellafield site.

---

[2] Defined as the fraction of the ground surface covered by the tree canopy ($m^2\ m^{-2}$).

[3] U*, a measure of momentum transfer from the atmosphere to an adjacent ground/vegetation surface.

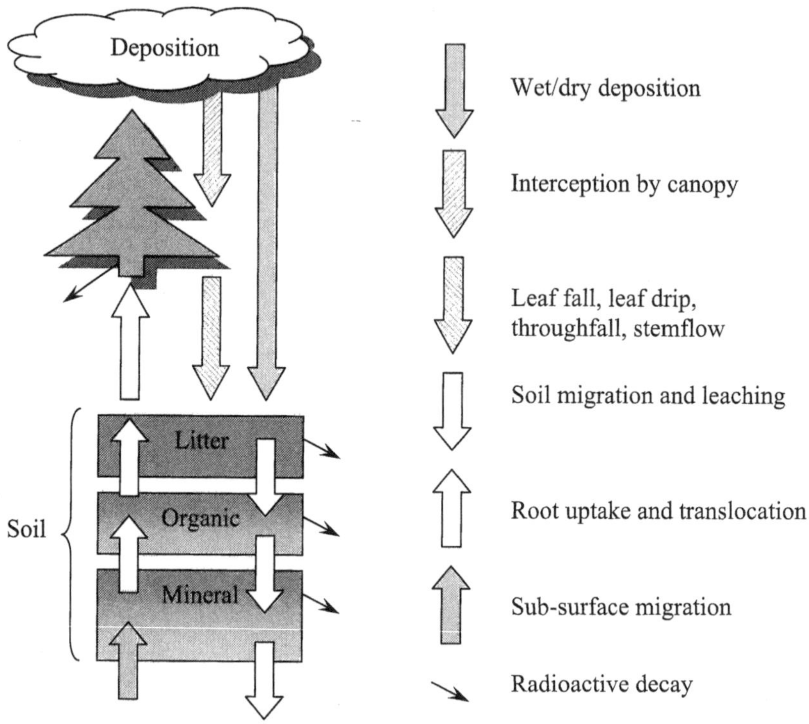


Fig. 2. Routes of contamination and recirculation of radionuclides within a forest ecosystem. The key to individual processes is given on the right.

Following interception by the tree canopy, pollutants are either absorbed directly by leaves and needles or they are lost by various processes. Weathering of leaf/needle surfaces results from wind and rainfall. Removal of deposits by wind is generally referred to as resuspension, since it results in deposited particles becoming airborne for a second time which can lead to redistribution and secondary contamination. Resuspension is difficult to measure under ambient conditions. Based on a series of wind tunnel studies, however, Ould-Dada and Baghini (2001) determined that resuspension rates of small (1.85 $\mu$m) particles from Norway spruce (*Picea abies*) canopies were in the range of $10^{-3}$ to $10^{-8}$ $s^{-1}$, resulting in an almost negligible fraction of deposited particles being removed by the wind. Washoff of leaf deposits by rainfall is much more significant and results in the transfer of contamination from the canopy to the ground surface by leaf drip and stem flow (i.e. water running down tree trunks). The chemistry of waters collected beneath tree canopies is significantly different to that of rainwater falling onto the canopy, which reflects the processes of leaching and weathering of a wide range of substances, including deposited pollutants, from the canopy (McCune and Boyce, 1992). Leaf and needle fall contributes to the removal of substances from the canopy and the redistribution of those substances to the forest floor. Patterns of distribution of radionuclides on the forest floor resulting from these various processes have been found to be heterogeneous: proximity to tree trunks and the species of tree both play a role in influencing this distribution (Förster et al., 1991; Fraiture, 1992).

Fig. 3. Upper horizons of a typical podzolic soil developed under Scots pine in southern England.

Figure 3 shows the uppermost portion of a typical podzolic soil obtained from beneath a 40 year old Scots Pine plantation overlying sandy parent material in southern England. Leaves and needles fall from the canopy to form a layer of litter on the forest floor. As this leaf/needle material accumulates it undergoes active microbial (predominantly fungal) decomposition within the fermentation (AoF) layer, with humified material accumulating in the AoH layer. The Ah layer, comprising an intimate mixture of highly humified organic matter and mineral material, represents the boundary between the organic surface horizons and the mineral soil beneath. Often, especially beneath coniferous trees, the uppermost mineral horizon (Ea) is subject to intense eluviation due to leaching by percolating rainfall. This removes iron and other materials, resulting in a 'bleached' appearance.

Radionuclides depositing to the forest floor are subject to downwards leaching through this series of discrete layers although the rate of movement is modified by sorption and complexation with organic and mineral components within the soil. The interaction of radionuclides with organic matter (OM) is also modified by biological turnover of the OM. Fungi play a dominant role in decomposing much of OM in forest soils and have been implicated in the retention of radionuclides such as $^{137}Cs$ within their mycelia (Olsen et al., 1990). Not only do fungi provide a significant driving force for the decomposition of organic matter, they also provide a functional connection between the soil and higher plants, including trees, growing within the soil (Smith and Read, 1997). Fungus-root associations, or mycorrhizas, are probably responsible for the bulk of uptake of ions such as K and Cs by tree roots. Once absorbed by roots, radionuclides are incorporated into the woody stems of trees, with a portion being translocated to the leaves/needles within the canopy, from which leaching and leaf/needle fall return a fraction of the biologically-incorporated radionuclides to the forest floor.

This brief sketch of the recirculation processes of elements/pollutants within a forest is highly generalised and the rates and quantitative importance of individual processes will vary from forest to forest and from one element to another. Nevertheless, it is within this general framework that the ecological behaviour of pollutants such as radionuclides can be evaluated.

## 4. Radionuclides in forests after Chernobyl

The area immediately surrounding the Chernobyl Nuclear Power Plant (ChNPP) is a mosaic of trees, agricultural land, rivers, lakes and villages which was contaminated by a wide variety of fission and activation products including $^{241}Am$, $^{95}Zr$, $^{95}Nb$, $^{154}Eu$, $^{155}Eu$, $^{144}Ce$, $^{125}Sb$, $^{106}Ru$, $^{134}Cs$, $^{137}Cs$, $^{60}Co$, $^{90}Sr$, $^{235}U$, $^{238}U$, $^{238}Pu$, $^{239}Pu$ and $^{240}Pu$. Forests occupy approximately 45% of the area surrounding the ChNPP which was evacuated in 1986. Of these forests, two thirds are artificially managed Scots pine (*Pinus sylvestris*) plantations although naturally occurring birch (*Betula* spp.) and oak (*Quercus* spp.) species are also widely distributed throughout the area (Davydchuk, 1997).

Contamination of the 30 km exclusion zone was unique (a) because of the variety and extremely high levels of radionuclides introduced and (b) because of the widespread introduction of reactor fuel particles to the environment. These so-called 'hot particles' (consisting of discrete fragments of uranium oxide reactor fuel with particle sizes ranging from 2–3 $\mu$m to hundreds of $\mu$m) were expelled from the reactor as a result of the initial explosion on 26$^{th}$ April 1986 (Chesser et al., 2004). The larger of these particles sedimented rapidly from the atmosphere and deposited within 5 to 10 km of the ChNPP along the so-called 'Western Trace'. The ecosystems immediately to the west of the ChNPP thus received significant inputs of fuel particles which gave rise to extremely high dose rates from short lived isotopes including $^{95}$Zr, $^{95}$Nb and $^{144}$Ce within the reactor fuel. Absorbed dose rates in excess of 5 mGy h$^{-1}$ resulted in the death of pine trees within a 500 ha area immediately to the west of the reactor and in another smaller area approximately 6 km to the north (Kozubov et al., 1991). The area of trees to the west of the reactor became known as the 'Red Forest' due to the red-brown needles of the dead pine canopy. The dead trees in this area were bulldozed into shallow trenches in 1987, which has since created problems of groundwater contamination in the sandy subsoil (Dewiere et al., 2004).

### 4.1. *Interception of deposition by tree canopies*

Shcheglov et al. (2001) have noted that the fallout from the Chernobyl accident was different from that experienced in any other major nuclear release to the atmosphere. These authors observed that, in the near-field, approximately 60 to 90% of the fallout was intercepted by stands of coniferous trees, with approximately 1.5 to 3 times less being captured by deciduous species, which were still without leaves at the time of the release. Of the radioactivity intercepted by coniferous tree canopies, approximately 50% was initially associated with the needles (Shcheglov et al., 2001). Even close to the ChNPP, where depositing particles were relatively large, the effects of the forest canopy on total deposition were evident. Shcheglov et al. (2001) reported that, at a site 6 km to the south of the ChNPP, the total deposition of radioactivity in the forest was 20% higher than deposition to an adjacent grassland, and as much as 40% higher at the forest edge.

Loss of these deposits was rapid and early observations showed that only 10–15% of the initial deposits remained in the canopies in August 1986. Interestingly, Shcheglov et al. (2001) found that natural decontamination of forest canopies was faster at sites closest to the ChNPP where deposited particles were largest, confirming a longstanding observation that rates of loss of particles from tree canopies are particle size-dependent (Witherspoon and Taylor, 1969).

Further afield, in the Höglwald forest in Bavaria (southern Germany) where depositing particle sizes were of the order of 1–2 $\mu$m, Schimmack et al. (1991) determined that approximately 20% of the deposited activity from Chernobyl was initially intercepted by the canopy and trunk/branch surfaces of beech (*Fagus sylvatica*). This compared with 70% interception of radiocaesium ($^{137}$Cs and $^{134}$Cs) and 60% interception of $^{106}$Ru by Norway spruce (*Picea abies*) at the same site. Figure 4 shows the dynamics of transfer of $^{134}$Cs from the spruce canopy at Höglwald to the forest floor over a period of approximately 600 days following deposition (30$^{\text{th}}$ April, 1986). In analysing these data, Bunzl et al. (1989) found that the rate of loss from the canopy was time-dependent. A decontamination half time of 90 days was applicable during the period 0–130 days, while a longer half time of 230 days applied to the 130–600 day data. The Höglwald data also demonstrate that deposition to the forest canopy was of the order of 20% higher than to adjacent grassland, which agrees with the observations of Shcheglov et al. (2001) despite the clear difference in the size range of particles deposited at both sites. Bunzl et al. (1989) also showed that the loss of radiocaesium from the canopy by litter fall was small compared with the loss due to weathering.

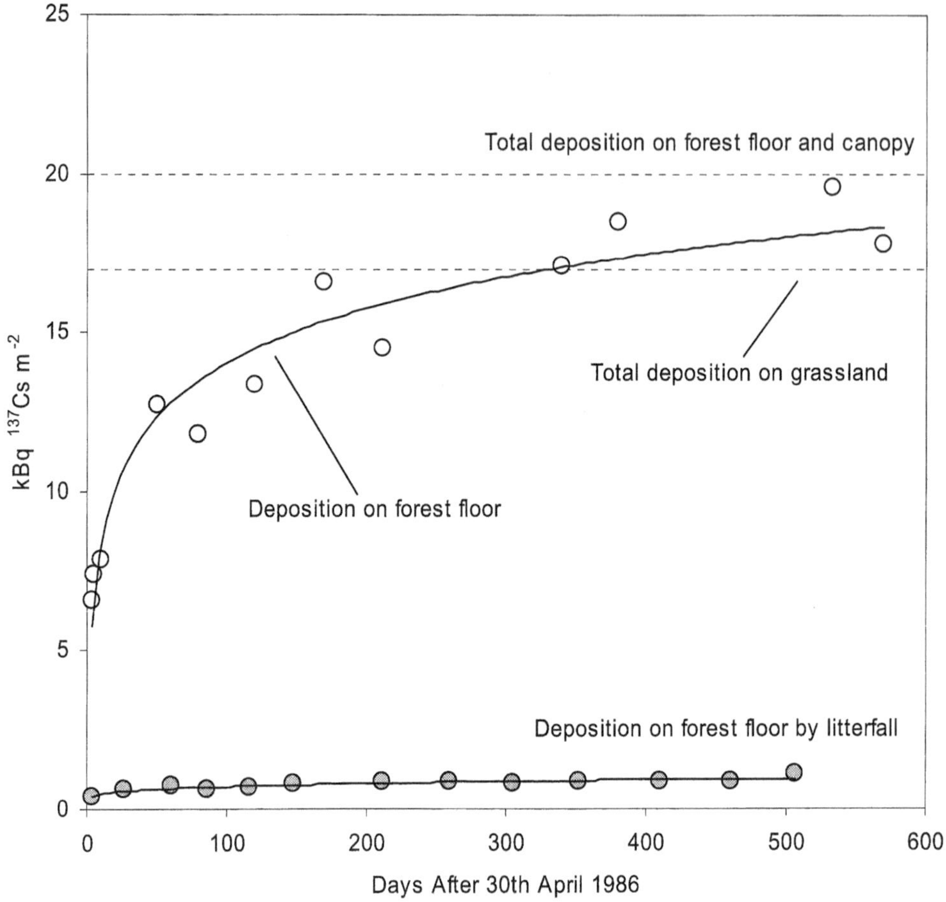


Fig. 4. Temporal redistribution of $^{134}$Cs from the canopy of a spruce forest in southern Germany to the forest floor beneath (redrawn from Bunzl et al., 1989, with permission from Elsevier).

## 4.2. *Forest soils*

A long term study by the University of Moscow, led by Tikhomirov and Shcheglov (Shcheglov et al., 2001), established monitoring sites at several forest sites within the Chernobyl 30 km exclusion zone. Shaw and Wang (1996) sampled soils at three of these sites which formed a transect perpendicular to the Western Trace, starting at Shepelichi, situated approximately 6 km west of the ChNPP, and running southwards for approximately 25 km to Dityatki. Details of the characteristics of these sites are given in Table 1. The dose rates reported by Shcheglov et al. (2001) for these sites (Figure 5) decline dramatically in the order Shepelichi > Kopachi > Dityatki, reflecting the declining deposition of radioactivity in a southerly direction from the ChNPP. As well as a reduction in total deposited activity along this transect, the nature of the radioactive deposits changed significantly. Figure 6 shows depth profiles of the $^{238}U/^{235}U$ ratios in soils at each of these sites. At Shepelichi, there was clear evidence of the presence of

Table 1
Sites sampled in the Chernobyl 30 km exclusion zone

| Site | Forest type | Location | Initial deposition ($MBq\,m^{-2}$) |
|---|---|---|---|
| Shepelichi | Pine, birch, oak forest on sandy podzol | ~6 km W of reactor | ~45 $^{137}Cs$ |
| Kopachi | Pure pine forest on sandy podzol | ~6 km S of reactor | ~2.5 $^{137}Cs$ |
| Dityatki | Pine, birch forest on sandy podzol | ~25 km S of reactor | ~0.4 $^{137}Cs$ |

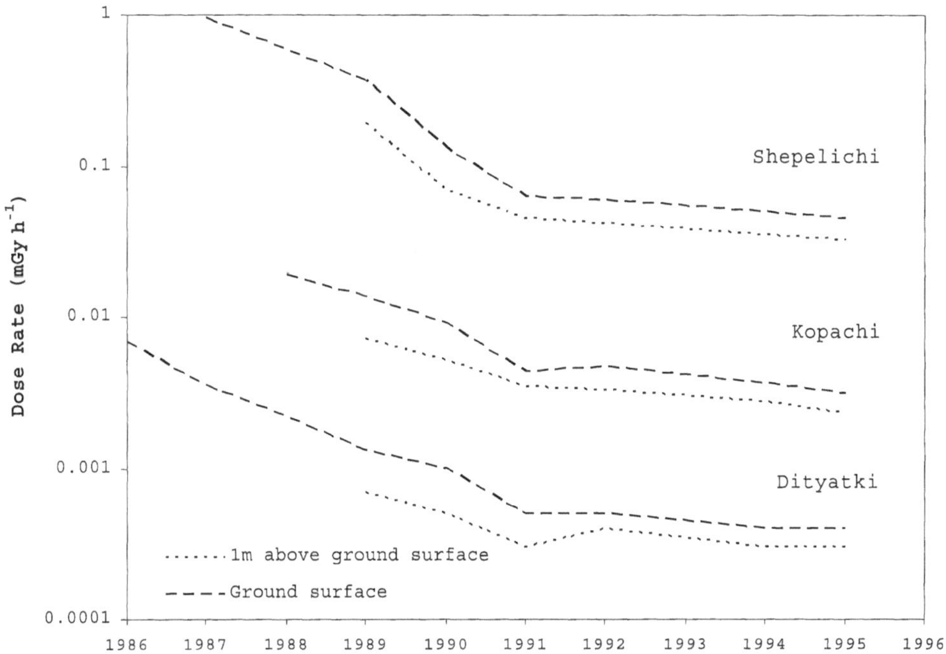


Fig. 5. Dose rates at selected forest sites in the Chernobyl 30 km exclusion zone plotted using data from Shcheglov et al. (2001).

enriched uranium in the three surface organic horizons (AoL, AoF and AoH). In the AoF and AoH horizons, the $^{238}U/^{235}U$ ratios were indicative of uranium enriched to 2% $^{235}U$, which is diagnostic for the uranium oxide fuel used in the RBMK reactor at Chernobyl (Borovoi et al., 1998). 6 km south of the ChNPP at Kopachi, there was evidence of significant enrichment of $^{235}U$ in the AoF layer, while 25 km south of the ChNPP the uranium appeared to be solely of natural origin. Thus, it can be concluded that Shepelichi, on the line of the Western Trace, received significant input of fuel particles, Kopachi received a smaller input of reactor fuel while Dityatki received no measurable fuel particle input.

Along with other fission and activation products, fuel particles from the ChNPP contained $^{238}Pu$, $^{239}Pu$ and $^{240}Pu$ at a combined specific activity of $1.62 \times 10^7$ $Bq\,g^{-1}$ and $^{137}Cs$ and $^{134}Cs$ at specific activities of $1.5 \times 10^9$ $Bq\,g^{-1}$ and $8.6 \times 10^8$ $Bq\,g^{-1}$, respectively (Borovoi

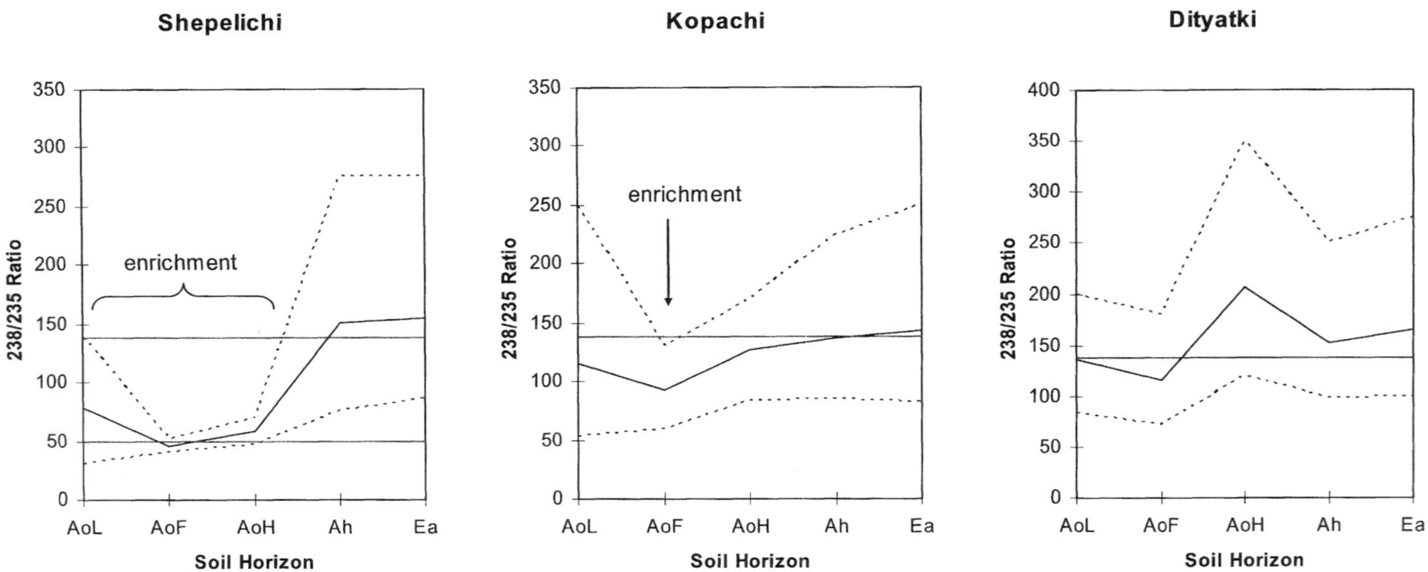


Fig. 6. $^{238}U/^{235}U$ ratios in forest soils within the Chernobyl 30 km exclusion zone (data adapted from Shaw and Wang, 1996). The ratio expected for natural uranium is 138 while a ratio of 49 is expected for the 2% enriched RBMK reactor fuel. The solid black line is the median measured ratio and the dotted lines are maximum and minimum ratios based on measurement uncertainties.

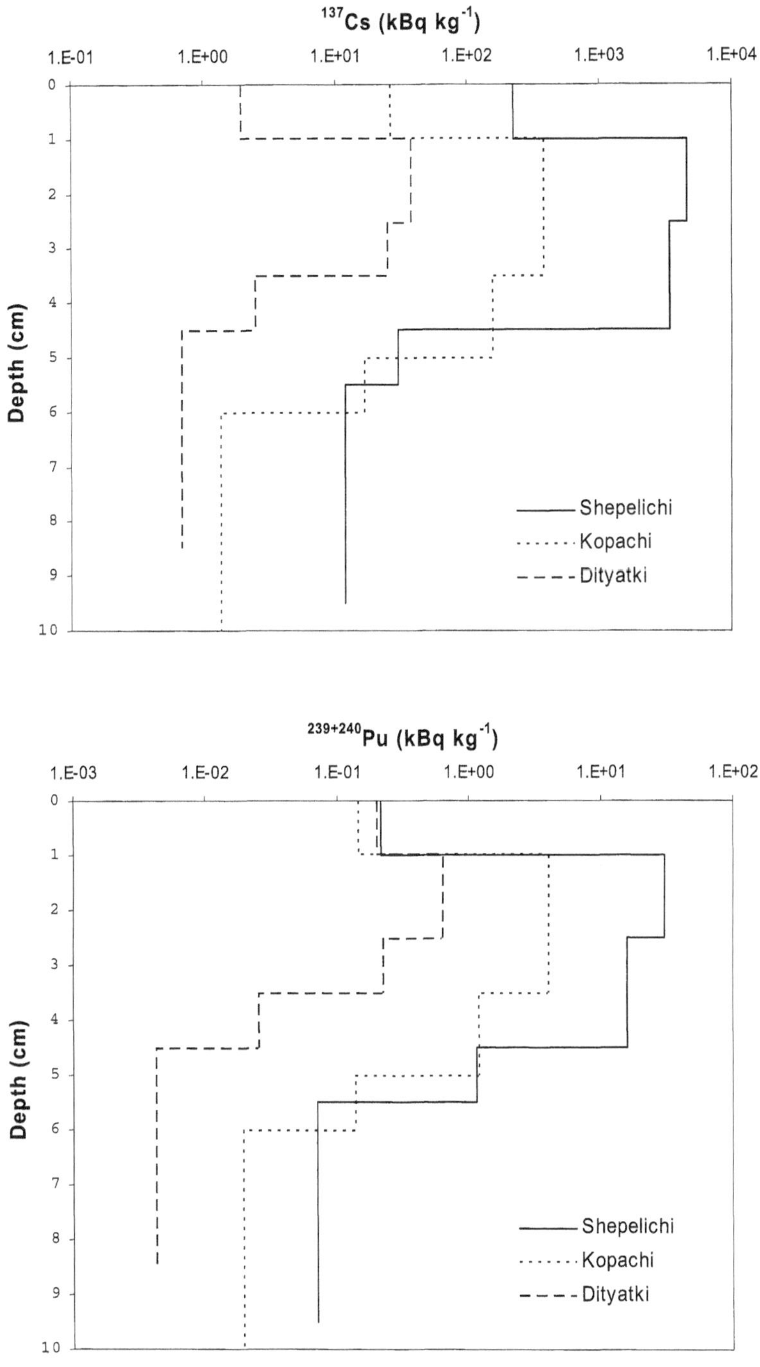


Fig. 7. $^{137}$Cs and Pu depth profiles in forest soils within the Chernobyl 30 km zone, 1993–1994 (redrawn from Shaw and Wang, 1996).

Table 2
Migration half times ($\tau$, years) estimated for $^{134}Cs$ and $^{137}Cs$ based on measurements made at three sites in the Chernobyl 30 km exclusion zone in 1993 and 1994

| | Shepelichi | | Kopachi | | Dityatki | |
|---|---|---|---|---|---|---|
| | $^{137}Cs$ | $^{239+240}Pu$ | $^{137}Cs$ | $^{239+240}Pu$ | $^{137}Cs$ | $^{239+240}Pu$ |
| AoL | 0.99 | 0.75 | 1.11 | 1.02 | 1.01 | 1.29 |
| AoF | 1.76 | 2.32 | 3.04 | 3.31 | 3.60 | 3.22 |
| AoH | 199 | 60.8 | 23.9 | 12.9 | 10.5 | 16.7 |
| Ah | 1.29 | 6.99 | 6.04 | 2.02 | 3.32 | 2.22 |
| Ea | 6.61 | 5.95 | 16.7 | 78.1 | 28.0 | 47.0 |

et al., 1998). Soil samples taken in 1993 and 1994 indicated that the majority of these radionuclides had migrated into the AoF and AoH soil layers at Shepelichi, Kopachi and Dityatki but not significantly deeper (Figure 7). Thus, the biologically active organic soil horizons of forest soils appear to represent an important sink for radionuclides and fuel particles, at least in the medium term. Interestingly, a similar distribution within forest soils has been observed for non-radioactive ferrimagnetic particles of anthropogenic origin (Strzyszcz, 1989), suggesting similar processes of migration and accumulation for both types of particle.

Uranium oxide is relatively insoluble although slow dissolution does occur under environmental conditions and leaching rates of fuel particles have been calculated. Konoplev et al. (1993) estimated a rate coefficient for leaching of $^{90}Sr$ and $^{137}Cs$ from fuel particles at Kopachi of $4 \times 10^{-4}$ day$^{-1}$, giving a half time of approximately 4.75 years. Thus, at the time the data in Figures 6 and 7 were obtained (1994), only 30% of $^{137}Cs$ is likely to have been retained in fuel particles at Shepelichi and Kopachi. The initial presence of radionuclides in fuel particles followed by slow dissolution to soluble and exchangeable forms influences both the migration potential of radionuclides and their availability for absorption by soil dwelling organisms. An analysis by Wang (1996), based on the depth profiles shown in Figure 7, quantified the leaching rates of Pu and Cs isotopes in the forest soils at Shepelichi, Kopachi and Dityatki (Table 2). For the AoH horizon, in which the highest accumulation of radionuclides occurred, the downwards migration of both caesium and plutonium isotopes was retarded substantially at Shepelichi ($\tau = 199$ years for Cs and 61 years for Pu) compared with the other sites at which the deposition of fuel particles was less important.

Table 3 shows estimates for migration half times of radiocaesium in soils at several forest sites across western Europe. Belli (2000) studied western European forests which had received significant deposition of $^{137}Cs$ from Chernobyl in the form of small aerosol particles (1–2 $\mu$m). Based on field measurements, estimates of 0.95, 10.7 and 168 years, respectively, were obtained for radiocaesium residence half times in litter, organic and mineral (0–5 cm) horizons. The geometric mean residence time within the organic horizon is close to that obtained for the Dityatki site which is only 25 km south of the ChNPP but apparently unaffected by fuel particles. Hence, for non-fuel particle fallout from Chernobyl the migration rates of radiocaesium in forests distributed over a wide area of the European continent appear similar. Schimmack and Bunzl (1992) obtained estimates of 4–6 years for Chernobyl-derived caesium within the organic soil horizon, and 15 years for Chernobyl caesium in the mineral soil (0–5 cm).

Table 3
Migration half times ($\tau$, years) estimated for $^{137}Cs$ based on measurements made at various western European forest sites

| Authors | Site | Country | Half time (years) | | | |
|---|---|---|---|---|---|---|
| | | | Litter | Organic | Mineral (5 cm) | |
| Belli (2000) | Novaggio | Switzerland | 2.95 | 19.8 | 170 | Chernobyl + fallout |
| Belli (2000) | Tarvisio | Italy | 0.46 | 4.22 | 27.8 | Chernobyl + fallout |
| Belli (2000) | Weinsberger | Austria | 1.24 | 13.6 | 385 | Chernobyl + fallout |
| Belli (2000) | Kobernausser | Austria | 0.77 | 15.6 | 278 | Chernobyl + fallout |
| Belli (2000) | Roundwood | Ireland | 0.28 | 23.1 | 174 | Chernobyl + fallout |
| Belli (2000) | Clogheen | Ireland | 1.16 | 6.93 | 217 | Chernobyl + fallout |
| Belli (2000) | Shanrahan | Ireland | 1.73 | 5.78 | 203 | Chernobyl + fallout |
| Belli (2000) | Geomean | | 0.95 | 10.73 | 168.36 | |
| Belli (2000) | GSD | | 2.22 | 1.93 | 2.11 | |
| Schimmack and Bunzl (1992) | Hoglwald spruce | Germany | | 4–6 | 15.0 | Chernobyl |
| Schimmack and Bunzl (1992) | Hoglwald spruce | Germany | | 10–15 | 50.0 | Fallout |
| Schimmack and Bunzl (1992) | Taxoldener pine | Germany | | | 4.0 | Chernobyl |
| Schimmack and Bunzl (1992) | Taxoldener pine | Germany | | | 11.0 | Fallout |

### 4.3. *Radiocaesium uptake by trees*

Just as the migration of radiocaesium in forest soils which received fuel particles from the ChNPP appears to be retarded, so the uptake of $^{137}Cs$ by trees appears to be reduced at sites at which radionuclides were initially deposited in the form of fuel particles. Figure 8 shows the percentage distributions of $^{137}Cs$ at three sites close to the ChNPP, indicating that, in each case, the soil contained the majority of contamination. Bourakovka lies 15 km to the west of the reactor and received 10× less deposition than Shepelichi (Myttenaere et al., 1993). However, its position on the 'Western Trace' means that it would certainly have received significant fuel particle deposition. As already described, Kopachi received a measurable degree of fuel particle deposition based on measured $^{238}U/^{235}U$ ratios in the soil. Kruki, on the other hand, received insignificant fuel particle deposits despite receiving high total deposition of radionuclides (5.5 MBq m$^{-2}$ $^{137}Cs$). Based on the observation that the significance of fuel particle deposition at each site was in the order:

Bourakovka > Kopachi > Kruki

two interesting trends emerge. First, it is evident from Figure 8 that the major reservoir for $^{137}Cs$ at Bourakovka and Kopachi is the organic soil layer. This is most likely due to the long residence half times of fuel-particle associated radionuclides, as indicated by Table 2. At Kruki, it is evident that more significant migration of $^{137}Cs$ into the mineral soil had taken place by 1992.

Secondly, the percentage of the total deposited $^{137}Cs$ absorbed into tree tissues at each site is negatively correlated with the significance of fuel particle deposition. Myttenaere et al. (1993) determined that approximately 1.5% of the total deposited $^{137}Cs$ at Bourakovka was located within pine tree tissues in 1990. Measurements by Belli et al. (1996) showed that pine

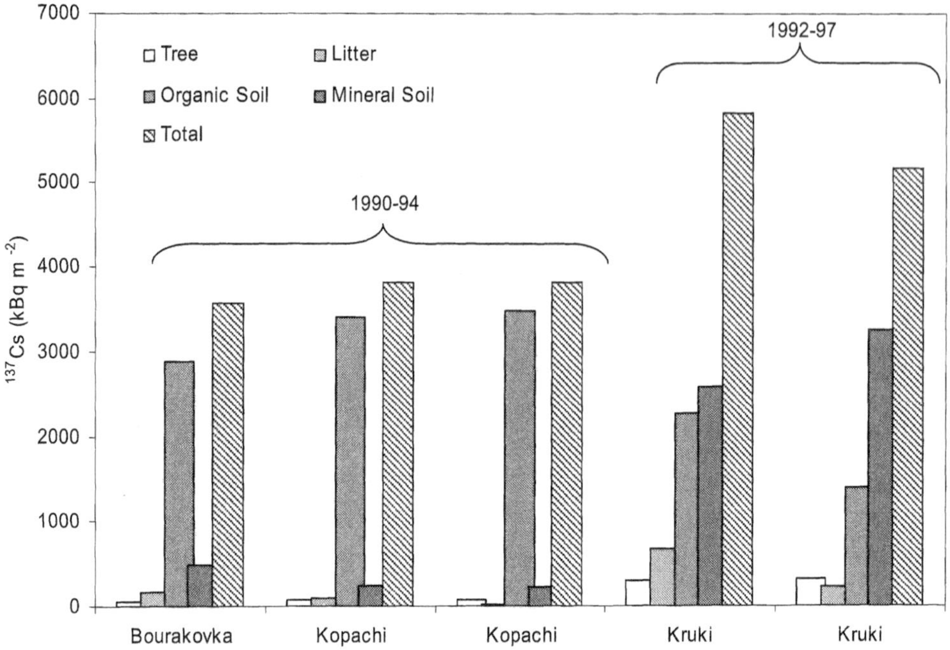


Fig. 8. Distributions of $^{137}$Cs in forest sites close to the Chernobyl reactor from 1990 to 1997.

trees at Kopachi contained 1.7% of total deposited $^{137}$Cs, while at Kruki, some 25 km to the north east of the ChNPP in Belarus, 4–5% of $^{137}$Cs was located within pine tree tissues from 1992 to 1997 (Dvornik and Zhuchenko, 1998, pers. comm.).

Thus, the behaviour of $^{137}$Cs in forests close to the ChNPP has been influenced by the initial presence or absence of fuel particles. Further afield from the ChNPP Belli (2000) studied western European forests which had received significant deposition of $^{137}$Cs from Chernobyl in the form of small aerosol particles. The distributions within the major forest compartments are shown in Figure 9, which indicates that, as for the sites close to the ChNPP, the major reservoir for $^{137}$Cs within the forests is the soil compartment. At five of the sites, however, the larger proportion of the $^{137}$Cs had migrated to the mineral soil suggesting that the physical mobility of the initial deposits was significantly greater than at sites, such as Shepelichi, Kopachi and Bourakovka, known to be contaminated with fuel particles. A trend of increasing bioavailability of $^{137}$Cs with increasing distance west from the ChNPP is indicated in Figure 10 which shows that, compared with the 'near-field' sites, trees in forests in the Irish Republic contained from 8 to 20% of the total deposited $^{137}$Cs some ten years after deposition. This observation is consistent with the measurements of Croom and Ragsdale (1980) who found 12–13% of $^{137}$Cs from nuclear weapons fallout in turkey oak trees in South Carolina in the USA after 25 years' residence time. Similarly, Melin et al. (1994) found 14% of $^{137}$Cs from nuclear weapons tests in tree tissues within a mixed pine/spruce forest in Sweden sampled in 1980.

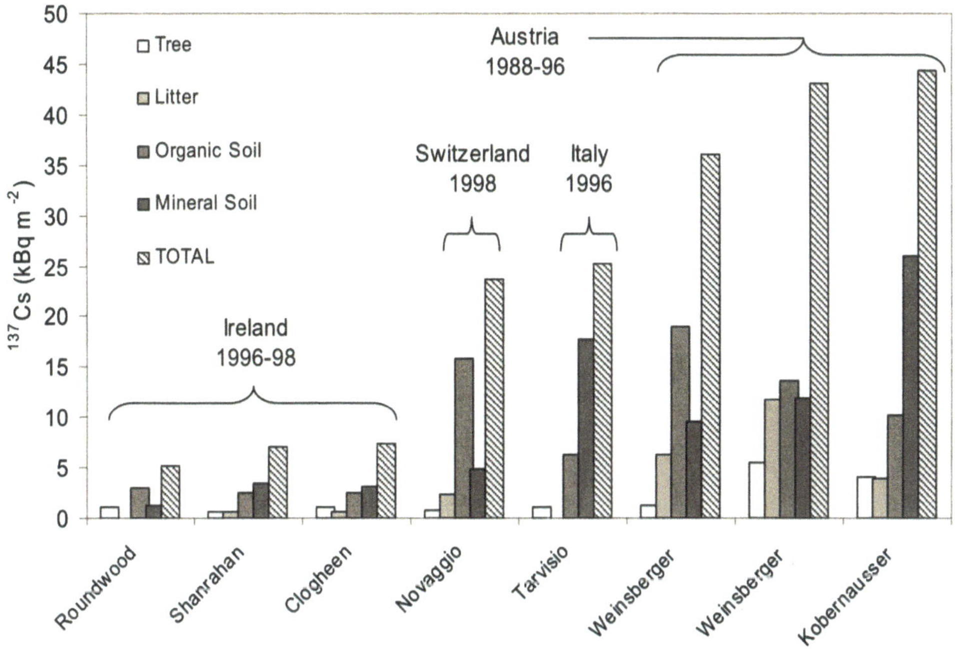


Fig. 9. Distributions of $^{137}Cs$ in forest sites across western Europe from 1996 to 1998 as measured by the SEMINAT project (Belli, 2000).

The uptake of $^{137}Cs$ is known to occur via the leaf/needle surface of trees as well as through the root system. The importance of foliar absorption following the Chernobyl deposition was extremely difficult to measure, however. Belli (2000) presented results of a model sensitivity analysis using a compartmental model which had been calibrated using data obtained at Kopachi and at a site in the Irish Republic. Using this model, root uptake rate coefficients were estimated for these sites (0.006 $y^{-1}$ and 0.12 $y^{-1}$, respectively) based on observed $^{137}Cs$ distributions in trees and soils at each site. The relatively low root absorption rate obtained for Kopachi was explained by the initial presence of $^{137}Cs$ in fuel particle matrices, leading to low bioavailability. A range of foliar absorption rate coefficients from 0 to 0.1 $y^{-1}$ was then assumed and model calculations carried out using each foliar and root absorption rate in combination. The results, shown in Figure 11, demonstrate that under circumstances in which $^{137}Cs$ availability in the forest soil is reduced due to the deposition of fuel particles, there is a potentially large contribution of foliar absorption of $^{137}Cs$ to the overall uptake by the tree. However, in conditions such as those observed in Irish forests after Chernobyl, and also in forests subject to global weapons fallout, in which the depositing activity is associated with small aerosols, even relatively high foliar absorption rates do not contribute significantly to the overall contamination of tree tissues. In conclusion, root uptake appears to be the most important biological assimilation mechanism in the long term recycling of radiocaesium in forests.

Dvornik and Zhuchenko (Shaw et al., 2002) have made 50 year predictions of $^{137}Cs$ uptake dynamics in trees in Belarus using the FORESTLIFE model. Figure 12 shows predicted

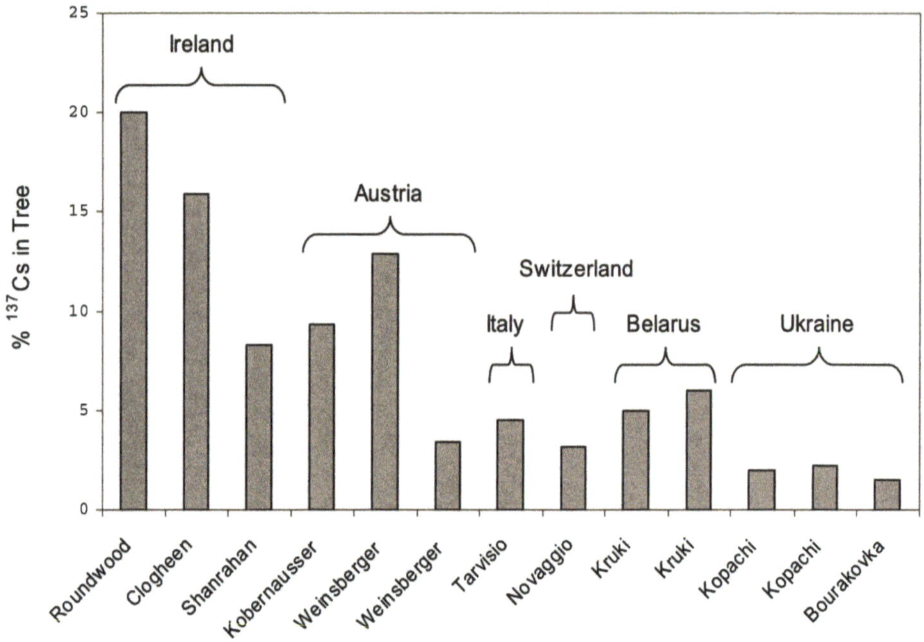


Fig. 10. Percentage distributions of $^{137}Cs$ in the tree compartment at forest sites ranging from as close as 15 km from the Chernobyl reactor to as far afield as the Irish Republic (measurements taken from 1990 to 1998).

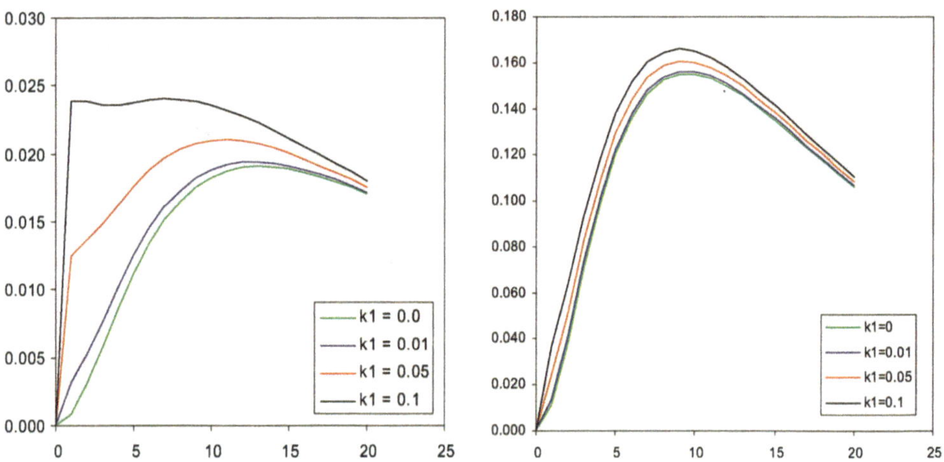


Fig. 11. Results of sensitivity analysis of foliar absorption rate on radiocaesium accumulation by trees (Belli, 2000). See section 4.3 for explanation.

dynamics for the aggregated transfer factor[4] of wood in trees of four distinct age classes. Measurements have shown that trees which are young at the time of initial contamination

[4] Aggregated transfer factor (Tagg) = $Bq\,kg^{-1}$ in tree tissue/$Bq\,m^{-2}$ deposited to the forest floor ($m^2\,kg^{-1}$).

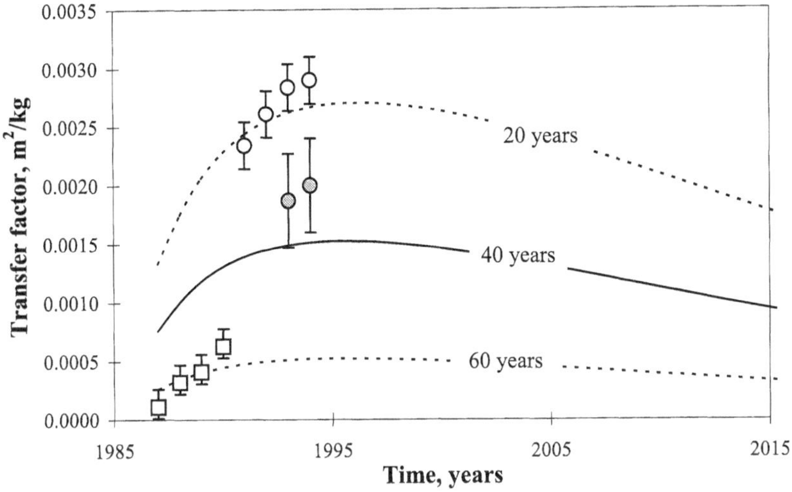


Fig. 12. FORESTLIFE predictions of soil–wood aggregated transfer factors for trees of different age classes compared with independent data from Belarus and the Russian Federation (Dvornik and Zhuchenko, in Shaw et al., 2002).

are more susceptible to contamination with radiocaesium than older trees, as a result of more active nutrient acquisition and a reduced dilution effect due to low standing biomass. The time dependent nature of the curve results from increasing root uptake with time, balanced by loss of Cs from the canopy due to weathering and leaf fall, as well as the physical decay of $^{137}$Cs over the time scale considered. Predictions for Belarus were for peak contamination in 1999–2001. The time to peak contamination in timber is important when considering the effects on the forest and related industries (section 6) although the available data may be sufficiently variable to make it difficult to determine when maximum activity concentrations have been reached (Shaw et al., 2005). Fesenko et al. (2001) concluded that contamination of trees with $^{137}$Cs was controlled by the relative distributions of $^{137}$Cs and tree roots within the soil profile, in combination with the bioavailability of $^{137}$Cs in different layers of the soil. Variations in these three factors from site to site can lead to substantial variability in the activity concentrations accumulated by individual trees, even within the same species.

### 4.4. *Fungi*

In the years after the Chernobyl accident, Basidiomycete fungi, which produce prominent fruiting bodies or 'mushrooms', have been observed consistently to play an important role in the migration and biological uptake of radiocaesium in forest soils. As described in section 3, some fungi are the principal decomposers (saprotrophs) within the organic horizons of forest soils but others provide functional connections between the soil and the roots of higher plants, notably trees, in the form of mycorrhizas (symbiotic fungi). Even within the two broad classes of saprotrophic and symbiotic fungi, individual species occupy a variety of distinct ecological niches which exposes them to a range of potential radionuclide concentrations within the soil. The main distinguishing feature which links many of these fungi, however, is the high

Table 4
Mean $^{137}Cs/^{134}Cs$ ratios and standard deviations for different species of fungi, normalised to 1 May 1986 (Rühm et al., 1997)

| Species | *n* | $^{137}Cs/^{134}Cs$ | Mycelium location |
|---|---|---|---|
| *Lepista nebularis* | 9 | $1.73 \pm 0.12$ | L/Of |
| *Collybia maculata* | 7 | $1.78 \pm 0.10$ | L/Of |
| *Collybia butyracea* | 11 | $1.81 \pm 0.16$ | L/Of/Oh |
| *Xerocomus badius* | 21 | $1.79 \pm 0.10$ | L/Of |
| *Russula ochrocleuca* | 15 | $1.82 \pm 0.08$ | Oh |
| *Lactarius deterrimus* | 6 | $1.83 \pm 0.06$ | Oh |
| *Clitocybe clavipes* | 8 | $1.83 \pm 0.07$ | Oh |
| *Hydnum repandum* | 15 | $1.91 \pm 0.12$ | Oh |
| *Paxillus atrotomentosus* | 9 | $1.92 \pm 0.17$ | Oh/Ah |
| *Lactarius scrobiculatus* | 8 | $1.91 \pm 0.15$ | Oh/Ah |
| *Macrolepiota rhacodes* | 11 | $1.85 \pm 0.18$ | Oh |
| *Armillariella spp.* | 12 | $1.81 \pm 0.18$ | LOf/Oh |
| *Sarcodon imbricatus* | 8 | $2.06 \pm 0.09$ | Ah/B |
| *Russula cyanoxantha* | 7 | $2.16 \pm 0.26$ | Ah/B |

degree of radionuclide contamination of their fruiting bodies after the Chernobyl accident (Mascanzoni, 1990).

Rühm et al. (1997) compared ratios of $^{137}Cs$:$^{134}Cs$ measured in discrete forest soil horizons and in mushroom fruiting bodies to locate the principal horizon(s) from which individual fungal species absorb radiocaesium. The characteristic $^{137}Cs$:$^{134}Cs$ ratio from the Chernobyl accident was approximately 1.74 at the time of deposition, before differential decay rates caused $^{134}Cs$ to become reduced in activity in comparison with $^{137}Cs$. $^{137}Cs$:$^{134}Cs$ ratios at the time of deposition to the Hochstadt forest near Munich (1$^{st}$ May 1986) were found to range from 1.74 in the litter layer to 2.41 in the deeper mineral horizons. Table 4 shows the results of this study which indicate that, of 14 species examined, only two (*Sarcodon imbricatus* and *Russula cyanoxantha*) absorbed radiocaesium exclusively from the mineral horizons. Most species, such as *Xerocomus badius* which has been found to be a particularly efficient accumulator of radiocaesium, contained $^{137}Cs$:$^{134}Cs$ ratios which suggested that absorption was occurring from the L, F and H horizons, or a combination of these horizons. Variations in the $^{137}Cs$:$^{134}Cs$ ratios within individual species in a particular year varied by less than 10% whereas the activity concentrations in individual species varied by up to a factor of 10. The high degree of concentration of radiocaesium in fungal fruiting bodies has led several radioecologists to estimate the potential of fungal mycelium to sequester and recycle radiocaesium within the organic soil layers of forests. Olsen et al. (1990) estimated that an average of 32% of radiocaesium within the soil could be associated with the fungal mycelium rather than with non-living mineral and organic matrices. More recently, Vinichuk et al. (2004) have suggested that the relative ease with which radiocaesium can be leached from fungal fruiting bodies with water indicates that the fungi may play an active role in the long term cycling of $^{137}Cs$ in forest soils. Certainly, the high $^{137}Cs$ transfer coefficients observed for edible mushrooms (IAEA, 1994) has led to their contributing significantly to dietary intake in both humans and animals exploiting forests for foodstuffs.

## 5. Strontium in forests

The Kyshtym accident in the southern Urals in 1957 resulted in a track of radioactive contamination within mixed birch and pine forests which extended 300 km to the NNE of the facility (Warner and Harrison, 1993). With a physical half life of 28 years, $^{90}$Sr was the most significant of the radionuclides released. An area of approximately 1200 km$^2$ was contaminated with $^{90}$Sr at levels ranging from 2 Ci km$^{-2}$ to >1000 Ci km$^{-2}$ within a small area of <20 km$^2$. Alexakhin et al. (1994) constructed a mathematical model of $^{90}$Sr cycling within forests of this area based on a data set which was collected over the first 11 years following the accident. One year after the initial deposition, $^{90}$Sr activity inventories (Bq m$^{-2}$) were highest in the litter layer (22%) and soil (87%) underlying birch trees. The soil under pine contained approximately 90% of $^{90}$Sr after one year. Soil inventories peaked approximately 4 years after deposition and thereafter declined year-on-year due principally to physical decay. Activity concentrations of $^{90}$Sr in wood were observed to increase during the first few years after initial contamination and reached a peak 6–7 years after deposition in birch and 10–12 years after deposition in pine. Tikhomirov (1993) presented comparative data on the migration of $^{137}$Cs and $^{90}$Sr in forest soils in the same area which indicated that, of these two radionuclides, $^{90}$Sr was significantly more mobile within the soil profile. After 6 years $^{137}$Cs remained predominantly at a depth of 1 cm within the soil profile while the peak $^{90}$Sr activity was located at almost 2 cm after 8 years, and at 3 cm after 13 years, suggesting that $^{90}$Sr migration was continuing while $^{137}$Cs migration had largely halted after the first few years.

Another notable study of $^{90}$Sr migration in a forest ecosystem was conducted by Jordan and Kline (1976) at a rain forest in Puerto Rico which received fallout from atmospheric weapons testing. Peak deposition to the site was from 1962 to 1967 and numerous data were collected during this period to determine $^{90}$Sr contamination within the forest canopy. Further samples were taken in 1974 of the soil, wood, canopy and litter in a study to validate a model which was based on the original samples some ten years before. The distribution of $^{90}$Sr within the ecosystem was significantly different to that observed for the Kyshtym forests, with less than 70% of the total $^{90}$Sr inventory present within the soil and 28% present within the wood of trees. The litter layer, which is a much less significant feature of tropical forests due to much higher decomposition rates compared with temperate and boreal forests, accounted for less than 2% of the total $^{90}$Sr inventory.

## 6. Plutonium in forests

Few studies have been conducted on the distribution and dynamics of plutonium isotopes in forest ecosystems. As an extremely heavy element, the biological uptake of Pu is generally considered to be very low and soil-plant transfer factors of the order of $10^{-4}$–$10^{-5}$ (IAEA, 1994) reflect this. Solid–liquid distribution coefficients ($K_d$) for Pu in soils tend to be high, of the order of $10^3$–$10^4$ (IAEA, 1994), implying that mobility of Pu in soils over the long term is likely to be low. In particular, Pu is known to form complexes with organic matter which suggests that the organic soil horizons in forest soils could be important in controlling the long term behaviour of Pu in forest ecosystems.

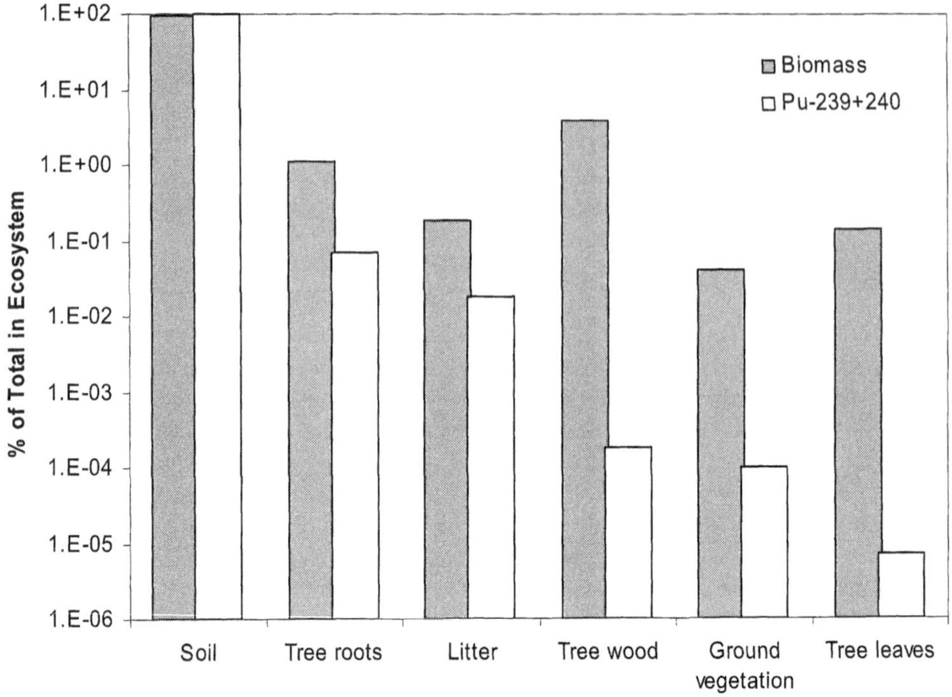


Fig. 13. Comparison of plutonium and biomass distributions within a deciduous forest at Oak Ridge, USA. Data recalculated from Garten et al. (1978).

Garten et al. (1978) studied the distribution and dynamics of $^{239,240}Pu$ in a forest at Oak Ridge (USA) which consisted of mixed deciduous species including white ash (*Fraxinus americana*) and sycamore (*Platanus occidentalis*). Figure 13 shows the percentage distributions of biomass and Pu activities determined by Garten and co-workers 30 years after the establishment of the forest on the Oak Creek floodplain. As expected from the low soil–plant transfer factors mentioned above, the proportion of the total Pu inventory associated with the standing biomass was very low at less than 0.25%. The soil accounted for more than 99% of the total Pu inventory, although reciprocal transfers between the soil, litter and tree roots were identified as the major processes affecting the redistribution of Pu in the ecosystem. Through the development of a compartment model based on field data, Garten et al. (1978) predicted that Pu uptake in deciduous trees would increase to a maximum over a period of approximately 100 years. Since the physical half life of $^{239}Pu$ is 24,600 years this period of 'equilibration' within an undisturbed forest ecosystem is relatively short, especially when considering the long term environmental impacts of radioactive waste disposal. One such waste disposal area is the Savannah River Burial Site in the USA which received low level transuranic wastes in unlined shallow trenches from 1953 to 1974. Murphy and Tuckfield (1994) investigated $^{238}Pu$ uptake and distributions in three tree species (loblolly pine, sweetgum and willow oak) planted in 1979 above the trenches. It was found that more $^{238}Pu$ was taken up by pine tree seedlings than by the other species and that uptake in pine trees increased with the trees' age, compared with seedlings. The activity concentrations of $^{238}$Pu measured in tree tissues were extremely small (maximum of 0.8 Bq kg$^{-1}$ in leaves) although this study demonstrates the potential for long term 'pumping' of $^{238}$Pu and other transuranics by growing trees from the sub-surface to the surface.

## 7. Technetium in forests

$^{99}$Tc, with a physical half life of 210,000 years, is another radionuclide which is of potential long-term significance in connection with radioactive waste disposal. In another study at the Oak Ridge Reservation (USA), Garten et al. (1986) investigated the cycling of $^{99}$Tc from waste trenches which received $1.6 \times 10^{16}$ Bq of beta activity during the period 1952–1959. $^{99}$Tc is relatively mobile within groundwater and measurements by Garten et al. revealed significant underground seepage of $^{99}$Tc away from waste disposal trenches in the form of the pertechnetate anion ($TcO_4^-$). The roots of several deciduous tree species had intercepted this sub-surface plume which resulted in the uptake and translocation of $^{99}$Tc into above ground tissues of trees. In estimating the overall distribution of $^{99}$Tc within the forest ecosystem Garten et al. (1986) found that almost 90% of the total inventory was located within the soil, with the remainder being mainly associated with the wood of trees (approximately 8% of the total). Measurements of the distribution of $^{99}$Tc in the various forest compartments and the annual fluxes of $^{99}$Tc between these compartments facilitated the development of a model of the long term recycling of $^{99}$Tc within the forest system at Oak Ridge (Garten, 1987). The model predicted a period of 30 years for $^{99}$Tc to reach a steady state distribution within the forest, although some uncertainty remains over the long term chemical behaviour of $^{99}$Tc within forest soils. Short term $K_d$ values for $^{99}$Tc in organic forest soil horizons are low (of the order of 3–4 l kg$^{-1}$) although interaction with humic substances occurs readily and 'ageing' over several years may result in $^{99}$Tc becoming immobilised to a greater degree than laboratory determined $K_d$ values suggest (Bostock, 2004).

Assessments of $^{99}$Tc behaviour in other forest ecosystems are rare because (a) technetium does not occur as a stable element and (b) the activity concentrations of $^{99}$Tc normally encountered are extremely low. Uchida et al. (1999) were able to determine $^{99}$Tc activity concentrations in forest soils of the Chernobyl 30 km exclusion zone using Inductively Coupled Plasma Mass Spectrometry (ICPMS). At the Kopachi site (Table 1) $^{99}$Tc was present at 8–14 Bq kg$^{-1}$ and at Dityatki activity concentrations of only 1.1–2.4 Bq kg$^{-1}$ were measured. As confirmation of the comparatively low $^{99}$Tc activities at these sites, the $^{99}$Tc:$^{137}$Cs ratio was determined to be in the range $0.5–1.25 \times 10^{-4}$.

## 8. Iodine and chlorine in forests

The halogens iodine and chlorine are both potentially important as contaminants emanating at low activities from radioactive waste disposal sites. $^{129}$I and $^{36}$Cl have physical half lives of $15.7 \times 10^6$ years and $3 \times 10^5$ years, respectively, and once introduced into environmental systems are likely to remain there for extremely long periods. Few direct measurements of either of these isotopes have been made in forest ecosystems. However, measurements of

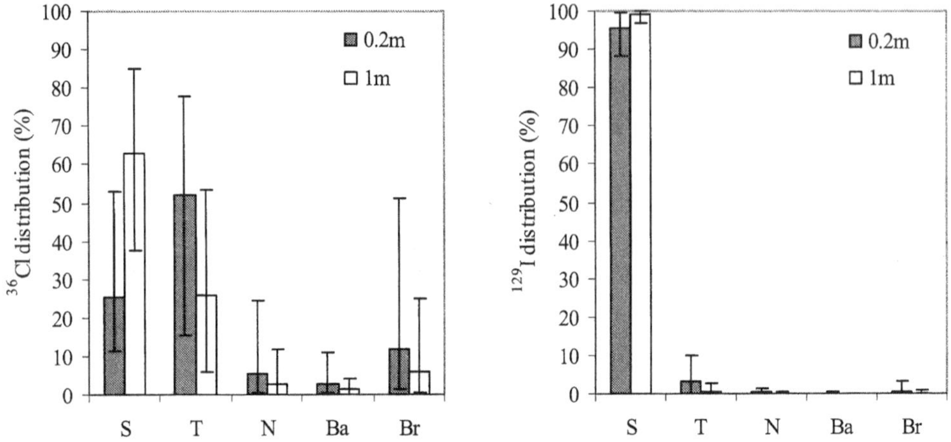


Fig. 14. Median percentages of chlorine (a) and iodine in 5 forest compartments soil (S), trunk wood (T), needles (N), bark (Ba) and branch wood (Br) derived from equilibrium mass balance modelling with an assumed soil depth of 0.2 m or 1 m. Error bars show minimum and maximum values (from Bostock, 2004).

stable (non-radioactive) iodine and chlorine have been made, from which the behaviour of the radionuclides can be inferred.

Stable iodine and chlorine concentrations in forest components have been measured by Bostock (2001) using Instrumental Neutron Activation Analysis (INAA). In the tissues of Scots pine (*Pinus sylvestris*) chlorine was found to be considerably more abundant than iodine. In needles, the Cl:I ratio was of the order of 1000:1, while in wood it ranged from approximately 100:1 to 230:1. Bostock (2004) used these measurements to calculate percentage distributions of Cl and I in a typical Scots pine plantation in southern England (Figure 14). Based on the high concentrations of Cl found in tree tissues, the soil and trunk wood represent approximately equal sinks for Cl, each compartment accounting for approximately 30–60% of the total Cl within the ecosystem depending on the assumed depth of the soil. In contrast, the soil is calculated to contain approximately 96 to 99% of stable iodine within the forest ecosystem. The higher concentrations of stable iodine found in forest soils are most likely due to the organophilic nature of iodine, which is known to have a very high affinity for humic substances and thus a high $K_d$ in organic soils (Sheppard and Thibault, 1990). Cl, on the other hand, is known to have a low $K_d$ within soils of all types and has a high soil–plant transfer factor when compared with most other elements. Measurements of stable Cl in a variety of vegetation types in boreal Canada were taken by Sheppard et al. (1999) who used the Cl concentrations obtained to calculate soil–plant transfer factors (concentration ratios, CR[5]) for several tree species including oak (*Quercus macrocarpa*), pine (*Pinus banksiana*), willow (*Salix* spp.), alder (*Alnus rugosa*) and aspen (*Populus tremuloides*). These CR values ranged from 0.8 for oak (leaves and stems) to 28 for aspen leaf. For the complex mixture of woody and herbaceous samples analysed, Sheppard et al. (1999) recommended a geometric mean CR for Cl of 7. This compares with transfer factors for radiocaesium in forest vegetation which are typically $\ll 1$.

[5] Concentration ratio (CR) = concentration in plant tissue/concentration in soil.

Chlorine is normally assumed to occur in soils as the chloride ($Cl^-$) ion which is extremely mobile, with a very low $K_d$ and short residence half time in soil horizons. However, observations of stable Cl in Swedish forest soils have shown that chlorination of soil organic matter occurs readily (Öberg and Grøn, 1998) and that this can lead to significant retention of chlorine within boreal forest soils (Rodstedth et al., 2003). Milton et al. (2003) used Accelerator Mass Spectrometry (AMS) to measure $^{36}Cl$ derived from global weapons testing in Canadian forests in Ontario. Measured activities in groundwater, soil and vegetation were summed and found to be of the order of 75% of the total $^{36}Cl$ activity known to have deposited to the area. Furthermore, sorption of $^{36}Cl$ to the organic soils was demonstrated in comparison with the movement of water, thus confirming the observations of Öberg and Grøn (1998) from Sweden. The 5–25% imbalance in the $^{36}Cl$ inventory was thought by Milton et al. (2003) to be the result of loss of volatile organohalides from forest vegetation. This emphasises the potential significance of volatilisation from soil/vegetation surfaces to the free atmosphere when considering the long term behaviour of radioactive chlorine and iodine within forests.

## 9. Radiation exposures associated with forest products

Radiation doses to humans in the forest environment can be incurred either *in situ*, in the form of direct external exposure to gamma and beta radiation (Figure 5) and the inhalation of radioactive particles, or *ex situ* as a result of exporting a variety of forest products away from the contaminated area. Most of the forests affected by deposition from the Chernobyl accident are artificially managed to produce timber crops, as havens for wildlife and as collecting areas for wild foodstuffs such as berries and mushrooms. The exploitation of forests for each of these resources leads to some surprising and often significant routes of radiation exposure.

### 9.1. *Timber and wood products*

Timber extracted from forests is used for a variety of domestic and industrial purposes, including the construction of buildings, the manufacture of paper on an industrial scale and the use of firewood on a domestic scale. Figure 15 shows a variety of possible uses of timber following its extraction from the forest: the routes of export, treatment and end-use of timber are often complex and may vary from country to country. Exposure routes may involve long distance, international transport of timber resulting in radiation doses of people hundreds or thousands of kilometres away from the forest which was originally contaminated.

The IAEA (2003) conducted a study to calculate the likely dose conversion coefficients (DCC) which could be applied to a wide variety of exposure pathways associated with timber and related products. Table 5 shows the summarised DCC values which represent the dose in mSv incurred per $Bq\,kg^{-1}$ in the original raw timber. For $^{137}Cs$, most significant exposure routes are related to industrial treatment of timber, in sawmills, in pulp manufacture and in the handling of wood ash. The burning of wood produces ash weighing approximately 1 to 5% of the original mass of wood and a concentration of the original radiocaesium content of 20–100 times higher than the original activity concentration. The use of wood contaminated with

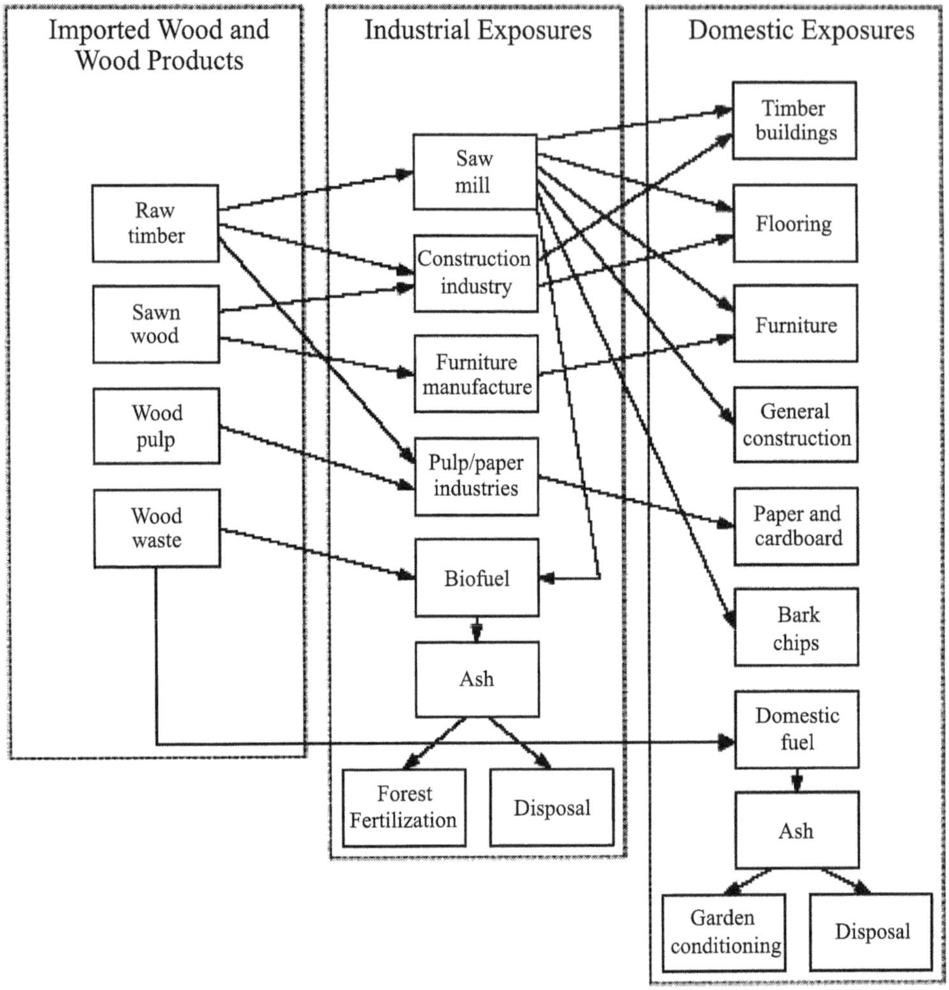


Fig. 15. Uses of wood in industrial and domestic products which may give rise to human radiation exposures. From IAEA (2003); reproduced by permission of the International Atomic Energy Agency.

radiocaesium at even modest levels can lead to significant accumulations of activity, especially if combustion is carried out on an industrial scale such as in biofuel burning (Hubbard, 1997). Naturally occurring radionuclides (NORM) within timber can also result in significant exposures in the pulp and paper industry (see section 11 of Chapter 10).

Domestic exposures are generally less significant than industrial exposures because of the much smaller masses of wood used in the home. Nevertheless, continued burning of firewood obtained from contaminated forests could eventually lead to significant doses being incurred, especially if wood ash is habitually disposed of in one location, such as in domestic gardens.

Table 5
Dose conversion coefficients for most exposed population groups (DCC, mSv per Bq/kg in wood) for different scenarios of treatment or application of wood contaminated with $^{137}Cs$ or $^{90}Sr$. (Source: IAEA, 2003)

| Category | Scenario | | DCC (mSv/Bq/kg in wood) $^{137}Cs$ | $^{90}Sr$ |
|---|---|---|---|---|
| Industrial | Sawmill | | $1.5 \times 10^{-4}$ | $5.0 \times 10^{-7}$ |
| | Pulp factory | | $1.1 \times 10^{-4}$ | 0 |
| | Ash disposal | Stockpile | $3.0 \times 10^{-4}$ | $5.0 \times 10^{-7}$ |
| | | Landfill | $2.0 \times 10^{-4}$ | $1.0 \times 10^{-7}$ |
| | | Forest fertilisation | $6.0 \times 10^{-6}$ | $5.0 \times 10^{-7}$ |
| Domestic | Timber house and sauna | | $2.0 \times 10^{-5}$ | $3.0 \times 10^{-8}$ |
| | Flooring | | $3.0 \times 10^{-6}$ | 0 |
| | Furniture | | $3.0 \times 10^{-5}$ | 0 |
| | Bark chips as a mulch | | $8.0 \times 10^{-7}$ | $2.0 \times 10^{-8}$ |
| | Domestic fuel | | $1.3 \times 10^{-6}$ | $1.7 \times 10^{-6}$ |

### 9.2. *Edible forest products*

The role of edible forest products in the dietary exposure of populations across Europe has been the subject of considerable study following the Chernobyl accident. Mushrooms, in particular, have been scrutinised because of their tendency to concentrate radiocaesium from forest soils, as described in section 4.4. However, edible berries from dwarf shrubs such as bilberry (*Vaccinium macrocarpon*), lingon berry (*Vaccinium vitis-idaea*) and cloudberry (*Rubus chamaemorus*) are often harvested in forests and can also contribute to dietary intake of radiocaesium. The importance of mushrooms and berries in contributing radiocaesium to the body burdens of large forest dwelling animals such as deer and moose can be seen clearly in data which have been collected since shortly after 1986 (Johanson and Bergström, 1994). A long term study by Zibold et al. (2001) has clearly shown the seasonal increase in $^{137}Cs$ activity concentrations in the muscle tissues of roe deer harvested in the September–October period during which the major appearance of fungal fruiting bodies occurs.

Just as grazing animals rely on mushrooms and berries as an important component of their diet at certain times of the year, so humans in many parts of the world collect forest foods on a seasonal basis. Johanson and Bergström (1994) calculated that the ingestion of meat from moose and roe deer hunted in Swedish forests contaminated with Chernobyl radiocaesium could lead to integrated collective doses ranging from 300 to 1500 man Sv. The seasonal consumption of mushrooms, in particular, is an extremely important dietary habit in many European countries. Fesenko et al. (2000) estimated that internal doses from wild mushrooms and berries obtained from forests in the Bryansk region of Russia were 10–15% of the dose derived from agricultural products in 1987, rising to 40–45% in 1996 as agricultural countermeasures reduced contamination of milk, meat and vegetables. These authors also estimated that, in the same area, the collective dose obtained from forest products during the decade following the Chernobyl accident was 213 man Sv, representing some 20% of the overall collective dose locally. Interestingly, however, Fesenko et al. (2000) also estimated that, due to the wider dispersal of forest products from the Bryansk region, the potential collective dose was greater than 650 man Sv. So, as in the case of exported timber products, the radiological impact of contaminated forests may be felt beyond the immediate boundaries of forested areas.

## 10. Conclusions

In terms of preparedness for future nuclear accidents and sustainable management of existing contaminated forests in Europe and the former Soviet Union it is clear that, in the aftermath of the Chernobyl accident, there was insufficient or piecemeal understanding of processes controlling radiocaesium recycling in forest ecosystems. Much has been learned of these processes since 1986, however, and major efforts have been made to integrate this information into computer-based models which can play a key role in consolidating our understanding of ecological processes as well as providing tools for the design of management and intervention strategies to be applied to contaminated forests (Fesenko et al., 2005, Shaw et al., 2005). Chernobyl provided the impetus for many new radioecological investigations in forests although several other contamination scenarios have been described in this chapter. One of the most challenging for future risk assessments is likely to be the issue of long term leakage of radionuclides from radioactive waste disposal facilities. The long time scales over which such risk assessments must be carried out (hundreds to thousands of years) are similar to those during which forests develop in the absence of human intervention. Changing climatic conditions over such time scales are highly likely to impact forest development and the degree of risk posed by long-lived radionuclides in the subsurface, providing a compelling reason for improving our understanding of the behaviour of as wide a variety of radionuclides in forest ecosystems as possible.

## References

Alexakhin, R.M., Ginsburg, L.R., Mednik, I.G., Prokhorov, V.M. (1994). Model of $^{90}$Sr cycling in a forest biogeocenosis. *Science of the Total Environment* **157**, 83–91.

Belli, M. (Ed.) (2000). *SEMINAT Long-Term Dynamics of Radionuclides in Semi-Natural Environments: Derivation of Parameters and Modelling*. Agenzia Nazionale per la Protezione dell'Ambiente, Rome.

Belli, M., et al. (1996). Dynamics of radionuclides in forest environments. In: *The Proceedings of the First International Conference of the European Commission, Belarus, Russian Federation & Ukraine on the Radiological Consequences of the Chernobyl Accident* (Minsk, 18–22 March, 1996), pp. 69–80. EUR 16544 EN.

Bergman, R. (1994). The distribution of radioactive caesium in boreal forest ecosystems. In: Dahlgaard, H. (Ed.), *Nordic Radioecology—The Transfer of Radionuclides Through Nordic Ecosystems to Man*. Elsevier, Amsterdam, pp. 335–379.

Borovoi, A.A., Dobrynin, Yu.L., Kulakov, V.M., Levina, L.A. (1998). Correlation calculations for the fuel of the 4$^{th}$ unit of the Chernobyl Nuclear Power Plant. Criteria of fuel release. In: *Abstract of All Union Conference on Problems of Liquidation of Consequences of Chernobyl Accident in Agricultural Production, Obninsk*, pp. 100–101 (in Russian).

Bostock, A.C. (2001). Volatilisation of I-129 from coniferous forest and grasland soils, MSc Thesis, University of London (Imperial College).

Bostock, A.C. (2004). Volatilisation and chemical speciation of chlorine, iodine and technetium in boreal ecosystems. PhD Thesis, University of London (Imperial College).

Bunzl, K., Kracke, W. (1988). Cumulative deposition of $^{137}Cs$, $^{238}Pu$, $^{239+240}Pu$ and $^{241}Am$ from global fallout in soils from forest, grassland and arable land in Bavaria (FRG). *Journal of Environmental Radioactivity* **8**, 1–14.

Bunzl, K., Schimmack, W., Kreutzer, K., Schierl, R. (1989). Interception and retention of Chernobyl-derived $^{134}Cs$, $^{137}Cs$ and $^{106}Ru$ in a spruce stand. *Science of the Total Environment* **78**, 77–87.

Chesser, R.K., Bondarkov, M., Baker, R.J., Wickliffe, J.K., Rodgers, B.E. (2004). Reconstruction of radioactive plume characteristics along Chernobyl's Western Trace. *Journal of Environmental Radioactivity* **71**, 147–157.

Croom, J.M., Ragsdale, H.L. (1980). A model of radiocesium cycling in a sand hills-Turkey Oak (Quercus laevis) ecosystem. *Ecological Modelling* **11**, 55–65.

Crossland, I.G., Pinedo, P., Kessler, J.H., Torres-Vidal, C., Walters, B. (2005). "Reference Biospheres" for solid radioactive waste disposal: the BIOMASS. *Journal of Environmental Radioactivity* **84**, 135–149.

Davydchuk, V. (1997). Ecosystem remediation in radioactively polluted areas: the Chernobyl experience. *Ecological Engineering* **8**, 325–336.

Dewiere, L., Bugai, D., Grenier, C., Kashparov, V., Ahamdach, N. (2004). $^{90}Sr$ migration to the geosphere from a waste burial in the Chernobyl exclusion zone. *Journal of Environmental Radioactivity* **74**, 139–150.

Dvornik, A., Zhuchenko, T. (1998). Belorussian Forestry Institute, Gomel. Personal communication to G. Shaw.

Fesenko, S.V., Voigt, G., Spiridonov, S.I., Sanzharova, N.I., Gontarenko, I.A., Belli, M., Sansone, U. (2000). Analysis of the contribution of forest pathways to the radiation exposure of different population groups in the Bryansk region of Russia. *Radiation and Environmental Biophysics* **39**, 291–300.

Fesenko, S.V., Soukhova, N.V., Sanzharova, N.I., Avila, R., Spiridonov, S.I., Klein, D., Lucot, E., Badot, P.M. (2001). Identification of processes governing long-term accumulation of $^{137}Cs$ by forest trees following the Chernobyl accident. *Radiation and Environmental Biophysics* **40**, 105–113.

Fesenko, S.V., Voigt, G., Spiridonov, S.I., Gontarenko, I.A. (2005). Decision making framework for application of forest countermeasures in the long term after the Chernobyl accident. *Journal of Environmental Radioactivity* **82**, 143–166.

Förster, H., Schimmack, W., Kreutzer, K.E. (1991). Horizontal distribution of radiocaesium in forest soils under spruce (*Picea abies* L. Karst.) and beech (*Fagus sylvatica* L.). *Z. Pflanzenernähr. Bodenk.* **154**, 87–92 (in German).

Fraiture, A. (1992). Introduction to the radioecology of forest ecosystems and survey of radioactive contamination in food products from forests. *Radiation Protection 57*. Commission of the European Communities, Brussels.

Garten, C.T. (1987). Technetium-99 cycling in deciduous forests: review and ecosystem model development. *Environment International* **13**, 311–321.

Garten, C.T., Gardner, R.H., Dahlman, R.C. (1978). A compartment model of plutonium dynamics in a deciduous forest ecosystem. *Health Physics* **34**, 611–619.

Garten, C.T., Tucker, C.S., Walton, B.T. (1986). Environmental Fate and Distribution of Technetium-99 in a Deciduous Forest Ecosystem. *Journal of Environmental Radioactivity* **3**, 163–188.

Hubbard, L.M. (1997). *Biofuel Burning and Earlier Radioactive Fallout.* Swedish Radiation Protection Institute, Stockholm. SSI News, No. 1, pp. 10–11.

IAEA (1994). *Handbook of Parameter Values for The Prediction of Radionuclides Transfer in Temperate Environments.* International Atomic Energy Agency, Vienna.

IAEA (2003). *Assessing Radiation Doses to The Public from Radionuclides in Timber and Wood Products.* International Atomic Energy Agency, Vienna. IAEA-TECDOC-1376.

Johanson, K.J., Bergström, R. (1994). Radiocaesium transfer to man from moose and roe deer in Sweden. *Science of the Total Environment* **157**, 309–316.

Jordan, C.F., Kline, J.R. (1976). Strontium-90 in a tropical rain forest: 12th year validation of a 32 year prediction. *Health Physics* **30**, 199–201.

Kinnersley, R.P., Farrington-Smith, J.G., Shaw, G., Minski, M.J. (1994). Aerodynamic characterisation of model tree canopies in a wind tunnel. *Science of the Total Environment* **157**, 29–33.

Konoplev, A.V., Viktorova, N.V., Virchenko, E.P., Popov, V.E., Bulgakov, A.A., Desmet, G.M. (1993). Influence of agricultural countermeasures on the ratio of different chemical forms of radionuclides in soil and soil solution. *The Science of the Total Environment* **137**, 147–162.

Kozubov, G.M., Taskaev, A.I., Fedotov, I.S., Arkhipov, N.P., Davydchuk, V.S., Abaturov, Y.D. (1991). *The Map-Scheme of Radiation Damage of Coniferous Forests in The Region of Accident on The Chernobyl Nuclear Power Station.* Institute of Biology, Syktyvkar, Russia.

Mascanzoni, D. (1990). Uptake of $^{90}Sr$ and $^{137}Cs$ by mushrooms following the Chernobyl accident. In: Desmet, G.M., Nassimbeni, P., Belli, M. (Eds.), *Transfer of Radionuclides in Natural and Semi-natural Environments*. Elsevier Applied Science, London and New York, p. 459.

McCune, D.C., Boyce, R.L. (1992). Precipitation and the transfer of water, nutrients and pollutants in tree canopies. *Trends in Ecology and Evolution* **7**, 4–7.

Melin, J., Wallberg, L., Suomela, J. (1994). Distribution and retention of cesium and strontium in Swedish boreal forest ecosystems. *Science of the Total Environment* **157**, 93–105.

Milton, G.M., Milton, J.C.D., Schiff, S., Cook, P., Kotzer, T.G., Cecil, L.D. (2003). Evidence for chlorine recycling—hydrosphere, biosphere, atmosphere—in a forested wet zone on the Canadian Shield. *Applied Geochemistry* **18**, 1027–1042.

Murphy, C.E., Tuckfield, R.C. (1994). Transuranic element uptake and cycling in a forest over an old burial ground. *Science of The Total Environment* **57**, 115–124.

Myttenaere, C., Sombre, C.L., Thiry, Y., de Brouwer, S., Ronneau, C. (1993). Cycling of radiocaesium in forest ecosystems. *Journal of Radioecology* **1**, 7–14.

Nylén, T., Grip, H. (1997). The origin and dynamics of $^{137}Cs$ discharge from a coniferous forest catchment. *Journal of Hydrology* **192**, 338–354.

Öberg, G., Grøn, C. (1998). Sources of organic halogens in spruce forest soil. *Environmental Science and Technology* **32**, 1573–1579.

Olsen, R.A., Joner, E., Bakken, L.R. (1990). Soil fungi and the fate of radiocaesium in the soil ecosystem. In: Desmet, G.M., Nassimbeni, P., Belli, M. (Eds.), *Transfer of Radionuclides in Natural and Semi-natural Environments*. Elsevier Applied Science, London and New York, p. 657.

Ould-Dada, Z., Baghini, N.M. (2001). Resuspension of small particles from tree surfaces. *Atmospheric Environment* **35**, 3799–3809.

Ould-Dada, Z., Copplestone, D., Toal, M., Shaw, G. (2002). Effect of forest edges on deposition of radioactive aerosols. *Atmospheric Environment* **36**, 5595–5606.

Pawlawski, L. (1997). Acidification: its impact on the environment and mitigation strategies. *Ecological Engineering* **8** (4), 271–288.

Rodstedth, M., Ståhlberg, C., Sandén, P., Öberg, G. (2003). Chloride imbalances in soil lysimeters. *Chemosphere* **52**, 381–389.

Rühm, W., Kammerer, L., Hiersche, L., Wirth, E. (1997). The $^{137}Cs/^{134}Cs$ ratio in fungi as an indicator of the major mycelium location in forest soil. *Journal of Environmental Radioactivity* **35**, 129–148.

Schimmack, W., Bunzl, K., Kreutzer, K., Rodenkirchen, E., Schierl, R. (1991). Effect of spruce (*Picea abies* L. Karst.) and beech (*Fagus sylvatica* L.) on the migration of radiocaesium in the soil. In: Kreutzr, K., Göttlein, A. (Eds.), *Ökosystemforschung Höglwald*. Verlag Paul Parey, Hamburg, Berlin, pp. 242–251.

Schimmack, W., Bunzl, K. (1992). Migration of radiocesium in two forest soils as obtained from field and column investigations. *Science of The Total Environment* **116**, 93–107.

Schuck, A., Paivinen, R., Hame, T., Van Brusselen, J., Kennedy, P., Folving, S. (2003). Compilation of a European forest map from Portugal to the Ural mountains based on earth observation data and forest statistics. *Forest Policy and Economics* **5**, 187–202.

Shaw, G., Wang, X. (1996). Caesium and plutonium migration in forest soils of the Chernobyl 30 km zone. In: *Ten Years Terrestrial Radioecological Research, Proceedings of an International Symposium on Radioecology*. Austrian Soil Science Society/Austrian Federal Environment Agency, Vienna, Austria, pp. 27–34.

Shaw, G., Avila, R., Fesenko, S., Dvornik, A., Zhuchenko, T. (2002). Modelling the behaviour of radiocaesium in forest ecosystems. In: Scott, E.M. (Ed.), *Modelling Radioactivity in the Environment*. Elsevier Science Publishers, Amsterdam. Chapter 11.

Shaw, G., Venter, A., Avila, R., Bergman, R., Bulgakov, A., Calmon, P., Fesenko, S., Frissel, M., Goor, F., Konoplev, A., Linkov, I., Mamikhin, S., Moberg, L., Orlov, A., Rantavaara, A., Spiridonov, S., Thiry, Y. (2005). Radionuclide Migration in Forest Ecosystems—Results of a Model Validation Study. *Journal of Environmental Radioactivity* **84**, 285–296 (special issue on the IAEA BIOMASS Programme).

Shcheglov, A.I., Tsvetnova, O.B., Klyashtorin, A.L. (2001). *Biogeochemical Migration of Technogenic Radionuclides in Forest Ecosystems*. Nauka, Moscow.

Sheppard, M.I., Thibault, D.H. (1990). Default soil solid/liquid partition coefficients, $K_d$s, for four major soil types: a compendium. *Health Physics* **59**, 471–482.

Sheppard, S.C., Evenden, W.G., Macdonald, C.R. (1999). Variation among chlorine concentration ratios for native and agronomic plants. *Journal of Environmental Radioactivity* **43**, 65–76.

Smith, J.T., Beresford, N.A. (2005). *Chernobyl: Catastrophe and Consequences*. Springer Praxis publishers, Chichester, UK.

Smith, S.E., Read, D.J. (1997). *Mycorrhizal Symbiosis*, 2nd edn. Academic Press, San Diego, California.

Sokolov, V.E., Krivolutski, D.A., Voronov, A.G., Shilova, S.A. (Eds.) (1993). *Ecological After-Effects of The Radioactive Contamination in The Southern Urals*. Nauka, Moscow (in Russian).

Strzyszcz, Z. (1989). Ferromagnetic properties of forest soils being under influence of industrial pollution. In: Bucher, J.B., Bucher-Wallin, I. (Eds.), *Air Pollution and Forest Decline*. EAFV, Birmensdorf, Switzerland, pp. 201–207.

Tikhomirov, F.A. (1993). Radionuclide migration and distribution in the EURT forests. In: Sokolov, V.E., Krivolutski, D.A., Voronov, A.G., Shilova, S.A. (Eds.), *Ecological After-Effects of The Radioactive Contamination in the Southern Urals*. Nauka, Moscow, pp. 21–39 (in Russian).

Uchida, S., Tagami, K., Wirth, E., Rühm, W. (1999). Concentration levels of technetium-99 in forest soils collected within the 30-km zone around the Chernobyl reactor. *Environmental Pollution* **105**, 75–77.

UNECE (1993). *UNECE Convention on Long-Range Transboundary Air Pollution: Manual for Integrated Monitoring*. Environment Data Centre, National Board of Waters and the Environment, Helsinki, Finland.

UNECE/FAO (2000). Forest resources of Europe, CIS, North America, Australia, Japan and New Zealand (industrialized temperate/boreal countries). UNECE/FAO Contribution to the Global Forest Resources Assessment (TBFRA) 2000, Main report. Geneva Timber and Forestry Study papers, No. 17. United Nations, New York and Geneva, p. 445.

Vinichuk, M.M., Johanson, K.J., Rosén, K., Nilsson, I. (2004). Role of the fungal mycelium in the retention of radiocaesium in forest soils. *Journal of Environmental Radioactivity* **78**, 77–92.

Wang, X. (1996). Physico-chemical behaviour of radionuclides in forest soils of the evacuated region of the Chernobyl nuclear power plant. PhD Thesis, University of London (Imperial College), UK.

Warner, F., Harrison, R.M. (1993). *Radioecology After Chernobyl*. John Wiley publishers, Chichester, UK. SCOPE 50.

Witherspoon, J.P., Taylor, F.G. (1969). Retention of simulated fallout containing $^{137}Cs$ by pine and oak trees. *Health Physics* **17**, 825–839.

Yoshida, S., Muramatsu, Y., Dvornik, A.M., Zhuchenko, T.A., Linkov, I. (2004). Equilibrium of radiocesium with stable cesium within the biological cycle of contaminated forest ecosystems. *Journal of Environmental Radioactivity* **75**, 301–313.

Zibold, G., Drissner, J., Kaminski, S., Klemt, E., Miller, R. (2001). Time-dependence of the radiocaesium contamination of roe deer: measurement and modelling. *Journal of Environmental Radioactivity* **55**, 5–27.

# Radioecology of terrestrial arctic ecosystems

Mark Dowdall[a], Justin P. Gwynn[a], George Shaw[b]

[a]*Statens Strålevern, Polar Environmental Centre, Tromsø, Norway*
[b]*Division of Agricultural & Environmental Sciences, University of Nottingham, Nottingham, NG7 2RD, UK*

## 1. Introduction

The common perception of the Arctic as a pristine wilderness and the vulnerability of its ecosystems have precipitated an increase in the amount of interest and attention focused on these regions. Much of this attention has been in relation to pollutant levels in the Arctic and its ecosystems, with radioactive contaminants one of the most often discussed suite of pollutants. Evidence of pollution of the Arctic environment with radionuclides has been collected since the 1960's, generating little in the way of public attention with the notable exception of contaminated reindeer. Two occurrences within the last twenty years, however, have increased public awareness of the contamination of the Arctic, these being the Chernobyl accident of 1986 and publication of the Yablokov report in 1993 (Yablokov, 1993). The effects of the Chernobyl accident were felt strongly in the Arctic, with some of the most affected areas lying in the Eurasian Arctic region. The media coverage of the aftermath and the plight of the reindeer herding population of this area highlighted for the public how contamination could travel quickly through a typical arctic food chain and the incident initiated a strong response from the scientific community. The momentum from this response continues and a large body of information continues to be gathered, relevant to the further understanding of radioactive contaminant behaviour in terrestrial ecosystems.

The Yablokov report, a Russian government white paper, outlined the extent of the radioactive threat posed to the Arctic some 7 years after the Chernobyl accident. The report provided, in some detail, information about the past dumping of submarine nuclear reactor elements and other radioactive waste materials into the seas adjacent to the Arctic. Publication of this report served to heighten concerns in arctic countries and efforts were implemented towards answering questions regarding the ultimate fate of the dumped materials and their implications for the Arctic. Public awareness of the issue of radioactive contamination of the Arctic has been sustained via media coverage of incidents such as the sinking of the nuclear submarine "Kursk" in 2000 and through the increased attention devoted by a number of countries to the activities of European nuclear reprocessing plants, the discharges of which have been detected in arctic waters.

RADIOACTIVITY IN THE ENVIRONMENT
VOLUME 10 ISSN 1569-4860/DOI: 10.1016/S1569-4860(06)10007-8

## 2. The arctic terrestrial environment

Although often defined using several criteria such as climate, flora and latitude, for the purpose of this chapter it is convenient to use the AMAP designation (AMAP, 1997) of the Arctic which includes parts of eight countries; Canada, Denmark (Greenland and the Faroe Islands), Finland, Iceland, Norway, Russia, Sweden and the United States (Alaska) and encompasses approximately 13.4 million square kilometres of land, some 40% of the entire region (Figure 1).

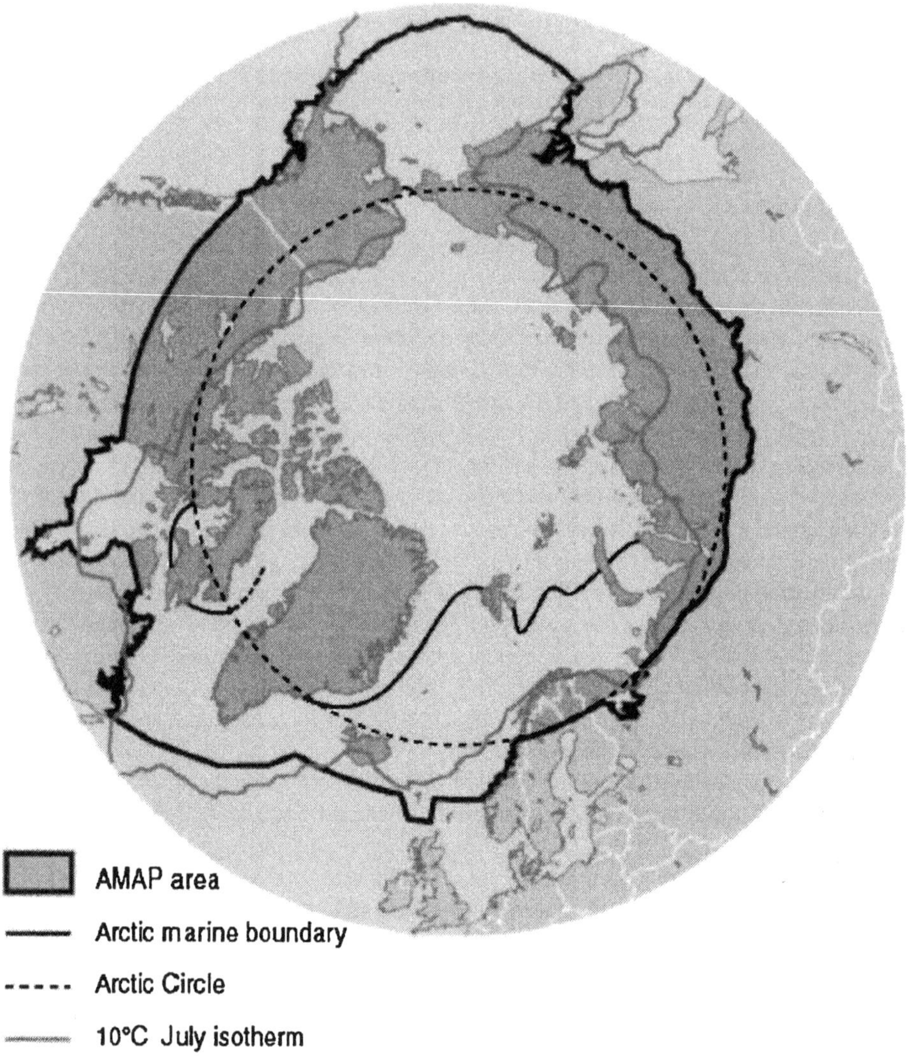


Fig. 1. The Arctic region. (Reproduced from AMAP, 1997, with permission).

Table 1
Arctic soil types and characteristics (after Goryachkin et al., 1998)

| Soil zone | Cold desert | High arctic | Tundra | | | |
|---|---|---|---|---|---|---|
| Subzone | | Soddy arctic | Northern | Subarctic | | |
| Typical soils | Primitive, stony, saline | Peaty organic soils, dwarf podsols, rendzinas | Tundra gleys, humus, podsols, peaty organic soils | Tundra gleys, podsols, peat and soddy soils | | |
| Typical features | | Sparse vegetation, aeolian inputs of salts and carbonates | Discrete soil cover, mull-moder horizon, sparse litter | Continuous cover, thick litter, peat and bog hummocks widespread | | |
| Distinguished zones and subzones | Arctic desert | Arctic tundra | | Typical tundra | Southern tundra | Forest tundra |

Arctic weather is often extreme although the climatological characteristics of the AMAP regions vary considerably. The High Arctic is a polar desert as cold air has little capacity to hold moisture, so precipitation is generally low and there is little open water. By comparison, the Low Arctic, dominated by tundra, is generally warmer and may exhibit greater precipitation. Annual temperature ranges in the Arctic are narrower over coastal regions that often exhibit a maritime climate but tend to be broader over continental regions such as Siberia.

The geology of the area covered by the AMAP designation ranges from the crystalline rocks of eastern Canada and Greenland and the younger sedimentary rocks of Russia to the volcanic lithologies of Iceland. Common to most arctic regions is the role ice has played in shaping the morphological features of landscape via the action of glaciers and permafrost. Permafrost, or ground that remains frozen for at least two consecutive summers, can reach as deep as 1500 m in the coldest arctic areas. The material above the permafrost that remains unfrozen during summer months is termed the active layer and can range in depth from a few centimetres to several metres. Water movement within areas affected by permafrost is different to environments from more temperate zones. The majority of freshwater input to the Arctic terrestrial environment occurs in spring when snowmelt occurs and water flows over the surface of the ground, rather than penetrating deeper layers, entering streams, rivers and lakes. The resulting saturation of the upper soil layers can cause the formation of wetlands, the saturated condition of which inhibits decomposition of organic material and results in the accumulation of peat and humic materials.

Recent years have seen a relative upsurge in the amount of published information pertaining to soils in the Arctic and, in particular, to the soil types present in poorly investigated regions such as Novaya Zemlya, Spitsbergen, northern Canada and the Eurasian Arctic (see, for example, Foscolos and Kodama, 1981; Dobrovol'skiy, 1990; Vasil'evskaya et al., 1993; Kalyakin, 1995). A brief summary of soil types prevalent in the Arctic is provided in Table 1.

### 2.1. *Terrestrial flora and fauna*

The terrestrial environment of the Arctic is conveniently divided into two biogeographical areas, the High Arctic and the Low Arctic as described in the section above. The growing season in the High Arctic is approximately two months long, with air temperatures rarely rising above 8°C. Soil cover may be discrete or nonexistent, whilst vegetation cover is sparse and consists mainly of lichens (*Neuropogon, Collema, Pertusaria* and *Tonina* spp.), mosses (*Pohlia, Myurella, Andreaea* and *Onchophorus* spp.) and in some isolated areas communities of cushion plants (*Dryas integrifolia, Saxifraga oppositifolia* and *Papaver* spp.), grasses (*Phippsia algidae* and *Poa abbreviata*) and shrubs (*Salix arctica*). Large herbivores graze intensively in the summer months and include musk ox (*Ovibus moschatos*) and Peary caribou (*Rangifer tarandus pearyi*) in Canada and reindeer subspecies (*Rangifer* species) in the Eurasian regions. Arctic hare (*Lepus timidus*) and arctic fox (*Alopex lagopus*) are smaller mammals that survive in the High Arctic whilst birds are represented by about 20 species that nest in the region. Wolves (*Canis lupus*) and the marine mammal, the polar bear (*Ursus maritimus*), are the main predators in the High Arctic.

Biological diversity is greater in the Low Arctic, due in part to the longer growing season (three to four months) and slightly warmer temperatures. The action of permafrost controls plant diversity with the thin, periodically un-frozen active layer restricting the occurrence of plants to shallow rooting species such as grasses (*Poa* and *Arctagrostis* spp.), sedges (*Eriophorum* spp.), heath shrubs (*Ledum, Cassiope* and *Empetrum* spp.) and a variety of lupins, buttercups and ferns as well as such lichens and mosses as may be found in the High Arctic. The permafrost layer prevents seepage of surface water, leading to vast waterlogged areas in spring and summer from the melting snow, resulting in the accumulation of substantial quantities of organic detritus. The ensuing anoxic conditions, coupled with low temperatures, result in the formation of widespread peat deposits.

Wetlands cover approximately 1.5 million square kilometres of the Arctic region and of the five typical wetland types (bogs, fens, swamps, marshes and shallow open waters), fens and bogs are the most common in the Arctic. Bogs are usually composed of *Sphagnum* mosses whilst fens consist of sedges and grasses. The rich vegetation attracts many bird species including waterbirds, swans, geese and a variety of others. Mammals are represented by foxes (*Vulpes vulpes*), wolves (*Canis lupus*), voles (*Microtus oeconomus, M. gregalis*), lemmings (*Lemmus sibericus*), hares (*Lepus timidus*), reindeer (*Rangifer tarandus*) and brown bears (*Ursus arctos*) as well as a variety of stoats, mink and rodents.

## **2.2.** *Arctic terrestrial food chains*

Terrestrial food chains in the Arctic tend to be short with slow-growing lichens and mosses as the primary producers followed by a small number of herbivores (reindeer, hares etc.) and ending with one or more of the carnivorous species, such as wolves or bears. Slow growth in arctic regions means that many plants are quite long-lived in contrast to temperate regions where a plant may only have been growing for a single season before it is consumed. The lack of species diversity in some regions of the Arctic also means that the fodder species of herbivores may be limited to one or two vegetation types, a prime example being the predominance of lichens in the diet of reindeer/caribou for much of the year. A second point to be considered is that in some parts of the Arctic, most notably High Arctic locations, nutrient input to the soil is provided mainly by the faecal material of birds, of which large populations exist in the Arctic. In such typically nutrient poor environments, plant distribution is closely linked to nutrient availability and plant numbers and diversity are often greatest where fertilisation by faecal material has occurred. Proportions of contaminants present in faecal material are available for uptake by vegetation growing at these locations and so, ultimately, may be introduced into terrestrial food chains. Therefore, marine contaminants may find their way into a terrestrial food chain in a manner that is not so prevalent or important in other regions.

In many cases in the Arctic, humans exist at the top of the food chain and it is this fact that has exacerbated concern in the transfer of contaminants within arctic ecosystems. Although fish and marine creatures constitute a proportion of the diet of indigenous arctic people, caribou/reindeer (either taken from the wild or from farmed populations) are the most important terrestrial source of food for these populations. The significance of this food source varies according to location: for example, reindeer meat is much less important in the diet of Faroe islanders than in the Saami diet.

### 2.3. *Arctic freshwater ecosystems*

The characteristics of lakes, streams and rivers vary extensively within the Arctic region. Freshwater ecosystems in the Low Arctic can be quite productive, remaining ice free for much of the year, a situation contrasted by the ice bound lakes and rivers of High Arctic areas where ice cover may attain thicknesses of up to 3 metres. Ice cover and the spring melt dominate the ecology of these systems, with productivity constrained by the low temperatures and lack of light whilst nutrient inputs originate largely from contributions from the spring snow melt. The formation of ice can result in the underlying water becoming enriched in contaminants by partial exclusion of solutes or the export of contaminants to distant locations by ice rafting. Prior to the spring melt, the snow pack accumulates such contaminants as may be deposited on it over the course of the previous winter. In the spring, the rise in temperature causes accumulation of these contaminants in the meltwater, with the initial 20–30% of the meltwater containing as much as 60% of the contaminant load of the snow pack. The nature of soils in the Arctic and the presence of permafrost results in the majority of this spring melt water entering the fresh water and marine ecosystems. In contrast to temperate regions, surface waters in arctic areas are more vulnerable to radioactive contamination, as the proportion of water in a freshwater arctic body that is derived from groundwater is substantially less than in a temperate waterbody.

Rivers and streams in the High Arctic are usually short and fed by snow or glacial melt water. They may provide habitats for some vegetation, insect larvae and fish, although strong, periodic flow can reduce the number of species present. Low Arctic rivers and streams may provide more stable environments with greater species diversity as in Fennoscandia. Rivers constitute an important vector for radionuclide contamination in the Arctic terrestrial environment. Industrial facilities are often located near rivers or within their catchments and rivers have historically been used as repositories for a variety of wastes. The long courses run by many rivers means that contamination originating from non-arctic regions can easily reach the Arctic, the situation concerning the Yenisey and Ob Rivers in Russia being a particularly topical example. River ice can incorporate bottom sediments and associated contaminants, and transport these contaminants long distances or remobilise the contamination back into the water column for uptake by flora and fauna. This movement of sediment is associated with the role of lakes as contaminant sinks via the accumulation of sediment and particles.

Terrestrial-aquatic food chains, which are typically a little more complex than purely terrestrial ones, are founded on free floating and benthic algae that are able to photosynthesise, depending on light conditions, between April and the end of the Arctic summer. These in turn are preyed upon by zooplankton, the number and diversity of which are largely a function of location and temperature. Insect larvae and crustaceans may play roles in the food chain but the final trophic level within the aquatic system is usually occupied by fish. Fish types in arctic freshwaters vary from region to region but may include arctic char (*Salvelinus alpinus*), Brown trout (*Salmo trutta*), Northern pike (*Esox lucius*) and Grayling (*Thymalls* spp.). Species feeding on freshwater fish include mink (*Mustela vison*), otters (*Lutra lutra*) and a variety of raptors.

## 3. Radionuclides in arctic terrestrial ecosystems

Radionuclides constitute one of a group of contaminants (the others including heavy metals, persistent organic pollutants, or POPs, and petroleum hydrocarbons) which are of special concern for arctic regions. To date, the focus of radioecological investigations has been largely directed at the marine environment. It is possible that this emphasis has overshadowed the fact that the Arctic terrestrial environment has been contaminated, continues to be contaminated and remains extremely vulnerable to contamination by radioactive materials. Due to the special processes and relationships that exist in the Arctic terrestrial environment, it is perhaps unwise to assess the impact of even the slightest radioactive contamination of this environment using the same concepts and procedures which are applied to radiological impact assessments in other terrestrial systems.

### 3.1. *Sources of contamination*

The primary mode of anthropogenic radioactive contamination in arctic terrestrial environments is via atmospheric deposition (Figure 2). The two main sources of contamination delivered in this way have been the atmospheric weapons tests carried out in the 1950s and 1960s and the accident at the Chernobyl power plant in 1986. Of the 543 atmospheric weapons tests conducted globally, 91 of these were carried out in the Arctic region by the Former Soviet Union (FSU) at Novaya Zemlya with a total yield of 239.6 Mt (UNSCEAR, 2000). Aarkrog (1993) estimated a level of fallout in the Arctic region of 20 PBq of $^{90}$Sr and 30 PBq of $^{137}$Cs from 87 of these tests alone. Levels of atmospheric contamination reaching the Arctic dropped significantly (Figure 3) in the years following the signing of the Partial Test Ban Treaty in 1963 (Barrie et al., 1992), with a resultant but slower reduction in radionuclide concentrations in the terrestrial environment.

Locations on Novaya Zemlya show elevated levels of radioactivity against generally low background levels as a result of these atmospheric tests, but it is difficult to assess the resultant levels of terrestrial contamination at greater distances in areas such as the Kola Peninsula and Arkhangelsk. Perhaps the most comprehensive attempt to assess the levels of fallout in the Arctic region as a result of nuclear weapon testing was conducted by AMAP (AMAP, 1998) using a precipitation-related deposition model to estimate terrestrial levels of $^{137}$Cs (Figure 4).

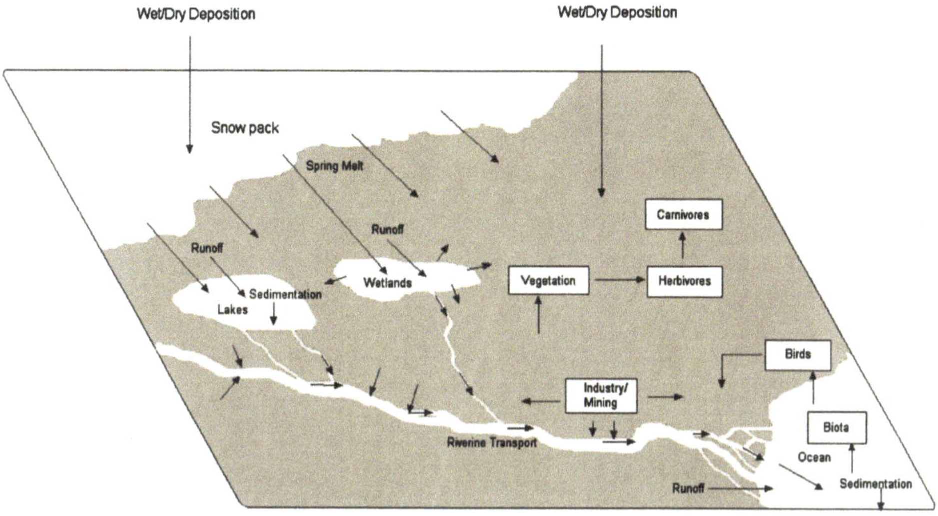


Fig. 2. Predominant contaminant pathways within the arctic terrestrial environment.

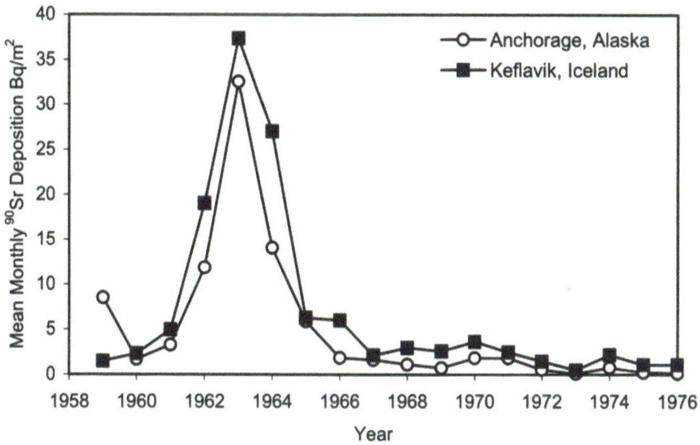


Fig. 3. $^{90}$Sr deposition at two Arctic locations between 1959 and 1976. Figure plotted using data obtained from www.eml.st.dhs.gov/database/fallout/.

As well as atmospheric tests, 39 subterranean tests were conducted at the Novaya Zemlya site (Mikhailov et al., 1996). The contamination patterns on Novaya Zemlya due to these tests are quite complex with respect to both the levels, locations and contaminants, but average terrestrial contamination is of the order of 3 kBq $m^{-2}$ ($^{137}$Cs) and 2 kBq $m^{-2}$ ($^{90}$Sr) (AMAP, 1998).

The FSU is not the only country to have carried out nuclear weapons testing and operations in the Arctic, with the United States conducting subterranean nuclear explosions on Amchitka Island, Alaska, the southernmost island of the Rat Island Group in the Aleutian Chain. These tests were conducted between 1965 and 1971 and consisted of 3 detonations, *Long Shot*

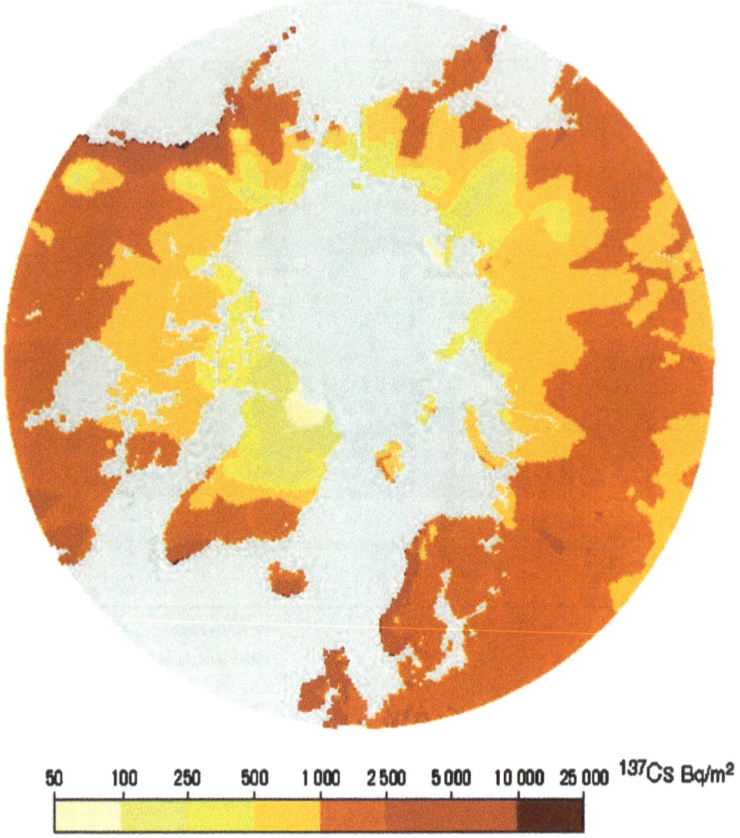


Fig. 4. Modelled $^{137}Cs$ contamination from nuclear weapons testing. (Reproduced from AMAP, 1998, with permission).

(80 kt), *Milrow* (1 Mt) and *Cannikin* (5 Mt). Contamination of the island chiefly comprised $^{131}I$, $^{85}Kr$ (trace amounts) and significant amounts of $^{3}H$ (Seymour and Nelson, 1977). Faller and Farmer (1997) concluded that maximum contamination at the *Long Shot* site occurred 100 m underground and in the immediate vicinity of the test site. Continued monitoring of the site by the United States Department of Energy (USDOE) is being carried out, but results to date do not indicate any significant contamination. Localised contamination has also occurred as a result of peaceful nuclear detonations, conducted by the FSU for mining and construction purposes, in the Russian Arctic (RCRA, 1997).

The Chernobyl accident of 1986 had a significant effect on the Arctic region, with radioactive fallout transported through the Baltic States and on to Sweden, Norway, Southern Finland, the Arkhangelsk region of Russia and the Kola Peninsula, with Eastern Siberia and Northern Canada being less affected (Barrie et al., 1992). In arctic regions of Norway and Sweden, deposition levels from Chernobyl releases of 2–10 $kBq\,m^{-2}$ were recorded (AMAP, 1997), dwarfing the previous inputs from global weapon test fallout (Backe et al., 1987). Despite the justifiable media attention devoted to this incident and its effects on the Arctic, outside of some limited (but nonetheless populated) areas in Fennoscandia and Western Russia, the most significant contribution to doses received by arctic residents from anthropogenic radioactive contamination continues to be derived from the atmospheric weapons test fallout (AMAP, 1998). Where dose rates from Chernobyl fallout are more significant, the application of effective countermeasures by the affected countries is serving to reduce this dose with time.

A variety of other sources or potential sources exist in the Arctic region ranging from power generation facilities to stored nuclear waste materials and, to emphasise the diversity of these sources, a number of examples are provided.

On January 24, 1978, the atmospheric burn-up of the Cosmos 954 satellite over the Canadian Northwest Territories spread radioactive material over a thousand kilometre long swathe to the northeast of the Great Slave Lake. Contamination occurred mostly as particles, one quarter of the radioactive inventory consisting of particles with diameters in the millimetre range being dispersed over some 124,000 $km^2$ and the remaining material being volatilised and dispersed as fine particles high in the atmosphere to be deposited over periods of years.

In 1962 a nuclear power plant was constructed at Fort Greely in Alaska by the US Army (Johnson, 1993). Decommissioning commenced in 1971 and although the fuel rods and other related materials were shipped off-site, highly radioactive waste was sealed in the inner vapour container and capped in concrete. A pipeline carrying high level waste ruptured at various points during operation of the plant and these leaks resulted in contamination of the surrounding soils. Remediation of the site commenced in 1997.

An accident at the Thule airbase in Greenland in January 1968 caused localised contamination when the detonation of conventional explosives in the nuclear payload of a crashed B-52 aeroplane released plutonium onto the snow and ice. Decontamination commenced and, although some plutonium was deposited in marine sediments, terrestrial contamination was minor. Other terrestrial contamination incidents of smaller size have occurred in the Arctic regions. During the 1960s, a plan was devised to excavate a harbour in Cape Thompson, Alaska using nuclear explosions. The United States Geological Survey conducted preliminary tests (Janzer and Beetem, 1963), which resulted in contamination of the area after burial of $^{137}Cs$-containing materials. Remediation of the area in 1993 showed little evidence of residual contamination (USDOE, 1994; O'Hara et al., 1999).

The remoteness of the Arctic and the difficulty of civil and military operations in the environment have led to nuclear materials being introduced as power sources for a variety of installations. Radioisotope Thermoelectric Generators (RTGs) are typically used as power supplies utilising the Seebeck thermoelectric effect, with a temperature gradient provided by the radioactive decay of a $^{90}Sr$ source. Based on data from 1994, around 165 RTGs are present in the Arctic, predominantly in Russia, with a combined inventory of over 2000 PBq (AMAP, 1997). Evaluation of RTG units at the US Air Force Burnt Mountain seismic observatory by the Office of Technology Assessment (1994) indicates low contamination risk in the event of tundra fire or similar accident, due to the physicochemical nature of the $^{90}Sr$ source.

The occurrence of Technologically Enhanced Naturally Occurring Radioactive Materials (TENORM—see Chapter 10) and their impacts on the Arctic have been largely uninvestigated. Materials in which the levels of naturally occurring radioactive materials have been enhanced by technological processes arise due to extraction industries such as oil and gas drilling, mining (both for uranium and other minerals), materials processing and geothermal energy. The activities of sea-based industries such as oil drilling in the North Sea would appear to pose little threat to the terrestrial arctic environment but, where such activities are conducted on land, the threat to the environment of contamination with resultant radioactive wastes warrants attention.

Despite the diversity of contamination incidents and sources and the location of nuclear weapons test sites in the Arctic region, current levels of terrestrial anthropogenic radioactive contamination in the wider Arctic area are generally considered to be low (Gilman et al., 1997). However, irrespective of the amount of contamination, the delicate nature of the Arctic environment means that any contamination is a source of concern. Although the contamination events which have impacted on the Arctic terrestrial environment are diverse, ranging from dissemination of weapons materials to power plant accidents, the suite of radionuclides contaminating the environment is quite narrow.

### **3.2.** *Radionuclides of concern*

*$^{137}Cs$ and $^{134}Cs$*

The occurrence of radioactive caesium in the Arctic is due to the fallout from atmospheric weapons testing and as a result of deposition after the Chernobyl accident. $^{134}Cs$ and $^{137}Cs$, with half-lives of approximately 2 and 30 years, respectively, are both present in arctic terrestrial matrices although $^{134}Cs$ levels have dropped rapidly due to its short physical half-life. $^{137}Cs$ is generally considered to be the radionuclide of greatest concern in arctic environments and provides a direct exposure to the human population of the areas via its accumulation in fish, in reindeer and caribou and its presence in berries, mushrooms and other edible plants. Caesium behaves in a similar manner to potassium and is almost completely absorbed by the intestine, becoming distributed in the muscle tissue of the body, its biological half-life in humans being approximately 3 months (Tracy, 1993).

*$^{90}Sr$*

There have been few studies of $^{90}Sr$ in arctic food chains, despite a physical half-life of 30 years and a relatively high mobility in soils. Contamination of the terrestrial environment of the Arctic with this isotope is due to the weapons tests of the 1950s and '60s. When ingested, 30% of strontium is absorbed and, as a chemical analogue of calcium, it is ultimately incorporated into bone tissue, where it may remain for many years.

*Plutonium*

Plutonium isotopes occur in the Arctic terrestrial environment primarily as a result of atmospheric weapons testing but also due to sundry other incidents such as satellite crashes and non-detonation incidents involving nuclear weapons. The isotopes of concern are $^{239}Pu$ and $^{240}Pu$ which originate from the volatilisation of unfissioned plutonium and such plutonium as may be produced by the neutron irradiation of $^{238}U$ during a nuclear explosion. Plutonium isotopes tend to bind tightly to soil constituents and the isotopes are regarded as relatively non-mobile in the terrestrial environment (Fisenne et al., 1980).

*$^{210}Pb$ and $^{210}Po$*

The greatest contribution to the radiation dose received by residents of the general Arctic region is from the two natural radioisotopes, $^{210}Pb$ and $^{210}Po$ (half-lives of 138 days and 22.3

years respectively). Both are naturally occurring members of the $^{238}U$ decay series and occur as airborne particles due to their progenitor, radon, being a gas which emanates from soils and other materials containing $^{238}U$ (see Chapter 10). Once airborne, the isotopes settle on, or are absorbed by, vegetation such as lichens and mosses from which point they may be ingested by herbivores. $^{210}Pb$ concentrates primarily in bone with levels in soft tissues of an order of magnitude less than for $^{210}Po$ (Tracy, 1993).

*Uranium and $^{226}Ra$*

A daughter of $^{238}U$, $^{226}Ra$ is present in all rocks and soils and is the immediate progenitor of the radioactive gas $^{222}Rn$. $^{226}Ra$ is of concern in the Arctic in TENORM in the waste products of uranium mining and processing. Veska and Eaton (1991) have shown that contamination of arctic areas with this isotope has occurred at abandoned mine tailings at Port Radium and Rayrock in the Canadian North West Territories as well as at points along the Mackenzie and Great Bear Rivers. However, $^{226}Ra$ is neither very mobile nor has it displayed evidence of bioaccumulation. It does, however, generate radon gas and the threat of elevated $^{210}Po$ and $^{210}Pb$ levels in the environment as a result of the presence of large amounts of $^{226}Ra$ should not be discounted.

*Other nuclides*

Other nuclides for consideration should include the activation product $^{55}Fe$ and the fission products $^{99}Tc$ and $^{129}I$, each resulting from weapons testing and discharges from nuclear facilities. All three nuclides are available for uptake by vegetation upon deposition with subsequent potential for trophic transfer. Elevated levels of $^{55}Fe$ have been found in Eskimo and Lapp blood due to its accumulation in fish and reindeer/caribou, whilst $^{99}Tc$ and $^{129}I$ may present problems despite their low specific activities due to their long physical half lives ($2.1 \times 10^5$ and $1.57 \times 10^7$ years respectively).

## 3.3. *Radionuclides in arctic soils*

Unlike more temperate ecosystems, the arctic soil compartment has not, with one exception, been the focus of much attention with respect to radioecological processes. The exception has been the large amount of effort devoted to the deposition of $^{137}Cs$ (see Figure 4) and, to a lesser extent, $^{134}Cs$ and $^{90}Sr$ and their consequential uptake into arctic food chains. In many arctic regions, $^{137}Cs$ penetration downwards into the soil profile is minimal, despite the main deposition outside of Chernobyl affected areas occurring in the 1950s and '60s. $^{137}Cs$ is detected predominantly within topsoil (upper 5 to 10 cm), in association with organic material rather than the underlying mineral horizons, as has been described for temperate soils (e.g. Baskaran et al., 1991; Schimmack and Bunzl, 1992; Strandberg, 1997). Recently reported levels of $^{137}Cs$ in arctic soils are scarce, with figures for West Greenland (0.6 to 1.77 $kBq\,m^{-2}$; Strandberg, 1997) and the Yamal Peninsula (1.77 to 2.02 $kBq\,m^{-2}$; Nifontova, 1995) the exception. Ultimately, penetration of caesium down the soil profile has been described as a function of latitude, the amount of precipitation at the time of deposition, the characteristics of the vegetative cover, the edaphic nature of the soil and the degree of bioturbation by soil organisms (Bergman, 1994; Strandberg, 1997). For other radionuclides, there is little information concerning their behaviour in arctic soils and a lack of contemporary data in general.

### 3.4. *Radionuclides in arctic vegetation*

Of all the components which constitute the terrestrial Arctic ecosystem, vegetation and the animal consumers of this vegetation have received the most attention. This is due to the close relationship between arctic peoples and herbivorous animals, notably reindeer/caribou, and the impact of the Chernobyl disaster on areas within the Arctic. The peculiarities of arctic plants result in their strong bioaccumulation of contaminants and, as a consequence, $^{137}Cs$ levels in some arctic plant species are amongst the highest in the world. To cope with the short growing season, arctic plants have evolved longer life spans in order to accumulate sufficient energy to reproduce. A long life span provides greater opportunity for the uptake of contaminants by arctic plants as compared with their temperate relatives. Arctic plants tend to draw nutrients directly from the air due to their physical characteristics and the shallow nature of the active soil layer, leading to greater susceptibility to airborne contamination. Furthermore, arctic plants tend to retain their leaves during the winter, so absorbed contamination is not removed by leaf fall, but rather stored within the tissues from season to season.

The accumulation of high levels of contaminants by arctic vegetation is an important factor in the consideration of the radioecology of, for example, caesium, in the arctic terrestrial environment. By 1961, attention was being focused on the problem of high $^{137}Cs$ body-burdens in arctic populations (see, for example Liden, 1961; Hanson et al., 1975) from weapons fallout and ultimately the spotlight was turned to the ability of arctic plants to accumulate these isotopes. The persistence of contaminant $^{137}Cs$ in the tissues of arctic plants long after the cessation of weapon testing was unexpected, given that the biological half-life of this isotope in higher plants could be measured in weeks as opposed to the years exhibited by arctic lichens. Lichens are thalloid organisms and obtain necessary water via the direct absorption of precipitation and its solute load which thus provides a direct route for deposited radionuclides to enter the tissues of the plant. The long life spans of these lichens and the ability of lichens to cycle this element internally (Martin and Koranda, 1971) was ultimately found to be the cause of the lichens' remarkable retentive capacity (Figure 5).

Fungi also accumulate long-lived radionuclides such as $^{137}Cs$ (Grueter, 1971) and, to a lesser extent, species such as $^{90}Sr$ (Seeger et al., 1982). The Chernobyl accident resulted in increased attention on fungi as accumulators of $^{137}Cs$ (see, for example, Heinrich, 1993) and their role in introducing radionuclide contamination to reindeer and humans. In addition, mushrooms have been shown to absorb $^{210}Pb$ selectively from soil, leading to concentrations which can dominate any potential committed effective dose (Kirchner and Daillant, 1998).

### 3.5. *Radionuclides in arctic terrestrial food chains*

The unique environment of the Arctic has produced many adaptations in the ecosystems which exist there. Many of these adaptations increase the vulnerability of the organisms involved, and the environment in general, to radioactive contamination. This is perhaps most evident when considering arctic terrestrial food chains. Although the Arctic can support a variety of mammals and birds in significant numbers the number of forage species is low and the food chains tend to be extremely short, as described in section 2.2. The most often studied food chain, from a radioecological aspect, has been the lichen/moss–reindeer/caribou–human chain.

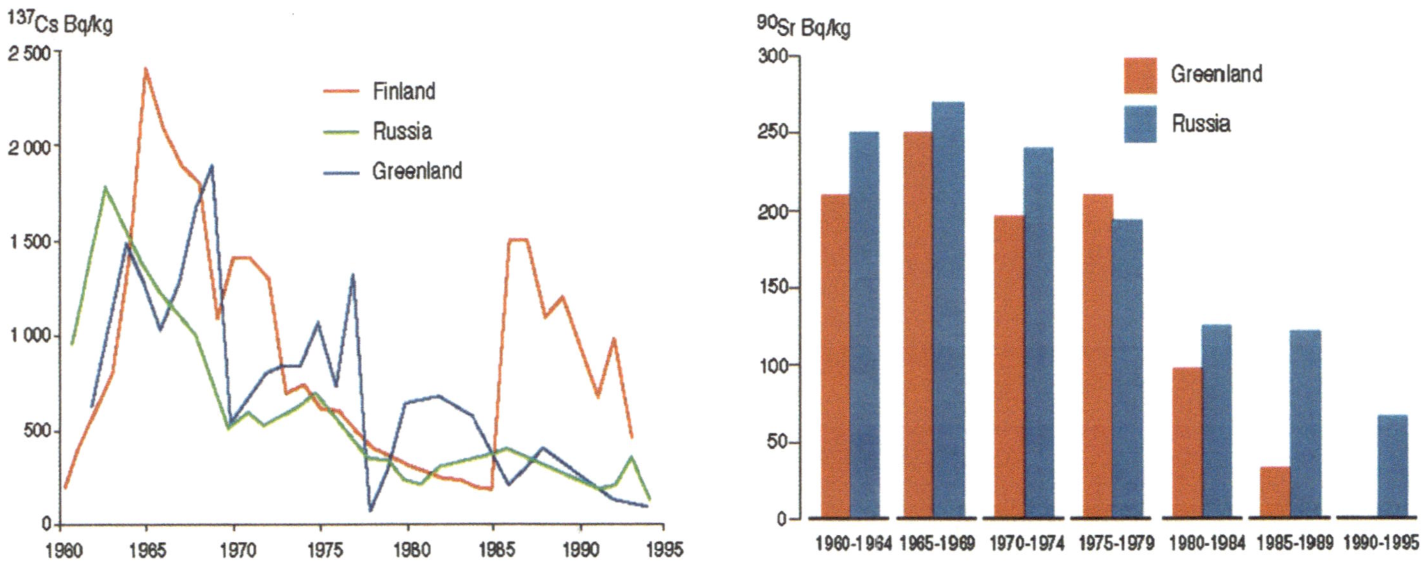


Fig. 5. $^{137}$Cs and $^{90}$Sr levels in lichens at selected arctic areas between 1960 and 1995. (Reproduced from AMAP, 1998, with permission).

For most of the year (Oct–May), depending on location, lichens constitute a large proportion (>60%) of the reindeer/caribou diet (Gaare and Staaland, 1994). The uptake and retention of radioisotopes by lichens therefore represents an efficient pathway for the introduction of radioactive contaminants to higher members of the food chain, ending in many cases with humans via their consumption of reindeer meat. Concentrations of $^{137}Cs$ peaked in the tissues of these herbivores in the 1960s, then decreased until 1986 when the fresh injection of this isotope into the biosphere from the Chernobyl accident resulted in a sudden rise in reindeer body burdens in Sweden, Norway and northwest Russia. In the period 1990 to 1995, average levels of $^{137}Cs$ in reindeer/caribou meat samples from across the Arctic were between 200 and 600 $Bq\,kg^{-1}$, levels being higher in the Fennoscandinavian region at between 600 and 2500 $Bq\,kg^{-1}$ (AMAP, 1997). The accumulation and retention of radioactive contaminants in forage species and the dietary constraints of herbivorous species in arctic environments allow for the efficient transfer of contamination. In temperate climes and agricultural ecosystems, concentrations of radionuclides in plants are reduced by factors such as relatively high biomass yields and agricultural practices such as ploughing and fertilisation of soils.

A hitherto unexplored transfer mechanism which may have significance for arctic radioecology is the potential of seabirds, through faecal deposits, to transfer radionuclides from the marine to the terrestrial environment. It has been shown that seabird faeces are the principal source of heavy metal input to arctic soils (Headley, 1996) and that transfer of these metals can occur to vegetation growing on faecally affected soils (Godzik, 1991). The role of arctic vegetation species in the accumulation and retention of radionuclide contaminants is closely linked to the overall nutrient status of the Arctic terrestrial environment. The enhanced nutrient status of faecally affected soils subsequently causes an increase in both the diversity and quantity of vegetation growing in these areas (Eurola and Hakala, 1977) and therefore provides enhanced grazing for the herbivores of the region. The consumption of such vegetation by herbivores and further trophic transfer may result in novel or increased exposure to radionuclide contaminants.

### 3.6. *Radionuclides in arctic freshwater systems*

Freshwater arctic systems typically accumulate radioactive contaminants from direct deposition of global fallout and from run-off containing previously deposited fallout and natural radionuclides leached from surrounding catchments. A regionally important third source is waterborne discharges from nuclear facilities, such as Mayak in the Urals from which major discharges of $^{90}Sr$ occurred in the 1950s. $^{90}Sr$ is one of the most mobile radionuclides entering freshwater systems as, unlike many radionuclides including $^{137}Cs$, it is not significantly retained by soils and sediments. The particle reactivity of $^{137}Cs$ accounts for its faster rate of elimination from freshwater systems as compared with $^{90}Sr$. Where drinking water is mostly derived from ice and snow (e.g. Greenland) a reservoir of fallout radionuclide contamination can develop contaminating water supplies for longer periods than water supplies derived solely from surface waters.

The accumulation of radionuclides in fish from freshwater systems depends on many factors. The uptake of $^{137}Cs$, for example, is largely determined by nutrient levels, the overall size of the catchment area of the watercourse and the total water volume. Radiocaesium levels in fish dwelling in highly biologically productive lakes tend to be lower than levels found in fish which reside in lakes whose productivity is lower (Håkanson et al., 1992). The levels in fish also tend to reflect their position in the food chain and their dietary habits. After the Chernobyl accident, the highest levels were first observed in plankton eating fish, with levels in predatory species such as pike rising after a longer period. Finnish studies after Chernobyl indicate that the maximum transfer of radionuclides to freshwater fish occurs within three years for most typical species (AMAP, 1997).

Levels of radiocaesium in freshwater fish can be similar to those in sheep and wild animals, but are typically lower than those found in reindeer meat and some mushroom species. Dietary studies conducted as part of AMAP (1997) indicated that freshwater fish from the Russian Arctic (1960 to 1994) contained levels of $^{137}Cs$ (15–40 $Bq\,kg^{-1}$) approximately 10 times greater than for saltwater fish, decreasing from 40 $Bq\,kg^{-1}$ in the 1960s to 15 $Bq\,kg^{-1}$, rising post-Chernobyl to 20 $Bq\,kg^{-1}$. A similar pattern of reduction and subsequent rise post-Chernobyl is observed for radiocaesium in Finnish rivers.

## 4. Arctic specific processes

When considering the behaviour of radionuclides within the Arctic terrestrial environment it is worthwhile bearing in mind a number of factors. The primary component of the Arctic terrestrial ecosystem is the soil which can exhibit some important characteristics that affect the behaviour of radionuclides within this matrix. Dobrovol'skiy (1996) concludes that although air temperature may be a defining criterion for the Arctic region, the level of precipitation largely determines the pedogenic character of soils in different parts of the Arctic. This factor governs the amount of vegetation present and the ratio of living organic material to dead material within the soil. In regions where appreciable precipitation is present, moisture is retained in the soil and decomposing plant material forms weakly condensed humic acids. The formation of humic acids and the nature of the humic materials have a significant bearing on the interaction of radionuclides with the soil. Dziadowiec et al. (1994), Flaig (1964) and Ross (1989) have shown that humic materials from an arctic tundra soil (Svalbard) bear more resemblance to fulvic acids than humic materials from other climatic areas and conclude that this is due to alternate freeze–thaw cycles and the low lignin content of the parent plant material. The role of these materials in the retention, mobilisation and behaviour of caesium and strontium isotopes has been well described elsewhere (Melin et al., 1994; Rafferty et al., 2000) but the peculiarities of radionuclide behaviour in arctic and high-arctic soils have received less attention and are therefore less well understood.

The formation of humic materials is not the only chemical property of soils that is affected by the arctic climate. Certain chemical reactions, crucial to radioecological considerations, are influenced by the extreme cold of the Arctic regions. Two of the most pertinent are precipitation–dissolution and cation exchange. As water freezes any solutes it contains are excluded from the solid phase and this can result in the formation of supersaturated solutions, promoting secondary mineral precipitation within the soil layers. This, in turn, can accelerate weathering of the underlying lithology with subsequent alteration of soil chemistry processes (Hallet, 1978; Sletten, 1988). Carbonate chemistry, which is of importance in the soil radiochemistry of U and Th series radionuclides among others (see, for example,

Ivanovich, 1994) is strongly affected by the freezing of arctic soils which leads to precipitation of insoluble $CaCO_3$ and an increase in the $Mg^{2+}/Ca^{2+}$ ratio within the solution phase (Vlasov and Pavlova, 1969). Of all the soil chemistry processes affected by the Arctic climate, perhaps the most important with respect to radionuclide behaviour are ion-exchange reactions. These reactions are responsible for processes such as pH buffering, ion transport and metal binding. Cation exchange capacity (CEC) appears to be unaffected by soil freezing (Hinman, 1970; Polubesova and Shirshova, 1992) although increases in exchangeable K and $NH_4$–N have been observed. Freeze–thaw cycles can liberate crystal lattice bound K (Graham and Lopez, 1969) and freezing has been observed to increase pH via adsorbed bases whilst thawing results in a lowering of pH (Federov and Bastistyi, 1974).

Furthermore, the mobility of radionuclides within arctic soils is subject to climatic effects, the results of which can often be contrary to what may be expected in the soils of more temperate climes. Supersaturated solutions formed by solute exclusion lead to freezing-point depression and increased quantities of unfrozen water in the soils. Liquid films on the surface of soil particles provide the dominant pathway for the flow of water and associated dissolved substances (Hoekstra, 1969; Murrman, 1973). Monovalent ions have been found to be more mobile within frozen soils (Czurda and Schababerle, 1988), due probably to a lesser attraction to charged surfaces within the soil column. A surprising aspect of the chemistry of metallic contaminants of arctic soil is the ability of metals to move upwards from the soil into overlying snow. This was demonstrated by Kadlec et al. (1988) and Ostroumov et al. (1992), the former establishing that freezing of wetlands and peat causes many solutes to move into the top soil layers and the latter establishing via laboratory experiments that solutes can move into snow layers in a process related to cation adsorption onto ice particles. Evidence for upward migration of radionuclides has been suggested by Schuller et al. (2002) describing the movement of $^{137}Cs$ in frozen soils. The importance of the movement of contaminants between water or snow and soil during freeze–thaw processes becomes apparent during thawing, with the release of solutes in meltwater. During winters with little or no snowmelt before spring, Johannessen and Henriksen (1978) found that 50–80% of the winter pollutant load is released with the first 30% of meltwater.

Soil mobility is a feature of the Arctic region that has a direct bearing on the distribution of deposited radionuclides within this matrix. Soil horizons in arctic soils can undergo severe physical distortion due to the action of processes such as frost heave. Tedrow and Walton (1977) have shown that differential movement of soil particles from cyclical freeze–thaw action can disturb soil profiles, an effect which has implications for the vertical redistribution of the radionuclide load. On a larger scale, soil movement can be produced via the high pressures generated by the freezing of soil that result in frost heave (see for example Nakano, 1990) or by the slipping of saturated layers (solifluction). Saturated surface soil layers arise in the spring and summer due to the inability of runoff water to penetrate permafrost vertically. The saturated soil slides downhill, even on shallow gradients, as the interface between the frozen and unfrozen layers acts as a lubricant.

Considering that the principal factors that govern the transport and biotic uptake of radionuclides in the terrestrial environment (persistence, solubility, nutrient competition and location/trophic level of biota within ecosystems) are affected by arctic specific processes, the role of these processes in arctic terrestrial radioecology has not received as much attention as is perhaps warranted. To date, the assumption among radioecologists appears to be that radionuclides in arctic terrestrial environments are subject to the same constraints as those in effect in temperate regions.

## 5. Conclusions

The status of radioactive contamination within the Arctic region remains the focus of much attention. Although contamination levels in the terrestrial environment remain quite low, the special status of the region necessitates ongoing monitoring and research to elucidate fully the impacts of current levels of radionuclides on the environment and how future contamination might affect the ecosystem and its inhabitants. The presence of potential sources of future nuclear contamination in locations that could impact on the Arctic places further emphasis on the necessity to be able to predict the behaviour and radioecology of contaminants within the Arctic environment. In addition, the possible exploitation of the Arctic region by natural resource exploration and extractive industries presents a relatively new threat to the radiological status of the region.

Despite the large amount of work that has been completed in the field of terrestrial radioecology over the past 20 years, the Arctic remains worthy of special consideration as it is subject to conditions that may not have been accounted for during these previous studies. If so, terrestrial arctic radioecology would appear to warrant further attention, with a view towards a better understanding of arctic radioecological processes, enabling greater protection of this special environment from future contamination events.

## References

Aarkrog, A. (1993). Radioactivity in polar regions: main sources. In: Strand, P., Holm, E. (Eds.), *Environmental Radioactivity in The Arctic and Antarctic: Proceedings, Kirkenes, Norway.*

AMAP (1997). Arctic Pollution Issues: A State of the Arctic Environment Report. Arctic Monitoring and Assessment Programme, Oslo, Norway, pp. 5–8.

AMAP (1998). AMAP Assessment Report: Arctic Pollution Issues. Arctic Monitoring and Assessment Programme (AMAP), Oslo, Norway, p. 859.

Backe, S., Bjerke, H., Rudjord, A.L., Ugletveit, F. (1987). Fallout pattern in Norway after the Chernobyl accident estimated from soil samples. *Radiation Protection Dosimetry* **18**, 105–107.

Barrie, L.A., Gregor, D., Hargrave, B., Lake, R., Muir, D., Shearer, R., Tracey, B., Bidleman, T.F. (1992). Arctic contaminants: sources, occurrence and pathways. *Science of the Total Environment* **122**, 1–74.

Baskaran, M., Kelley, J.J., Naidu, A.S., Holleman, D.F. (1991). Environmental $^{137}Cs$ in subarctic and arctic Alaska following Chernobyl. *Arctic* **44**, 346–350.

Bergman, R. (1994). The distribution of radioactive caesium in boreal forest ecosystems. In: Dahlgaard, H. (Ed.), *Nordic Radioecology*. Elsevier Science Publishers, Amsterdam, pp. 335–379.

Czurda, K.A., Schababerle, R. (1988). Influence of freezing and thawing on the physical and chemical properties of swelling clays. In: Jones, R.H., Holden, J.T. (Eds.), *Proceedings, 5th International Symposium on Ground Freezing, 26–28 July, Nottingham, England*. A.A. Balkema, Rotterdam, Netherlands, pp. 51–88.

Dobrovol'skiy, V.V. (1990). Geochemistry of Spitsbergen soils. *Pochvovedenie* **2**, 5–20.

Dobrovol'skiy, V.V. (1996). Basic features of the geochemistry of Arctic soil formation. *Eurasian Soil Science* **28**, 218–230.

Dziadowiec, H., Gonet, S., Plichta, W. (1994). Properties of humic acids of arctic tundra soils in Spitsbergen. *Polish Polar Research* **15** (1–2), 71–81.

Eurola, S., Hakala, U.K. (1977). The bird cliff vegetation of Svalbard. *Aquilo Ser. Bot.* **15**, 1–18.

Faller, S.H., Farmer, D.E. (1997). Long Term Hydrological Monitoring Program, Amchitka, Alaska 1997. Draft Report, U.S. Environmental Protection Agency, Radiation and Indoor Environments National Laboratory, Las Vegas, NV.

Federov, A.A., Bastistyi, V.P. (1974). Winter freezing and chemical properties of meadow brown soils. *Sib. Vestn. Skh. Nauki* **4** (6), 8–12.

Fisenne, I.M., Cohen, N., Neton, J.W., Perry, P. (1980). Fallout plutonium in human tissues from New York city. *Radiation Research* **83**, 162–168.

Flaig, W. (1964). Chemische Untersuchungen an Humusstoffen. *Z. Chem.* **4** (7), 253–265.

Foscolos, A.E., Kodama, H. (1981). Mineralogy and chemistry of Arctic desert soils on Effef Ringnes Island, Arctic Canada. *Soil Science Society of America Journal* **45**, 987–993.

Gaare, E., Staaland, H. (1994). Pathways of fallout radiocaesium via reindeer to man. In: Dahlgaard, H. (Ed.), *Nordic Radioecology: The Transfer of Radionuclides Through Nordic Ecosystems to Man.* Elsevier, The Netherlands, pp. 303–334.

Gilman, A., Dewailly, E., Feeley, M., Jerome, V., Kuhnlein, H., Kwavnik, B., Neve, S., Tracy, B., Usher, P., Van Oostdam, J., Walker, J., Wheatley, B. (1997). Chapter 4: Human health. In: Jensen, J., Adare, K., Shearer, R. (Eds.), *Canadian Arctic Contaminants Assessment Report.* Indian and Northern Affairs, Ottawa, Canada, pp. 337–341.

Godzik, B. (1991). Heavy metals and macroelements in the tundra of southern Spitsbergen: the effect of little auk Alle alle (L.) colonies. *Polar Research* **9** (2), 121–131.

Goryachkin, S.V., Karavaeva, N.A., Targulian, V.O. (1998). Geography of Arctic soils: current problems. *Eurasian Soil Science* **31** (5), 467–476.

Graham, E.R., Lopez, P.L. (1969). Freezing and thawing as a factor in the release and fixation of soil potassium as demonstrated by isotopic exchange and calcium exchange equilibria. *Soil Science* **108**, 143–147.

Grueter, H. (1971). Radioactive fission product $^{137}Cs$ in mushrooms in W. Germany during 1963–1970. *Health Physics* **20**, 655–656.

Håkanson, L., Andersson, T., Nilsson, Å. (1992). Radioactive caesium in fish in Swedish lakes 1986–1988—general pattern related to fallout and lake characteristics. *Journal of Environmental Radioactivity* **15**, 207–229.

Hallet, B. (1978). *Solute redistribution in freezing ground.* In: *Proceedings, 3rd International Conference on Permafrost, 10–13th July, Edmonton, Alberta*, vol. 1. National Research Council of Canada, Ottawa, pp. 86–91.

Hanson, W.C., Whicker, F.W., Lipscomb, J.F. (1975). Lichen forage ingestion rates of free roaming caribou estimated with fallout cesium-137. In: Luick, J.R. (Ed.), *Proceedings of The First International Reindeer and Caribou Symposium.* Fairbanks, Alaska, University of Alaska, pp. 71–79. Special Report No. 1.

Headley, A.D. (1996). Heavy metal concentrations in peat profiles from the high Arctic. *Science of the Total Environment* **177**, 105–111.

Heinrich, G. (1993). Distribution of radiocaesium in the different parts of mushrooms. *Journal of Environmental Radioactivity* **18**, 229–245.

Hinman, W.C. (1970). Effects of freezing and thawing on some chemical properties of three soils. *Canadian Journal of Soil Science* **50**, 179–182.

Hoekstra, P. (1969). The Physics and Chemistry of Frozen Soils. National Research Council, Highway Research Board, Special Report 103.

Ivanovich, M. (1994). Uranium series disequilibrium: concepts and applications. *Radiochimica Acta* **64**, 81–94.

Janzer, V.J., Beetem W.A. (1963). Hydrologic tracer studies conducted August 20–25, 1962 near Cape Thompson, Alaska. US Dept. Interior-Geological Survey Report, Anchorage, AK.

Johannessen, M., Henriksen, A. (1978). Chemistry of snow meltwater: changes in concentration during meltwater. *Water Resources Research* **14**, 615–619.

Johnson, W. (1993). Testing Nuclear Power in Alaska: The Reactor at Fort Greely. M.A. Thesis, University of Alaska, Fairbanks.

Kadlec, R.H., Li, X.-M., Cotton, G.B. (1988). Modelling solute segregation during freezing of peatland waters. *Water Resources Research* **24**, 219–224.

Kalyakin, V.N. (1995). New data on the unique biogeographical properties of Novaya Zemlya archipelago. *Dokl. Akad. Nauk* **343** (1), 139–141.

Kirchner, G., Daillant, O. (1998). Accumulation of $^{210}Pb$, $^{226}Ra$ and radioactive cesium by fungi. *The Science of the Total Environment* **222**, 63–70.

Liden, K. (1961). Cesium-137 burdens in Swedish laplanders and reindeer. *Acta Radiologica* **56**, 237–240.

Martin, J.R., Koranda, J.J. (1971). Recent Measurements of Cesium-137 Residence Time in Alaskan Vegetation. In: Nelson, D.J. (Ed.), *Radionuclides in Ecosystems.* National Technical Information Service, Springfield, VA, pp. 108–115. CONF-710501-P1.

Melin, J., Wallberg, L., Suomela, J. (1994). Distribution and retention of cesium and strontium in Swedish boreal forest ecosystems. *Science of the Total Environent* **157**, 93–105.

Mikhailov, V.N., Andryshin, I.A., Bogdan, V.V., Vashchinkin, S.A., Zelentsov, S.A., Zolotukhni, G.E., Karimov, V.M., Kirichenko, V.V., Matuschenko, A.M., Silkin, Y.A., Strukov, V.G., Kharitnov, K.V., Tchernyshev, A.K., Tsyrkov, G.A., Shumaev, M.P. (1996). *USSR Nuclear Weapons Tests and Peaceful Nuclear Explosions, 1949 through 1990.* RFNC-VNIIEF, Sarov, 63 pp.

Murrman, R.P. (1973). Ionic mobility in permafrost. In: *Permafrost: The North American Contribution to The 2nd International Conference on Permafrost, Yakuysk, 13–28 July.* National Academy of Sciences, Washington, DC, pp. 352–359.

Nakano, Y. (1990). Quasi-steady problems in freezing soils: I. Analysis on the steady growth of an ice layer. *Cold Regions Science and Technology* **17**, 207–226.

Nifontova, M. (1995). Radionuclides in the moss-lichen cover of tundra communities in the Yamal peninsula. *Science of the Total Environment* **160/161**, 749–752.

Office of Technology Assessment (1994). Power Sources for Remote Arctic Applications. OTA-BP-ETI 129. U.S. Congress, Washington, DC, 47 pp.

O'Hara, T.M., Dasher, D., George, J.C., Woshner, V. (1999). Radionuclide levels in caribou of northern Alaska in 1995–96. *Arctic* **52** (3), 279–288.

Ostroumov, V.E., Demidov, V.V., Gershevich, V.D. (1992). The transfer of ions within the soil–snow system and the role of ice surface in their migration in frozen soils. In: Gilichinsky, D.A. (Ed.), *Proceedings of The First International Conference on Cyropedology.* Russian Academy of Sciences, Pushchino, Russia, pp. 127–135.

Polubesova, T.A., Shirshova, L.T. (1992). Effect of prolonged freezing on chemical properties of soils and clays. In: Gilichinsky, D.A. (Ed.), *Proceedings of The First International Conference on Cyropedology.* Russian Academy of Sciences, Pushchino, Russia, pp. 140–145.

Rafferty, B., Brennan, M., Dawson, D., Dowding, D. (2000). Mechanisms of $^{137}$Cs migration in coniferous forest soils. *Journal of Environmental Radioactivity* **48**, 131–143.

RCRA (1997). Radioactive contamination in the Russian Arctic. Balonov, M., Tsaturov, Y., Howard, B., Strand, P. (Eds.) Report of Russian Experts for AMAP, Oslo.

Ross, S. (1989). *Soil Processes: Systematic Approach.* Routledge, London, New York. 444 pp.

Schimmack, W., Bunzl, K. (1992). Migration of radiocesium in two forest soils as obtained from field and column investigations. *Science of the Total Environment* **116**, 93–107.

Schuller, P., Bunzl, K., Voigt, G., Handl, J., Castillo, A. (2002). 137Cs accumulation and downward migration in selected soils from the South Shetland Islands. In: Strand, P., Jolle, T., Sand, A. (Eds.), *Proceedings of The 5th International Conference on Environmental Radioactivity in The Arctic and Antarctic, St Petersburg, Russia, 16–20 June*, pp. 90–92.

Seeger, R., Orth, H., Schweinshaut, P. (1982). Strontiumvorkommen in Pilzen. *Z. Lebensm Unters Forsch.* **174**, 381–389. (in German).

Seymour, A.H., Nelson, V.A. (1977). Radionuclides in air, water, and biota. In: Merritt, M.L., Fuller, R.G. (Eds.), *The Environment of Amchitka Island, Alaska.* NTIS, Springfield, VA, pp. 579–613. USERDA Report TID-26712.

Sletten, R.S. (1988). The formation of pedogenic carbonates on Svalbard: the influence of cold temperatures and freezing. In: *Proceedings, 5th International Conference on Permafrost, 2–5 August, Trondheim, Norway*, vol. 1. Tapir Publishers, Norway, pp. 467–472.

Strandberg, M. (1997). Distribution of $^{137}$Cs in a low Arctic ecosystem in West Greenland. *Arctic* **50**, 216–223.

Tedrow, J.C.F., Walton, G.F. (1977). Rendzina formation on Bathurst Island. *Journal of Soil Science* **28**, 519–525.

Tracy, B.L. (1993). Exposure of northern Canadians to natural and artificial radioactivity in caribou meat. In: Strand, P. (Ed.), *Proceedings of the International Conference on Environmental Radioactivity in the Arctic and Antarctic, Kirkenes, Norway, August 23–27.*

UNSCEAR (2000). *Sources and Effects of Ionizing Radiation. volume 1: Sources.* United Nations Scientific Committee on the Effects of Atomic Radiation, New York, 654 pp.

U.S. Department of Energy (USDOE) (1994). Project Chariot Site Assessment and Remedial Action Report. USDOE Report DOE/NV-368. USDOE, Las Vegas, NV.

Vasil'evskaya, V.D., Karavaeva, N.A., Naumov, E.M. (1993). Formation of soil cover structure in polar regions. *Pochvovdenie* **7**, 44–45.

Veska, E., Eaton, R.S. (1991). Abandoned rayrock uranium mill tailings in the Northwest Territories: environmental conditions and radiological impact. *Health Physics* **60** (3), 399–409.

Vlasov, N.A., Pavlova, L.I. (1969). Effect of cryogenic processes on the formation of bicarbonated waters. *Gidrochimicheskie Materialy* **51**, 76–82.

Yablokov, A. (1993). *Facts and Problems Related to Radioactive Waste Disposal in Seas Adjacent to The Territory of The Russian Federation*. Office of the President of the Russian Federation, Moscow.

# The distribution and transport of radionuclides in dryland ecosystems

Tom E. Hakonson

*Environmental Evaluation Services, LLC Daniel, WY, USA*

## 1. Introduction

This chapter addresses radionuclide behaviour in arid and semiarid ecosystems (collectively known as drylands) which comprise about one-third of the world's surface area. The majority of the above and below ground nuclear tests that were conducted by nations of the world were located in dryland regions. Furthermore, many of the research and development (R&D), as well as production facilities, to support nuclear weapons programs are located in dryland regions of their respective countries. As such, dryland regions often contain radionuclides at concentrations over and above worldwide fallout levels. This chapter discusses characteristics of dryland climates, nuclear activities that have resulted in dispersal of radionuclides, and processes which govern the distribution and transport of radionuclides in arid and semi-arid regions.

## 2. Characteristics of dryland ecosystems

Approximately one-third of the Earth's land surface is dryland that is characterized by low annual precipitation which supports only sparse vegetation and a limited population of humans and animals (Evenari, 1985, Figure 1). Most classifications of dryland regions rely on some combination of the number of days of rainfall, the total amount of annual rainfall, temperature, humidity, or other factors. Meigs (1953) divided desert regions into three categories according to the amount of precipitation they received. Extremely arid lands were defined to receive at least 12 consecutive months without rainfall, arid lands receive less than 250 millimeters of annual rainfall, and semiarid lands receive mean annual precipitation of between 250 and 500 millimeters. Following the classification of Potter (1992), arid and semiarid regions are those subject to precipitation to evaporation ratios (P/E ratio) smaller than 0.5, and between 0.5 and 1.0, respectively. Arid and extremely arid lands are deserts, and semiarid grasslands are generally referred to as steppes.

RADIOACTIVITY IN THE ENVIRONMENT
VOLUME 10 ISSN 1569-4860/DOI: 10.1016/S1569-4860(06)10008-X

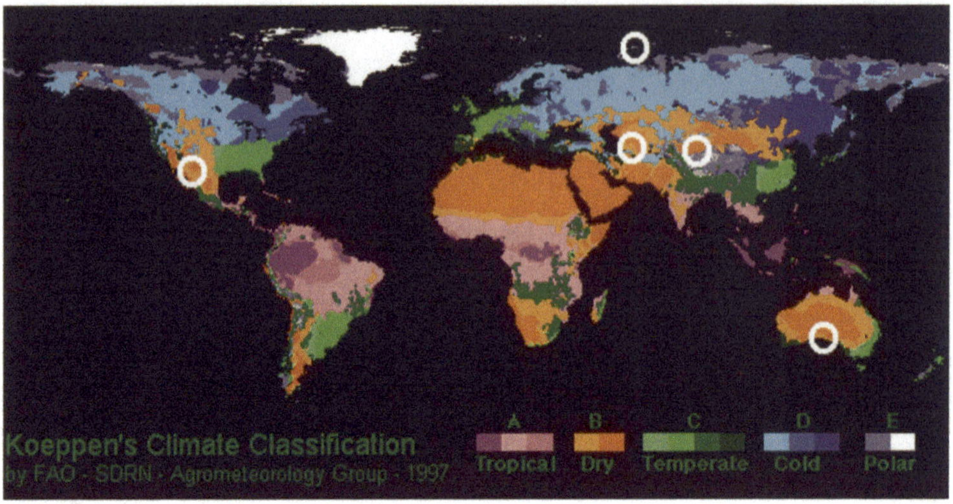


Fig. 1. Climatic zones of the world. Circles represent locations where atmospheric testing of nuclear weapons in dryland areas has occurred. Testing countries represented are the USA (Nevada Test Site), USSR (Kazakhstan and Russia), China (Lop Nur), and UK (Australia). Reproduced by permission of the Food and Agriculture Organization of the UN.

Many dryland watersheds contain little or no permanent surface water. Precipitation is received as intermittent high intensity rainstorms that bring a major portion of the annual rainfall to the surface during a very short period of time. Flash floods and the high soil erosion rates associated with these events are a direct result of these short duration, high intensity storms.

Because of the nature of these precipitation events, the limits on rainfall infiltration rates into the soil surface are easily exceeded which results in the generation of runoff, characterized as flash floods, that propagates rapidly through the watershed. Erosion processes on upland and stream channel areas, associated with these events, can contribute large amounts of sediment to downstream areas.

Once on the surface, water in an arid region is subject to a high evaporative demand from the low humidity and high temperature environment. In some cases, the surface flows never reach the valley bottom (Lavee et al., 1997) while in others, flows carry tremendous sediment loads to river systems and oceans.

The thickness of the unsaturated zone is another distinguishing feature in dryland watersheds (Scanlon et al., 1997). The presence of a deep water table allows unsaturated water or water vapor to move in either vertical direction depending on the surface conditions. This may have the effect of preventing water from percolating into the underlying aquifer. Coupled with the reduced availability of surface water, the various loss mechanisms in the unsaturated zone (evaporation from the soil surface, transpiration via plant roots and shoots, upwards vertical vapor and water flux, etc.) reduce the probability of water reaching the underlying deep aquifer. Evapotranspiration losses of soil moisture usually remove nearly 100% of available soil moisture. This means that aquifer recharge in an arid environment is usually very low under most circumstances.

Wind erosion also can be an important mechanism in dryland areas depending on soil type, soil moisture, vegetation cover, and physical disturbances resulting from catastrophic events such as fire, drought, or from human activities such as agriculture. Wind erosion physically removes the lighter, less dense soil particles such as organic matter, clays and silts. While transport of these constituents reduces soil fertility and productivity (Lyles, 1975), it also can lead to the transport of many soil contaminants including most radionuclides. Wind erosion in the USA is most widespread on agricultural land in the Great Plains states and is a major cause of soil degradation in arid and semiarid areas world wide.

Both wind and water erosion are major transport mechanisms for radionuclides in dryland regions. This is because most radionuclides bind tightly to soil particles so that, as soil particles move during soil erosion, so do the attached radionuclides.

## 3. Sources of radionuclides in dryland environments

Major sources of man-made radionuclides currently in dryland ecosystems are fallout from nuclear weapons testing, releases from weapons R&D and production facilities, and accidents such as Chernobyl. During 1945–98, an estimated 2051 atmospheric and underground nuclear tests were conducted by seven nations, with the United States and Soviet Union accounting for 85% of the tests (Table 1). About 26% of the tests (528) were conducted in the atmosphere with total yields amounting to 428 megatons.

About 85% of these tests were conducted in dryland regions of North America, Europe, Asia, and Australia (Figure 1 and Table 2). Test sites in these regions have large tracts of land that have low human population densities, very little surface water, and sparse fauna and flora. Why dryland regions were chosen for most of the world's nuclear weapons testing is unknown but probably relates primarily to the low human population densities and to facilitating physical security. Nuclear R&D facilities also are often located in dryland regions for the same reasons. In the USA, of the 12 major Department of Energy (DOE) facilities, 7 are located in dryland regions (Table 3). Additionally, nearly all of the nuclear testing in the continental USA has occurred in dryland areas and especially at the Nevada Test Site. Radioactive materials from R&D and nuclear production facilities include most elements listed in the Periodic Table, including most radionuclides. However, the major radionuclides which are subjects of DOE cleanup programs are the actinides and fission products such as $^{137}Cs$, $^{90}Sr$, $^{60}Co$ and tritium ($^{3}H$).

Table 1
Known nuclear tests worldwide, 1945–98

| | U.S. | | U.S.S.R. | | U.K. | | France | | China | | Total |
|---|---|---|---|---|---|---|---|---|---|---|---|
| | A | U | A | U | A | U | A | U | A | U | |
| Number of tests | 215 | 815 | 219 | 496 | 21 | 24 | 50 | 160 | 23 | 22 | 2051 |
| Total yield (megatons) | 141 | 38 | 247 | 38 | 8 | 0.9 | 10 | 4 | 22 | 1.5 | 510 |

Atmospheric (A) megatons = 428; underground (U) megatons = 82.4.

Table 2
Nuclear tests by location (dryland sites are shown in bold type and include both hot and cold regions)

| Tests by location | Number of tests | Climate region |
|---|---|---|
| **Nevada** | **935** | **Dryland** |
| **Kazakhstan** | **496** | **Dryland** |
| **Russia** | **214** | **Dryland** |
| Mururoa Atoll | 175 | Humid |
| Enewetak | 43 | Humid |
| **China (Lop Nur)** | **41** | **Dryland** |
| Christmas Island | 30 | Humid |
| Bikini | 23 | Humid |
| **Algeria** | **17** | **Dryland** |
| Johnston Island | 12 | Humid |
| **Australia** | **12** | **Dryland** |
| Fangataufa Atoll | 4 | Humid |
| Pacific Ocean | 4 | Humid |
| Malden Island | 3 | Humid |
| So. Atlantic Ocean | 3 | Humid |
| **Alaska** | **3** | **Dryland** |
| **New Mexico** | **3** | **Dryland** |
| **Pakistan** | **2** | **Dryland** |
| Mississippi | 2 | Humid |
| **Colorado** | **2** | **Dryland** |
| **Ukraine** | **2** | **Dryland** |
| **Uzbekistan** | **2** | **Dryland** |
| **Turkmenistan** | **1** | **Dryland** |
| Total | 2051 | |
| Dryland sites total | 1730 | |

Table 3
Climates of major U.S. Department of Energy nuclear weapons R&D and production facilities

| Facility | Climate |
|---|---|
| LLL | Semi-arid |
| HANFORD | Arid |
| NTS | Arid |
| RF | Semi-arid |
| LANL | Semi-arid |
| INEL | Arid |
| BNL | Humid |
| MOUND | Humid |
| SRP | Humid |
| ORNL | Humid |
| ANL | Humid |
| PANTEX | Arid |

## 4. Radioecology in dryland ecosystems

Studies of the ecological behaviour of man-made radionuclides in arid ecosystems began in the early 1940's with the development and testing of the world's first atomic bomb at the Trinity Site in south-central New Mexico. Intensive radioecological studies on plutonium and other radionuclides were conducted soon after the test in the desert surrounding the Trinity Site by the University of California at Los Angeles (Larson et al., 1951; Leitch, 1951).

In the 30 years following this historic nuclear weapons test, a very large quantity of data was gathered on the distribution and transport of fission and activation products deposited as fallout as a consequence of measurable accumulation and concentration of some fallout radionuclides in natural and human food webs (Pendleton et al., 1964; Hanson, 1967). It was not until the mid 1970's and especially the early 1980's that intensive radioecological studies on plutonium and other actinides began in the USA under the auspices of the DOE, Office of Health and Environmental Research (U.S. DOE, 1980) and Nevada Applied Ecology Group (U.S. DOE, 1985) research programs.

Most of the radionuclide data summarized in this section were collected in the USA at the Nevada Test Site, Trinity Site (New Mexico), Los Alamos National Laboratory (New Mexico), Rocky Flats Plant (Colorado), Idaho National Engineering Laboratory and Hanford Reservation (Washington State). All of these sites average less than 500 mm of annual precipitation. Several, including Nevada Test Site, Trinity Site, Hanford Reservation and Idaho National Engineering Laboratory, average less than 220 mm per year.

## 5. Radionuclide distribution in arid ecosystems

### 5.1. *Spatial and temporal relationships*

Most sources of radionuclides present in arid ecosystems in the USA were initially deposited on the ground surface decades ago as single contamination events or as chronic events over long periods of time. Results from field studies in arid and semiarid areas in the western USA show quite convincingly that soil will eventually be the primary repository of plutonium, cesium, strontium and most other radionuclides with long physical half-lives (i.e. years to millennia) (Hakonson and Nyhan, 1980; Little, 1980; Watters et al., 1983; Gilbert et al., 1987; Romney et al., 1987; Ritchie and McHenry, 1990).

Furthermore, the importance of soil as a reservoir for many radionuclides is often independent of source term (e.g. weapons fallout, effluents, accidents) and the type of terrestrial ecosystem (e.g. deserts, woodlands, shrublands, grasslands) (Watters et al., 1983). For example, on a unit area basis, the percentage of the total plutonium inventory associated with the soil component of contaminated ecosystems was over 99% at four different arid sites contaminated with different sources of plutonium (Watters et al., 1983). Less than 1% of the inventory was associated with biological components of these ecosystems.

After the contaminating events have ceased, the concentration of radionuclides on the soil surface have usually decreased, as shown by the plutonium data in Trinity Site soils (Table 4), due to horizontal and vertical redistribution and, in the case of shorter lived $^{137}Cs$ and $^{90}Sr$, also to radioactive decay. Note that the concentrations of plutonium in the 0–2.5 cm soil profile

Table 4
Comparison of plutonium concentrations in surface (0 to 2.5 cm) soils from Chupadera Mesa as a function of time after the atomic bomb test at Trinity Site in 1945 (from Hakonson and Nyhan, 1980)

| Plutonium concentration (kBq m$^{-2}$) | | |
|---|---|---|
| 1950 | 1951 | 1973 |
| 27.6 (0.01)[1] | 12.6 (0.03) | 0.67 (0.02) |
| $n = 6$ | $n = 3$ | $n = 8$ |

[1]Parenthetic value is coefficient of variation (standard deviation/mean).

Table 5
Percentage of plutonium inventory as a function of depth in alluvial soils at Los Alamos, New Mexico (adapted from Hakonson and Nyhan, 1980)

| Depth (cm) | Mortandad Canyon[1] | Acid-Pueblo Canyon[2] |
|---|---|---|
| 0–2.5 | 20 (0.44)[3] | 4 (0.76) |
| 2.5–7.5 | 36 (0.23) | 10 (0.48) |
| 7.5–12.5 | 22 (0.55) | 20 (1.3) |
| 12.5–30 | 22 (0.79) | 66 (0.18) |

[1]Actively receiving treated radioactive liquid waste.
[2]Had not received waste for 10 years.
[3]Mean percent ($n = 10$) and coefficient of variation.

at Trinity Site (where average annual precipitation is about 200 mm) decreased by a factor of 40 over a 23 year period.

Corresponding data on $^{90}$Sr and $^{137}$Cs in surface soils from Nevada Test Site (about 130 mm annual average precipitation) have shown less dramatic reductions (i.e. up to a factor of two) (Romney et al., 1983). However, the soil surface in many areas of the test site is protected with erosion pavement, a natural stone covering (resulting from deflation of the soil surface) which greatly reduces the potential for horizontal movement of radionuclides by erosion processes.

As radionuclides move into the soil they can become less available for transport by wind and water erosion. For example, in the stream channels used for treated radioactive liquid waste disposal at Los Alamos, only 4–20% of the plutonium in channel alluvium was in the 0–2.5 cm depth profile where it could be readily transported to downstream areas by storm runoff (Table 5). Barring major down-cutting through the alluvium during extreme runoff events, the plutonium that was deeper in the profile would likely continue to remain unavailable for transport by runoff processes.

In another example from the fallout areas of the Nevada Test Site, 30–60% of the $^{90}$Sr and $^{137}$Cs inventories in soil, 18 years after atmospheric testing ceased, was located at depths below 2.5 cm (Romney et al., 1983). In the absence of major disturbances of these soil profiles at the test site, both $^{90}$Sr and $^{137}$Cs will probably remain in place below the ground surface until radioactive decay has occurred.

## 5.2. *Particle size relationships*

An especially important distributional relationship which influences radionuclide transport in dryland sites is that between contaminant concentration and soil particle size distribution. The highest concentrations of many radionuclides in soil are usually associated with the smaller particle size fractions (Tamura, 1975; Nyhan et al., 1976; Little, 1980; Watters et al., 1983; Essington and Romney, 1986) as illustrated in Table 6. For example, in areas receiving liquid effluent at Los Alamos, the highest concentrations of plutonium and $^{137}Cs$ in intermittent stream bed alluvium are invariably in the silt-clay (<53 $\mu$m) fraction.

There are exceptions to this relationship as shown for plutonium in surface soil from ground zero (Area GZ) at Trinity Site (Table 6). The highest plutonium concentrations (0.2 $Bq\,g^{-1}$) were found in the sand size fraction (1–2 mm) while very low concentrations (<0.004 $Bq\,g^{-1}$) were measured in the <53 $\mu$m fraction. Similar results were also obtained from nuclear ground zero sites at Nevada Test Site (Lee and Tamura, 1981) and have been attributed to the deposition of larger plutonium particles closer to ground zero than further along the fall-out pathway.

Even though concentrations of radionuclides may be much higher in a particular soil size fraction, the total inventory of radioactivity in that fraction will depend on its mass in relation to the whole soil. For example, despite the higher concentrations of plutonium in the silt-clay fraction of Los Alamos alluvial soils, less than 15% of the total plutonium inventory in the whole soil was in this fraction (Table 3). In contrast, the silt-clay fraction in Area 21 at Trinity Site (a rangeland site) not only exhibited the highest plutonium concentrations, but it also contained over 70% of the total plutonium inventory in the soil (Table 6).

Soil serves as the primary reservoir of many long-lived radionuclides in arid environments and complex spatial, chemical, and physical relationships determine the mechanisms of transport and their relative importance. Wind and water sort soil particles during the detachment, transport and deposition phases of erosion. Because radionuclide concentrations in soils can be closely related to soil particle size, there is a potential to enrich the concentrations of radionuclides in eroding soil. This potential for enrichment can affect the long term redistribution of radionuclides and their transport to biological components of ecosystems.

Table 6
Comparative distribution of plutonium in silt-clay fraction (<53 m) of surface soil from New Mexico (adapted from Hakonson and Nyhan, 1980)

| | Los Alamos | | Trinity Site | |
|---|---|---|---|---|
| | Mortandad Canyon | Acid-Pueblo Canyon | Area GZ | Area 21[1] |
| $Bq\,g^{-1}$ | 55.5 | 3.15 | 0.003 | 0.14 |
| Soil weight (%) | 2.2 | 3.0 | 8.9 | 36.0 |
| Pu in fraction (%) | 14.0 | 7.0 | 0.78 | 73.0 |

[1] Plutonium concentration in 0–2.5 cm depth profile.

### 5.3. *Abiotic processes*

The importance of wind and water in transporting many radionuclides in arid environments results from the fact that most radionuclides deposit near quantitatively in soil and are tightly bound to this matrix (Hanson, 1975; Whicker and Schultz, 1982; Ritchie and McHenry, 1990). Consequently, processes which transport soil also transport soil-associated contaminants. In many cases, both wind and water preferentially detach and transport the finer size fractions which often contain the highest concentrations of the radionuclide. Moreover, the finer soil fractions are carried farther (and deposited later) than coarser fractions of eroding soil.

Advances in the ability to predict plutonium transport by hydrological processes are discussed here to illustrate relationships that govern wind and water transport of radionuclides. More detail on wind erosion and radionuclide transport by wind can be found in Bagnold (1941); Graf (1971); Marshall (1973); Gallegos (1978); Sehmel (1980); and Anspaugh et al. (1974).

Hydrological processes of particular importance in the physical transport of soil-associated radionuclides include soil detachment by raindrop splash (Ellison, 1945; Mutchler and Young, 1975; Martinez-Menez, 1979) and soil detachment and transport by overland flow (Meyer and Wischmeier, 1966; Foster and Meyer, 1972; Wischmeier and Smith, 1978). Soil particles detached by raindrop impact are important because they can be deposited on vegetation surfaces and thus provide a pathway for movement of soil-associated radionuclides to animals (see section 4). Sediment transported by overland flow is important because it can redistribute contaminants within a watershed and also deliver them to stream channels for subsequent transport to downstream areas.

The combined processes of runoff, soil erosion, sediment transport and deposition of sediment on upland areas and in stream channels usually result in enrichment of smaller soil particles and organic matter in transported sediment (Graf, 1971) including concentration of sediment associated contaminants (Massey and Jackson, 1952; Menzel, 1980; Lane and Hakonson, 1982). This enrichment is often expressed as an enrichment ratio, defined as the concentration of contaminant in the transported sediment divided by its concentration in the residual or uneroded soil. Enrichment ratios have been related to sediment concentration, sediment discharge rate and sediment yield (Massey and Jackson, 1952; Menzel, 1980). Lane and Hakonson (1982) analyzed sediment transport rates by particle size classes in alluvial channels and derived the following expression:

$$\mathrm{ER} = \frac{q[C_\mathrm{s}(d_i) \times Q_\mathrm{s}(d_i)]}{C_\mathrm{s} q[Q_\mathrm{s}(d_i)]}, \qquad (1)$$

where: ER = alluvial channel enrichment ratio, $C_\mathrm{s}(d_i)$ = concentration of contaminant in sediment particles of size class $i$, with representative diameter, $d_i$, in millimeters, $Q_\mathrm{s}(d_i)$ = sediment transport (mass/time) for particles in size class $i$, with representative diameter, $d_i$, in millimeters, and $C_\mathrm{s}$ = mean concentration of contaminant in soil over all particle size classes.

This equation supports the empirical observation that the enrichment ratio increases with decreasing sediment discharge rates. For example, at very low sediment discharge rates (those associated with low runoff velocities) the bedload discharge rate for coarse sediment particles is low and most of the transported sediment is in the smaller particle size classes. Under such conditions, ER would approach the ratio of concentrations in the finest size classes ($C_\mathrm{s}(d_i)$)

Table 7
Approximate enrichment ratios for nutrients and plutonium associated with locations in the United States

| Land use and location | Approximate enrichment ratios | | Comments |
|---|---|---|---|
| | Mean | Range | |
| Cropland, USA[a] | 4.5 | 2.5–7.4 | Nitrogen |
| | 3.6 | 2.6–6.0 | Phosphorus |
| Rangeland, USA[a] | 2.6 | 1.1–6.7 | Nitrogen |
| | 7.1 | 2.7–17 | Phosphorus |
| Cropland, USA[b] | 1.6 | 1.1–2.5 | Fallout plutonium |
| Pasture, USA[b] | 2.3 | 0.8–4.0 | Fallout plutonium |
| Mixed cropland, USA[c] | 2.5 | 1.2–4.0 | Fallout plutonium, transport in Perennial River |
| Semiarid, USA[d] | 5.5 | 1.4–13.3 | Waste effluent plutonium, transport in Ephemeral Streams |

[a]Small agricultural watersheds (5.2–18 ha) at Chickasha, Oklahoma.
[b]Small agricultural watersheds (2.6–2.9 ha) near Lebanon, Ohio.
[c]Great Miami River (drainage area = 1401 km) at Sidney, Ohio.
[d]Los Alamos Watersheds (176–15,000 ha) near Los Alamos, New Mexico.

to the mean concentration over all size classes ($C_s$). At high sediment discharge rates (those associated with high runoff velocities) more of the bed sediments are in transport. At the extreme, if all of the bed sediments were in transport in the same proportion as they exist in the bed material, ER in equation (1) would be unity.

Field measurements of enrichment ratios for nutrients and plutonium at several locations in the United States are given in Table 7. The first four entries are for soil nutrients in runoff from small agricultural areas and mean values vary from 2.6 to 7.1. The next three entries represent enrichment of fallout plutonium in runoff from agricultural watersheds; mean values range from about 1.6 to 2.5. The last entry represents enrichment of plutonium in runoff in ephemeral stream channels at Los Alamos, New Mexico. Field derived values for Los Alamos ranged from 1.4 to 13.3 with a mean of 5.5. Predicted enrichment ratios for Los Alamos stream channels, obtained using the above equation, ranged from 2.9 to 7.0 with a mean of 5.2 (Lane and Hakonson, 1982). Although other factors undoubtedly influence the observed enrichment ratios, the close agreement between observed and predicted ratios in addition to several analyses (Menzel, 1980; Foster et al., 1981; Lane and Hakonson, 1982) suggest that particle sorting alone can account for the observed ratios at Los Alamos. Furthermore, in spite of wide differences in watershed size, hydrologic regime and chemical characteristics, the enrichment ratios resulting from sediment transport given in Table 7 are quite similar for several sediment associated chemicals. Particle sorting is clearly one of the important factors involved in transport of soil and sediment associated radionuclides.

Recent studies were conducted at the Waste Isolation Pilot Plant, Rocky Flats Environmental Technology Site, and Los Alamos National Laboratory to evaluate the effects of fire on transport of fallout $^{137}Cs$ by wind and water erosion (Johansen et al., 2001a, 2001b). These studies, which used a large rainfall simulator to generate runoff and applied controlled or natural fires as study treatments, found that soil erosion and transport of soil associated $^{137}Cs$ increased by a factor of 4–25 on burned plots when compared with unburned plots, depending on the study area. The results demonstrated that erosion and soil radionuclide transport rates determined under undisturbed conditions do not reflect possible changes in rates due to a catastrophic disturbance such as fire. Without intense post fire management, such as reseeding, the effects of fire on water erosion can persist for years (Simanton and Emmerich, 1994).

### 5.4. *Biotic Processes*

#### 5.4.1. *Transport to vegetation*

Two mechanisms for the transport of soil contaminants to terrestrial plants are absorption by roots and deposition of contaminated soil particles on foliage surfaces, with or without subsequent absorption into plant tissues. Despite the host of chemical, biological and physical factors which can modify the physiological availability of radionuclides and subsequent transport to internal plant tissues (Adriano et al., 1980; Cataldo and Vaughn, 1980; Wildung and Garland, 1980), field studies suggest that contamination of foliage surfaces with soil particles contaminated with radionuclides is a major transport mechanism under many arid site and radionuclide source conditions (Romney and Wallace, 1976; Romney et al., 1987; White et al., 1981; Arthur and Alldredge, 1982). For example, comparative studies of plant uptake of plutonium under both field and laboratory conditions generally yield results of the type shown in Table 8. Laboratory studies focused on root uptake of plutonium from soils yield concentration ratios which are at least one order of magnitude (and often 2–3 orders of magnitude) lower than ratios observed under comparable conditions at field sites. The differences in concentration ratios between laboratory and field studies imply that a mechanism(s) exists in arid environments for delivering at least 10 times more plutonium to vegetation than would be predicted based on greenhouse studies.

The higher concentration ratios observed at field sites are generally attributed to the presence of surficial contamination on field site vegetation (Hakonson and Nyhan, 1980; Little et al., 1980; Romney et al., 1987). That conclusion is supported by the obvious presence of soil on foliage surfaces in the field and by the ability to remove up to 90% of the plutonium contamination from vegetation by washing (White et al., 1981; Arthur and Alldredge, 1982).

Studies at Los Alamos demonstrated that rain-splash of soil particles with subsequent deposition on foliage surfaces can easily contribute all of the plutonium measured in field-site

Table 8
Comparison of plutonium concentration ratios for field and glasshouse conditions (Romney and Wallace, 1976)

| Soil source | Field | Glasshouse |
|---|---|---|
| NTS Area 11B[a] | $1.3 \times 10^{-2}$ to $1.6 \times 10^{-1}$ | $1.5 \times 10^{-4}$ |
| NTS Area 11C | $4.5 \times 10^{-2}$ to $3.4 \times 10^{-1}$ | $1.8 \times 10^{-4}$ |
| NTS Area 13 | $7.8 \times 10^{-2}$ to $4.4 \times 10^{-1}$ | $1.1 \times 10^{-4}$ |

[a]NTS (Nevada Test Site).

vegetation (Dreicer et al., 1984). More importantly, these studies, which employed a labeled-soil particle technique and scanning electron microscope, have shown that relationships that govern translational movement of plutonium by soil erosion processes also govern transport of plutonium to foliage surfaces. For example, the energy of impacting raindrops caused an enrichment of the smaller soil particles (up to 105 $\mu$m in the study by Dreicer et al., 1984) on foliage surfaces. The amount of soil deposited on the plants was also related to height from the ground surface and the characteristics of individual rainstorms. Calculations based on the mass and plutonium content of soil measured on the plants demonstrated that the rainsplash mechanism could easily account for the observed plant/soil concentration ratios of $5 \times 10^{-2}$ (White et al., 1981; Foster et al., 1985).

While absorption of plutonium through leaf surfaces has been demonstrated (Cataldo and Vaughn, 1980) it is likely to be of limited importance in arid field sites, particularly for annual or deciduous vegetation. Studies on the uptake of plutonium by vegetable crops grown in field sites at Los Alamos show that as much as 50% of the plutonium in crop samples was surface contamination that could be removed by washing (White et al., 1981). Plutonium that cannot be removed from vegetable crop surfaces in arid environments does not necessarily reflect contamination of plant tissues. Cataldo and Vaughn (1980) showed that submicron particles on foliage surfaces are difficult to remove by either simulated wind or rain.

### **5.4.2.** *Transport to animals*

As with vegetation, soil can be a major source of radionuclides to animals in arid ecosystems. Soil particles can be transported to animals in association with exterior surfaces of food and by direct transfer of soil to the animal via inhalation, ingestion and contamination of the pelt.

Plutonium is the best example of a radionuclide whose transport to animals in arid ecosystems is dominated by physical processes. Data from many field sites and source conditions show that gut availability of plutonium in a variety of animals including rodents, deer and cattle is very low leading to very low concentrations of plutonium in internal tissues and organs (Moore et al., 1977; Smith, 1977; Hakonson and Nyhan, 1980; Arthur et al., 1987). The highest concentrations are generally found in tissues exposed to the external environment including the pelt, gastro-intestinal tract and lungs. At Los Alamos, about 96% of the plutonium body burden in rodents from the canyon liquid waste disposal areas was in the pelt and gastro-intestinal tract (Hakonson and Nyhan, 1980).

Because soil passes through the gastro-intestinal tract of free-ranging animals on a daily basis, there is a potential for these animals to redistribute soil radionuclides across the landscape. Studies at Nevada Test Site with cattle (Moore et al., 1977), at Rocky Flats Plant with mule deer and small mammals (Arthur and Alldredge, 1979; Little, 1980), and at Idaho National Engineering Laboratory with small mammals and coyotes (Arthur and Markham, 1982, 1983) demonstrated that horizontal (and vertical, in the case of burrowing animals) redistribution of soil plutonium occurs as animals move within and outside contaminated areas. While the magnitude of this transport is small over the short-term, there are circumstances in which it assumes much more importance. For example, in a nuclear waste burial site at Hanford, defecation of $^{90}$Sr on the ground surface and surrounding area by jackrabbits burrowing into the waste required remedial action to prevent further migration of the strontium (O'Farrell and Gilbert, 1975).

Strontium and cesium are potentially more soluble than plutonium and transport of these elements to animals in arid ecosystems involves a combination of physical and physiological processes. The more tightly bound $^{90}$Sr and $^{137}$Cs are to soil (related to the clay content of soil and local climate), the more their transport will be governed by soil particle transport. Data on $^{90}$Sr and $^{137}$Cs in small mammals from Nevada Test Site (Romney et al., 1983) and at a burial ground at Idaho National Engineering Laboratory (Arthur et al., 1987) show relatively high concentrations of these radionuclides in lung, pelt and gastro-intestinal tract similar to plutonium. This suggests that physical transport of these more bioavailable radionuclides is also important as it is for plutonium. The bioavailability of radionuclides such as $^{90}$Sr and $^{137}$Cs will depend on chemical form, local environmental conditions, and the structure and function of the relevant food webs (Whicker and Pinder, 2002).

## 6. Conclusions

In arid ecosystems within the USA, contaminated with radionuclides several decades ago, nearly 100% of the residual radioactivity is sorbed to the soils and sediments of affected areas. For radionuclides that are tightly bound to soils, such as plutonium, erosion processes, driven by wind and water, control their horizontal movement as well as their entry into biological systems and pathways.

Radionuclide concentrations in arid soils are often strongly related to soil particle size with higher concentrations appearing in the smaller size fractions including silts and clays. This is an especially important distribution relationship because wind and water erosion processes preferentially sort and transport the smaller soil particles compared with coarser fractions. This leads to an enrichment of fine particles, often containing the highest radionuclide concentrations, in eroding soil. The transport of fine soil particles, with higher radionuclide concentrations, followed by deposition on plant and animal surfaces, inhalation and ingestion, serve as primary pathways for radionuclide entry into biota in these dusty, dry environments.

The continued importance of physical transport of long-lived radionuclides in arid ecosystems will ultimately depend on the changes that occur in the distribution of the radionuclides in soil and sediment. At present, variable but significant amounts of the residual radionuclide inventory in most contaminated sites still occur near the ground surface where erosion processes operate. Ultimately, the fate of long-lived radionuclides tightly bound to soil in arid environments will depend on the fate of the soils and sediments themselves.

## References

Adriano, D.C., Wallace, A., Romney, E.M. (1980). Uptake of transuranic nuclides from soil by plants grown under controlled environmental conditions. In: Hanson, W.C. (Ed.), *Transuranic Elements in The Environment*. U.S. Department of Energy, NTIS, Springfield, VA, pp. 336–360. DOE/TIC-22800.

Anspaugh, L.R., Shinn, J.H., Wilson, D.W. (1974). Evaluation of the resuspension pathway toward protective guidelines for soil contamination with radioactivity. In: *Population Dose Evaluation and Standards for Man and His Environment*, pp. 513–524. STI/PUB/375, Vienna.

Arthur, W.J., III, Alldredge, A.W. (1979). Soil ingestion by mule deer in northcentral Colorado. *J. Range Mgmt.* **32** (1), 67–71.

Arthur, W.J., III, Alldredge, A.W. (1982). Importance of plutonium contamination on vegetation surfaces at Rocky Flats Colorado. *Environ. and Experimental Botany* **22** (1), 33–38.

Arthur, W.J., Markham, O.D. (1982). Radionuclide export and elimination by coyotes at two radioactive waste disposal areas in southeastern Idaho. *Health Physics* **43**, 493–500.

Arthur, W.J., III, Markham, O.D. (1983). Small mammal soil burrowing as a radionuclide transport vector at a radioactive waste disposal area in southeastern Idaho. *J. of Environ. Qual.* **12** (1), 117–122.

Arthur, W.J., Markham, O.D., Groves, C.R., Keller, B.L. (1987). Radionuclide export by deer mice at a solid radioactive waste disposal area in southeastern Idaho. *Health Physics* **52** (1), 45–54.

Bagnold, R.A. (1941). *The Physics of Blown Sand and Desert Dunes*. Methuen, London, 265 pp.

Cataldo, D.A., Vaughn, B.E. (1980). Interaction of airborne plutonium with plant foliage. In: Hanson, W.C. (Ed.), *Transuranic Elements in The Environment*. U.S. Department of Energy, NTIS, Springfield, VA, pp. 288–299. DOE/TIC-22800.

Dreicer, M., Hakonson, T.E., White, G.C., Whicker, F.W. (1984). Rainsplash as a mechanism for soil contamination of plant surfaces. *Health Physics* **46**, 177–188.

Ellison, W.D. (1945). Some effects of raindrops and surface flow on soil erosion and infiltration. *Trans. AGU* **26**, 415–429.

Essington, E.H., Romney, E.M. (1986). Mobilization of $^{137}$Cs during rainfall simulation studies at the Nevada Test Site. In: Lane, L.J. (Ed.), *Erosion on Rangelands: Emerging Technology and Data Base, Proc. Rainfall Simulation Workshop, Jan 14–15, 1985, Tucson, AZ*. Soc. Range Mgmt., Denver, CO, ISBN 0-9603692-4-4, pp. 35–38.

Evenari, M. (1985). The desert environment. In: Evenari, M., Noy-Meir, I., Goodall, D.W. (Eds.), *Ecosystems of The World. Hot Deserts and Arid Shrublands, A.*, vol. 12A. Elsevier, Amsterdam, pp. 1–22.

Foster, G.R., Meyer, L.D. (1972). A closed form soil erosion equation from upland areas. In: Shen, H.W. (Ed.), *Sedimentation: Symposium to Honor Professor H.H. Einstein*. Fort Collins, CO, pp. 121–129.

Foster, G.R., Lane, L.J., Nowlin, J.D., Laflen, J.M., Young, R.A. (1981). Estimating erosion and sediment yield on field-sized areas. *Trans. ASAE* **24**, 1253–1262.

Foster, G.R., White, G.C., Hakonson, T.E., Dreicer, M. (1985). A model for splash retention of sediment and soil-borne contaminants of plants. *Trans. ASAE* **28** (5), 1511–1520.

Gallegos, A.F. (1978). Preliminary model of plutonium transport by wind at Trinity Site. In: White, M.E., Dunaway, P.B. (Eds.), *Selected Environmental Plutonium Research Reports of The NAEG*, pp. 681–695. U.S. DOE Report NVO-192 (Vol. 2), Nevada Operations Office, NTIS.

Gilbert, R.O., Simpson, J.C., Engel, D.W., Kinnison, R.R. (1987). Estimating isotopic ratios, spatial distribution, and inventory of radionuclides at nuclear sites 201, 209 and 221. In: Howard, W.A., Fuller, R.G. (Eds.), *The Dynamics of Transuranics and Other Radionuclides in Natural Environments*. U.S. Department of Energy, NTIS, Springfield, VA, pp. 381–430. U.S. DOE Report NVO-272.

Graf, W.H. (1971). *Hydraulics of sediment transport*. McGraw-Hill Book Co., New York, 513 pp.

Hakonson, T.E., Nyhan, J.W. (1980). Ecological relationships of plutonium in Southwest ecosystems. In: Hansen, W.C. (Ed.), *Transuranic Elements in The Environment*. U.S. Department of Energy, NTIS, Springfield, VA, pp. 403–419. DOE/TIC-22800.

Hanson, W.C. (1967). Radioecological concentration processes characterizing arctic ecosystems. In: Aberg, B., Hungate, F.P. (Eds.), *Radioecological Concentration Processes*. Pergamon Press, Oxford, UK, pp. 183–192.

Hanson, W.C. (1975). Ecological considerations of the behavior of plutonium in the environment. *Health Phys.* **28**, 529–537.

Johansen, M.P., Hakonson, T.E., Breshears, D.D. (2001a). Post-fire runoff and erosion from rainfall simulation: Contrasting forests with shrublands and grasslands. *Hydrol. Processes* **15**, 2953–2965.

Johansen, M.P., Hakonson, T.E., Whicker, F.W., Simanton, J.R., Stone, J.J. (2001b). Hydrologic response and radionuclide transport following fire at semiarid sites. *J. Environ. Quality* **30**, 2010–2017.

Lane, L.J., Hakonson, T.E. (1982). Influence of particle sorting in transport of sediment associated contaminants. In: Post, R.G. (Ed.), *Proc. Waste Mgmt. '82 Symp.*, vol. 2. University of Arizona Press, Tucson, Arizona, pp. 543–557.

Larson, L.H., Olafson, J.H., Neel, J.W., Dunn, W.F., Gordon, S.H., Gillooly, B. (1951). The 1949 and 1950 radiological soil survey of fission product contamination and some soil-plant interrelationships of areas in New Mexico affected by the first atomic bomb detonation. USAEC Report, UCLA-140.

Lavee, H., Poesen, J., Yair, A. (1997). Evidence of high efficiency water harvesting by ancient farmers in the Negev desert, Israel. *Journal of Arid Environments* **35**, 341–348.

Lee, S.Y., Tamura, T. (1981). Distribution and characterization of radionuclides in soils from Nevada Test Site. *J. Environ. Qual.* **10**, 234–239.

Leitch, J.L. (1951). Summary of the radiological findings in animals from biological surveys of 1947, 1948, 1949, and 1950. USAEC Report, UCLA-111.

Little, C.A. (1980). Plutonium in a grassland ecosystem. In: Hanson, W.C. (Ed.), *Transuranic Elements in The Environment*. U.S. Department of Energy, NTIS, Springfield, VA, pp. 420–440. DOE/TIC-22800.

Little, C.A., Whicker, F.W., Winsor, T.F. (1980). Plutonium in a grassland ecosystem. *J. Environ. Qual.* **9** (3), 350–354.

Lyles, L. (1975). Possible effects of wind erosion on soil productivity. *J. Soil and Water Cons.* **30**, 279–283.

Marshall, J.K. (1973). In: Lovett, J.V. (Ed.), *The Environmental, Economic, and Social Significance of Drought*. Angus and Robertson, London.

Martinez-Menez, M.R. (1979). Erosion modeling for upland areas, Ph.D. Diss., University of Arizona, Tucson, AZ, pp. 145.

Massey, H.F., Jackson, M.L. (1952). Selective erosion of soil fertility constituents. In: *Proc. SSSA 16*, pp. 353–356.

Meigs, P. (1953). World distribution of arid and semi-arid homoclimates. In: *Reviews of Research on Arid Zone Hydrology*. United Nations Educational, Scientific, and Cultural Organization, Arid Zone Programme-1, Paris, pp. 203–209.

Menzel, R.G. (1980). Enrichment ratios for water quality modeling. In: Knisel, W.G. (Ed.), *CREAMS-MA Field Scale Model for Chemicals, Runoff, and Erosion from Agricultural Management Systems*, pp. 486–492. USDA, Conservation Research Report No. 26, III, 12.

Meyer, L.D., Wischmeier, W.H. (1966). Mathematical simulation of the process of soil erosion by water. *Trans. ASAE* **12** (6), 754–758.

Moore, K.S., Naegle, S.R., Bradley, W.G. (1977). Plutonium-239 and Americum-241 contamination of small vertebrates in NAEG study areas at NTS and TTR. In: White, M.G., Dunaway, P.B., Howard, W.A. (Eds.), *Environmental Plutonium on the Nevada Test Site and Environs*. U.S. Department of Energy, NTIS, Springfield, VA, pp. 193–217. U.S. DOE report NVO-171UC-2.

Mutchler, C.K., Young, R.A. (1975). Soil detachment by raindrops. In: *Present and Prospective Technology for Predicting Sediment Yields and Sources*. USDA, Agric. Res. Serv., Wash, DC, p. 285. ARS-S-40.

Nyhan, J.W., Miera, F.R., Peters, R.J. (1976). Distribution of plutonium in soil particle size fractions of liquid effluent-receiving areas at Los Alamos. *J. Environ. Qual.* **5**, 50–56.

O'Farrell, T.P., Gilbert, R.O. (1975). Transport of radioactive materials by Jack Rabbits on the Hanford Reservation. *Health Physics* **29**, 9–15.

Pendleton, R.C., Lloyd, R.D., Mays, C.W., Church, B.W. (1964). Trophic level effect on the accumulation of cesium-137 in cougars feeding on mule deer. *Nature* **204**, 708–709.

Potter, L.D. (1992). Desert characteristics as related to waste disposal. In: Reith, C.C., Thomson, B.M. (Eds.), *Deserts as Dumps? The Disposal of Hazardous Materials in Arid Ecosystems*. Univ. of N.M. Press, Albuquerque, NM, pp. 21–56.

Ritchie, J.C., McHenry, J.R. (1990). Application of radioactive fallout cesium-137 for measuring soil erosion and sediment accumulation ratios and patterns: A review. *J. Environ. Qual.* **19** (2), 215–233.

Romney, E.M., Wallace, A. (1976). Plutonium contamination of vegetation in dusty environments. In: White, M.G., Dunaway, P.B. (Eds.), *Transuranics in Natural Environments*. NTIS, Springfield, VA, pp. 287–302. USERDA Rept. NVO-178.

Romney, E.M., Lindberg, R.G., Kinnear, J.E., Wood, R.A. (1983). $^{90}$Sr and $^{137}$Cs in soil and biota of fallout areas in southern Nevada and Utah. *Health Physics* **45** (3), 643–650.

Romney, E.M., Hunter, R.B., Wallace, A. (1987). Distribution of $^{239,240}$Pu, $^{241}$Am, $^{137}$Cs, and $^{90}$Sr on vegetation at nuclear sites 201, 219 and 221. In: Howard, W.A., Fuller, R.G. (Eds.), *The Dynamics of Transuranics and Other Radionuclides in Natural Environments*. U.S. Department of Energy, NTIS, Springfield, VA, pp. 69–78. U.S. DOE Report NVO-272.

Scanlon, B.R., Tyler, S.W., Wierenga, P.J. (1997). Hydrologic issues in arid, unsaturated systems and implications for contaminant transport. *Reviews of Geophysics* **35** (4), 461–490.

Sehmel, G.A. (1980). Transuranic and tracer simulant resuspension. In: Hanson, W.C. (Ed.), *Transuranic Elements in The Environment*. U.S. Department of Energy, NTIS, Springfield, VA, pp. 236–287. DOE/TIC-22800.

Simanton, J.R., Emmerich, W.E. (1994). Temporal variability in rangeland erosion processes. In: *Variability of Rangeland Water Erosion Processes*, pp. 51–65. SSSA Special Publication 38.

Smith, D.D. (1977). Review of grazing studies on plutonium contaminated rangelands. In: White, M.G., Dunaway, P.B. (Eds.), *Transuranics in Natural Environments*. Nevada Operations Office, NTIS, Springfield, VA, pp. 407–417. ERDA Report NVO-178.

Tamura, T. (1975). Distribution and characterization of plutonium in soils from Nevada Test Site. *J. Environ. Qual.* **4**, 350–354.

U.S. DOE (1980). In: Hanson, W.C. (Ed.), *Transuranic Elements in The Environment*. U.S. Department of Energy, NTIS, Springfield, VA. DOE/TIC-22800. 725 pp.

U.S. DOE (1985). In: Howard, W.A., Fuller, R.G. (Eds.), *The Dynamics of Transuranics and Other Radionuclides in Natural Environments*. U.S. Department of Energy, NTIS, Springfield, VA. NVO-272, DE87014456. 565 pp.

Watters, R.L., Hakonson, T.E., Lane, L.J. (1983). The behavior of actinides in the environment. *Radiochimica Acta* **32**, 89–103.

Whicker, F.W., Schultz, V. (1982). *Radioecology: Nuclear Energy and The Environment*. CRC Press, Boca Raton, Florida. Volumes I (212 pp.) and II (228 pp.).

Whicker, F.W., Pinder, J.E. (2002). Food chains and biogeochemical pathways: contributions of fallout and other radiotracers. *Health Physics* **82**, 680–689.

White, G.C., Hakonson, T.E., Ahlquist, A.J. (1981). Factors affecting radionuclide availability to vegetables grown at Los Alamos. *J. of Environ. Qual.* **10**, 294–299.

Wildung, R.E., Garland, T.R. (1980). The relationship of microbial processes to the fate and behavior of transuranic elements in soils, plants, and animals. In: Hanson, W.C. (Ed.), *Transuranic Elements in The Environment*. U.S. Department of Energy, NTIS, Springfield, VA, pp. 300–335. DOE/TIC-22800.

Wischmeier, W.H., Smith, D.D. (1978). *Predicting Rainfall Erosion Losses—A Guide to Conservation Planning*. USDA. Agriculture Handbook No. 537, 58 pp.

# Radionuclides in tropical and subtropical ecosystems

Shigeo Uchida

*Research Center for Radiation Protection, National Institute of Radiological Sciences, Anagawa 4-9-1, Inage, Chiba 263-8555, Japan*

## 1. Introduction

Tropical areas lie between the two latitudes of 23.5°N and 23.5°S of the equator while sub-tropical regions lie immediately to the north and south of this tropical belt. Many of the countries in the tropical and sub-tropical areas are heavily populated. Daily life and activities in tropical and sub-tropical areas differ from those in temperate areas; there are also differences in agricultural products, food consumption habits and lifestyles which affect potential human exposures to radionuclides. Consequently, dose assessments for radionuclides deposited in these areas are likely to be significantly different from those usually carried out in temperate zones. In this chapter the behavior of radionuclides in both tropical and sub-tropical areas is discussed.

Since the Chernobyl accident, the International Atomic Energy Agency [IAEA], the Food and Agriculture Organization [FAO] and the International Union of Radioecology [IUR] have each paid special attention to safety assessments for the people who live in tropical and sub-tropical regions. An accident at a nuclear power plant potentially affects not only the temperate regions where most nuclear power stations are currently located but, as a result of long-range atmospheric transport, it could also contaminate tropical and sub-tropical regions. Furthermore, the potential risk of a local nuclear power plant accident will become higher in these equatorial regions, as several nuclear power plants are planned to be constructed within them in the near future. Thus, in 1993, IAEA, FAO and IUR established a Coordinated Research Programme to collect transfer parameters in (sub-)tropical regions.

In this chapter, transfers of radionuclides from soils to crops under tropical conditions are described. The focus is especially on the behavior of radionuclides in rice paddy fields, because of the importance of rice as a staple crop in the (sub-)tropical regions (Karube et al., 1995). Rice paddy fields represent a unique ecosystem for which data on radionuclide behavior and transfers have been obtained. A notable recent development in this area has been the establishment by the IUR of a taskgroup focusing on the Radioecology of

RADIOACTIVITY IN THE ENVIRONMENT
VOLUME 10 ISSN 1569-4860/DOI: 10.1016/S1569-4860(06)10009-1

Rice (Uchida and Tsukada, 2004). Another unique tropical ecosystem is represented by the coral atolls of the Marshall Islands for which radioecological data have been obtained by the Lawrence Livermore National Laboratory (USA). These data are also reviewed in this chapter.

## 2. Input of fallout radionuclides in tropical areas and critical foods for human intake of radionuclides

In general, the deposition rates of global fallout radionuclides in (sub-)tropical areas are not high. Although radionuclides produced by nuclear weapons testing and released into the atmosphere have been dispersed globally, they have not deposited uniformly on the earth's surface. The highest deposition rate was observed between 40°N to 50°N latitudes (i.e., temperate regions), while tropical and subtropical areas received lower deposition rates of global fallout radionuclides. UNSCEAR (1982) reported that 76% of the $^{137}$Cs produced in above-ground nuclear weapons tests (the total $^{137}$Cs activity released was estimated to be $9.6 \times 10^{17}$ Bq) was deposited in the northern hemisphere and 24% in the southern hemisphere: this means that less fallout is expected in the tropical and subtropical areas than in the temperate area of the northern hemisphere. Also, after the Chernobyl accident, only low activities of radionuclides were deposited onto the ground surface in tropical and subtropical areas. Thus, because of these low deposition rates, radioecological studies in these (sub-)tropical regions have had a much lower priority than in the temperate regions.

However, even in (sub-)tropical areas, there are some locations where high contamination levels are found. Bikini and Enewetak Atolls, in the Marshall Islands, were US nuclear test sites from 1946 to 1958. A total of 66 nuclear devices were detonated on these atolls, as a consequence of which huge amounts of radionuclides were deposited on the Marshall Islands. The total $^{137}$Cs deposit on Bikini Atoll was about 80 $\text{kBq}\,\text{m}^{-2}$. This is about two orders of magnitude higher than the $^{137}$Cs global fallout deposited on the Marshall Islands (Simon and Graham, 1997).

In addition to these sites, there is the potential risk caused by a nuclear power plant accident within (sub-)tropical regions. Thus it is necessary to determine the critical pathways for human dose assessment, one of which is the ingestion of locally produced foods. Because agricultural products and food consumption habits in the tropics are different from those in temperate regions, it is expected that the critical foods will differ as well. Figure 1 shows the relative contributions of eleven food groups to the total dietary intake in various countries (Ministry of Agriculture, Forestry and Fisheries [Japan], 1988). In European and North American countries, livestock products including meat, eggs and milk make a major contribution to diet while, in Asian and South American countries, agricultural products including cereals and vegetables are the main contributors.

In order to identify foods in Japan which are critical for the intake of fallout $^{137}$Cs and $^{90}$Sr to the human body, radiochemical analyses of foodstuffs in nine categories (root vegetables, leafy vegetables, fish and shellfish, meat, eggs, milk and milk products, potatoes, beans, and cereals) were conducted in the 1970s (Ueda et al., 1974). As can be seen from Figure 2, the contribution of cereals to the intake of $^{137}$Cs is the highest (about 40%) followed by beans, milk and milk products, leafy vegetables, fish and shellfish. It should be noted that rice is

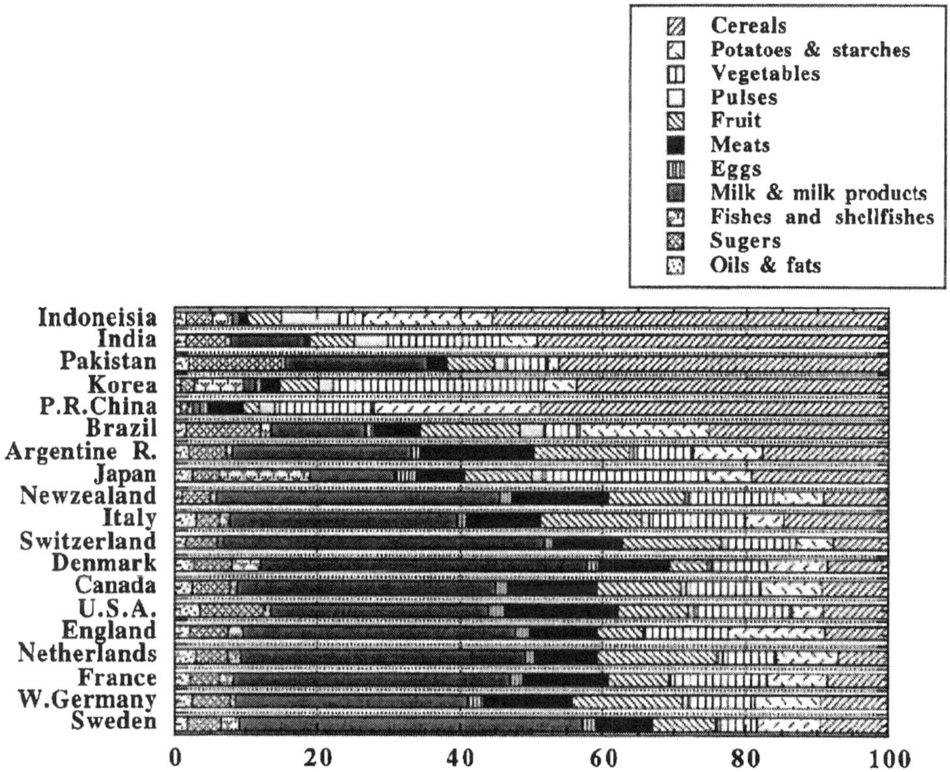


Fig. 1. Percentage contributions of eleven food groups to total annual dietary intake per capita in selected countries. Figure drawn from data compiled by the Ministry of Agriculture, Forestry and Fisheries [Japan] (1988).

the main cereal consumed in Japan. In the case of fallout $^{90}Sr$, the highest contribution to dietary intake is made by leafy vegetables, followed by beans and rice. From these results it is to be expected that cereals and leafy vegetables are likely to be the critical foodstuffs when assessing dietary intake of radionuclides. The results found in Japan are equally applicable to monsoon Asian regions, which are classified as (sub-)tropical areas because the Japanese food culture is quite close to that of the (sub-)tropical regions.

When considering the dietary exposure of human consumers of crops, the activity concentrations of radionuclides within the crops' edible tissues are obviously important. Crops are mainly contaminated with radionuclides via two routes: direct deposition from the atmosphere to plant surfaces and root uptake from the soil (see Chapter 2). After radionuclides are deposited to the ground surface, they may remain in the surface soil layer for a long time and, consequently, root uptake from the soil often represents the most important long-term dietary exposure route for radiological dose assessment.

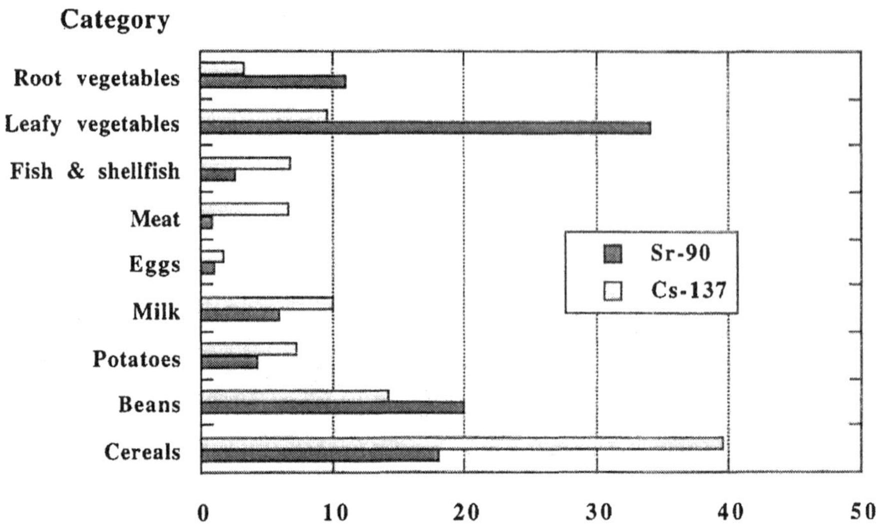


Fig. 2. Percentage contributions of nine food groups to total annual dietary intake of $^{90}$Sr and $^{137}$Cs. Each value is a mean during the period from 1966 to 1971 in three prefectures of Japan. Figure drawn from data presented by Ueda et al. (1974).

## 3. Soil-to-crop transfers of radionuclides in (sub-)tropical areas

As described above, the IAEA and FAO in cooperation with the IUR established a Co-ordinated Research Programme [CRP] in 1993 to collect radioecological transfer parameters for (sub-)tropical environments. In this section, preliminary results on soil to plant transfer factors [TFs—see Chapter 2] obtained in this CRP are described. The TF is defined as the ratio of radionuclide activity concentrations in vegetation and soil (Bq kg$^{-1}$ dry weight plant: Bq kg$^{-1}$ dry weight soil). It should be noted that TF values are generally determined for the edible parts of crops. These transfer data can be significantly affected by environmental conditions such as the mobilities of specific radionuclides in specific soil types, as well as the crop type considered. The key question asked in the CRP was if there are any systematic differences between data obtained in temperate and (sub-)tropical environments (Frissel, 1997).

Because human beings habitually consume a wide variety of crops, it is not feasible to compare the TF of all crops grown under both temperate and (sub-)tropical conditions. Thus, Uchida and Okabayashi (1988) proposed that TF values for crops typically consumed in Japan should be obtained for the edible parts of several generic crop types viz. rice, cereals except rice, tubers, root vegetables, leafy vegetables, fruit and seeds. In the CRP, though the classification method was slightly different from that proposed by Uchida and Okabayashi, crops were pragmatically divided into crop classes as cereals, green vegetables, tubers, root crops, fruits, pods, etc.

Distributions of TF values of Cs and Sr for four crop groups (pods and seeds of beans, cereals, vegetables, and others) are shown in Figures 3 and 4, respectively. In Figure 3, the values obtained in the Marshall Islands are not included, but these results are described later. The TF values for both Cs and Sr show wide distributions. In fact, the TF values obtained in

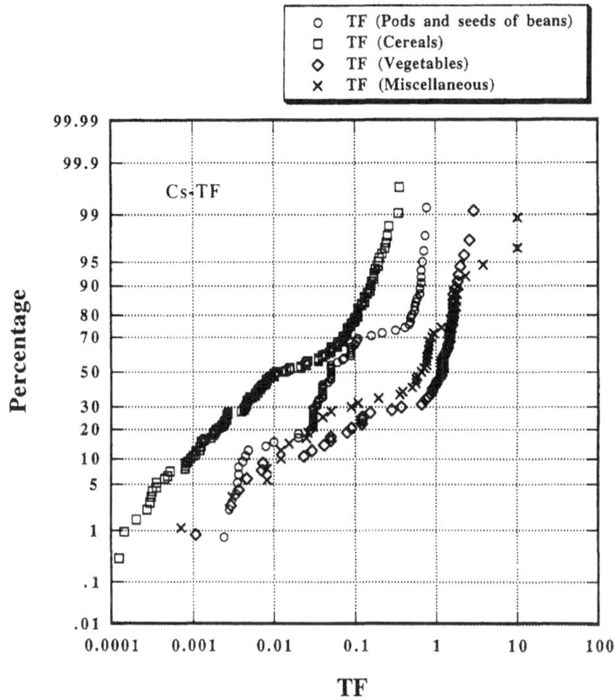


Fig. 3. Probability distribution of soil–plant transfer factors (TFs) for Cs obtained in tropical and subtropical areas (previously unpublished data; M.J. Frissel, pers. comm.).

the (sub-)tropical systems deviate more from the mean value than those in the temperate areas do, a result which is mainly due to the wider range of soil and vegetation types and weather conditions within temperate and (sub-)tropical zones. In addition to these major factors, soil, water and fertilizer managements are also likely to affect the uptake of radionuclides. The TF results are summarized in Table 1. TFs for vegetables are the highest among the four crop groups for both Cs and Sr, although the TFs for Sr are generally higher than those for Cs in all crop groups except 'miscellaneous'. Comparison of the TF data obtained by the CRP for (sub-)tropical environments with IUR data determined in temperate environments showed close agreement between the mean values. This result implies that no systematic differences exist between TFs in temperate and (sub-)tropical environments (Frissel, 1997).

However, attention should be paid to the results obtained by the CRP, because it is clear that the TF values for (sub-)tropical crops are very widely distributed. This large degree of uncertainty is important and could potentially pose a problem when applying 'expected' TF values in dose assessments for these regions. Frissel et al. (2002) proposed the use of cereals and cabbage as indicator crops for screening purposes. This is because the TFs of $^{90}Sr$ for cereals and cabbage varied by a factor of 100 overall, but the ratios of $TF_{cabbage}/TF_{cereals}$ varied by no more than a factor of 20. Similarly, the TFs of $^{137}Cs$ for cereals and cabbage varied by a factor of 1000 overall, but the ratios of $TF_{cabbage}/TF_{cereals}$ varied by less than a factor of 30.

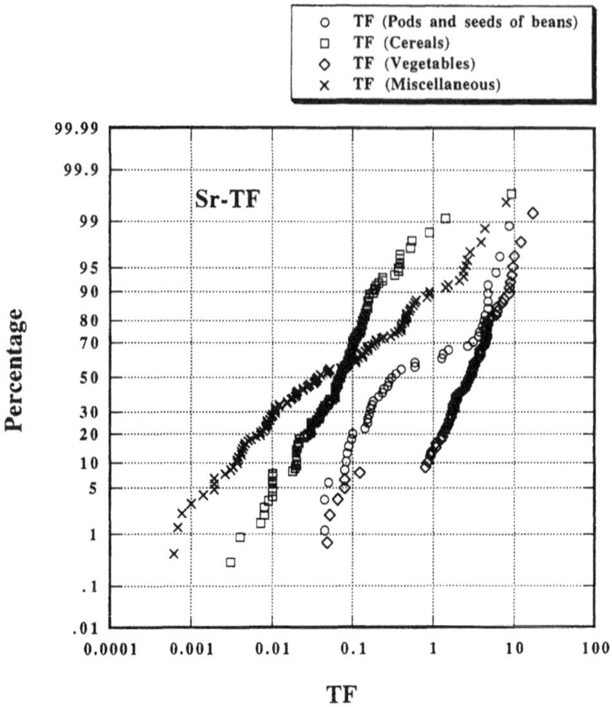


Fig. 4. Probability distribution of soil–plant transfer factors (TFs) for Sr obtained in tropical and subtropical areas (previously unpublished data; M.J. Frissel, pers. comm.).

Table 1
Statistical characteristics of soil–plant TF values for Cs and Sr for four generic crop groups (previously unpublished data; M.J. Frissel, pers. comm.)

| Nuclide | Crop type | Number of samples | TF value | | | | |
|---|---|---|---|---|---|---|---|
| | | | Min | Median | Max | Arithmetic mean | Geometric mean |
| Cs | Pods and seeds of beans | 69 | 0.0024 | 0.050 | 0.769 | 0.19 | 0.066 |
| | Cereals | 164 | 0.00012 | 0.012 | 0.35 | 0.053 | 0.013 |
| | Vegetables | 61 | 0.0011 | 1.2 | 2.9 | 1.0 | 0.42 |
| | Miscellaneous | 45 | 0.0010 | 0.665 | 10.2 | 1.2 | 0.27 |
| Sr | Pods and seeds of beans | 43 | 0.045 | 0.30 | 8.5 | 1.7 | 0.54 |
| | Cereals | 170 | 0.0030 | 0.070 | 9.3 | 0.16 | 0.065 |
| | Vegetables | 72 | 0.048 | 2.7 | 17.2 | 3.6 | 2.2 |
| | Miscellaneous | 116 | 0.00060 | 0.037 | 7.9 | 0.41 | 0.050 |

## 4. Paddy field ecosystems

### 4.1. *Characteristics of a paddy field and its water budget*

Agricultural practices and systems of cultivation are important factors which are likely to modify crop uptake of radionuclides from soil. As described in section 2, one of the critical foods for the intake of radionuclides by humans, especially in the monsoon Asian regions, is rice (*Oryza sativa* L.). Rice is usually planted under flooded conditions, which provide the best growth conditions for the rice plants. No crop rotation is necessary and, consequently, rice has been grown in the same paddy fields for hundreds of years. Furthermore, rice crops can be grown more or less continuously. Two rice crops per year are commonly obtained in the same paddy fields in tropical or subtropical areas and even in some parts of the temperate zones. In tropical areas, five crops in two years or triple cropping in one year is possible, if sufficient irrigation water is available (Karube et al., 1995). Under flooded conditions, the soil pH increases with soil reduction and, consequently, phosphate and iron sorbed onto the soil particles are gradually modified to soluble forms which are easily taken up by rice plants. In addition, under reducing conditions nitrogen is present as the positively charged ammonium ion which is absorbed efficiently by the roots of rice plants, unlike the situation in upland fields.

For realistic estimations of radionuclide transfer in the food-chain, a good knowledge of the physical transfer of radionuclides in media such as groundwater and irrigation water is required. Figure 5 shows the characteristics of the water budget in Japanese paddy fields (Watanabe, 1992). Under normal hydrological conditions, the water budget in a paddy field during a rice planting season consists of precipitation onto the field, introduction of irrigation water into the field, evapotranspiration from the leaves of rice plant and the surface water,

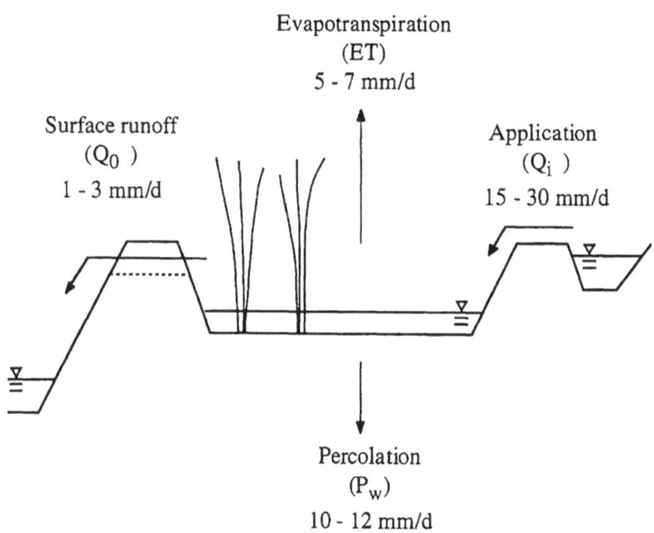


Fig. 5. Diagram of the water budget in a typical Japanese paddy field (Watanabe, 1992; reproduced with permission).

percolation into the soil, and surface runoff. A paddy field includes both the land area on which rice plants grow and the embankments (bunds) that surround the planting area to retain the irrigation water.

The water budget varies with field conditions and water management practices, area by area. The evapotranspiration rate changes from a minimum of 0 to a maximum of 8 $mm\,d^{-1}$ during different stages in the growth of the rice crop, and is also strongly influenced by weather conditions. Because the percolation rate is affected by soil type and soil structure, its rate also varies widely with a range from 5 to 30 $mm\,d^{-1}$. The normal ranges of each factor in the water budget are shown in Figure 5. These factors provide a rough estimate of water volume per unit area for rainless days in summer under stable water management.

## 4.2. *Distributions of radionuclides in paddy fields*

### 4.2.1. *$^{137}Cs$, $^{237}Np$ and $^{239,240}Pu$*

Yamamoto et al. (1994) measured the global fallout of $^{137}Cs$, $^{237}Np$ and $^{239,240}Pu$ in paddy soils collected periodically from 1959 to 1989 in Japan. The $^{137}Cs$, $^{237}Np$ and $^{239,240}Pu$ activity concentrations ranged from 12 to 150 $Bq\,kg^{-1}$, 0.39 to 4.55 $mBq\,kg^{-1}$ and 0.20 to 1.41 $Bq\,kg^{-1}$, respectively, on a dry soil basis. The highest values for each of the three radionuclides were observed in 1967 at each sampling site.

Figure 6 shows the activity concentrations of these radionuclides in paddy soils in Akita Prefecture as a function of time. Most of the radionuclides derived from global fallout were still retained within the surface soils over 25 years later. This result is very interesting, be-

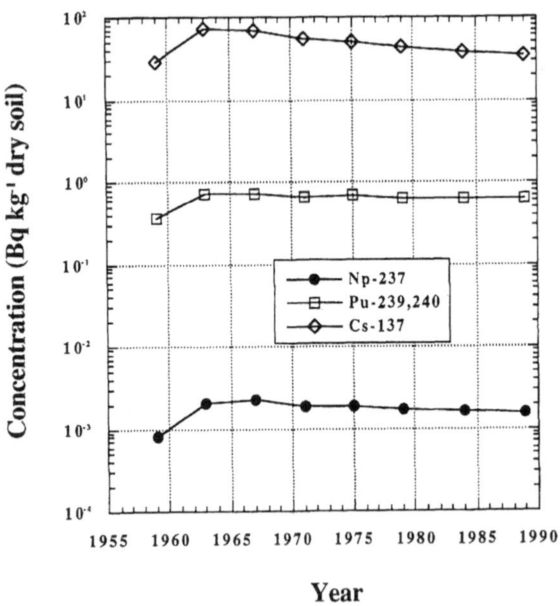


Fig. 6. Concentrations of fallout $^{237}Np$, $^{239,240}Pu$ and $^{137}Cs$ in paddy fields in Akita Prefecture, Japan (Yamamoto et al., 1994; reproduced with permission).

cause, as described above, paddy fields have a greater infiltration rate of percolating water than upland or uncultivated fields. However, in spite of this, these radionuclides have remained tightly sorbed on the paddy soil solid phase for a long time period after atmospheric nuclear weapons testing ceased.

### 4.2.2. *Iodine*

In radiological safety assessments, the root uptake pathway to rice grains is likely to be the most important for $^{129}$I (physical half-life = $1.57 \times 10^7$ y). It is known that radioiodine deposited on farm land which is generally under oxidizing conditions without water saturation, is retained within the surface soil layer and hardly any iodine dissolves into the soil solution. By contrast, under flooded conditions, radioiodine is soluble in soil solutions and is possibly taken up by crops through their roots.

In order to clarify the behavior of iodine in two paddy soil samples (Andosol and Gray lowland soils) under flooded conditions, Muramatsu et al. (1991) carried out radiotracer experiments. Radioiodine ($^{125}I^-$) concentrations in the soil solutions are shown in Figure 7 as a function of time after it was added once to each soil sample at the start of the experiment (day 0). Most of the added $^{125}$I was easily adsorbed by the soil solid phase in the first growing stage and thus the solution concentrations in both soils are low. In spite of these initially high adsorption capacities of both soils, the adsorbed $^{125}$I was gradually desorbed from the soil solid phase into the soil solutions about one month after transplanting. For both soil samples, the highest $^{125}$I concentrations in the soil solutions were observed around the flowering pe-

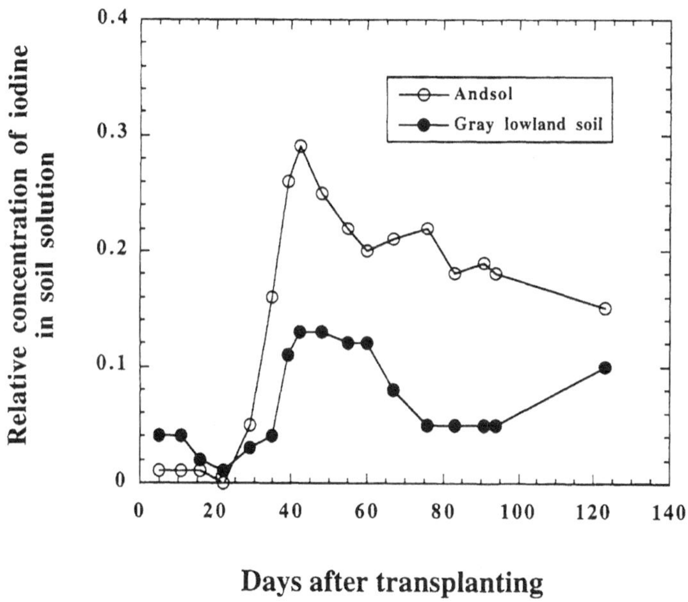


Fig. 7. Relative concentration of iodine in soil solution after transplanting rice plants as a function of time (Muramatsu et al., 1993). Relative concentration is defined as the activity of $^{125}$I present in the soil solution divided by the total activity added to the soil. Reproduced with permission of the Japan Radioisotope Association.

riod, 40 days after transplanting. At this stage about 30% and 13% of the $^{125}$I was present in the soil solutions of the flooded Andosol and Gray lowland soils, respectively. After reaching these maximum values, the concentrations in the solutions gradually decreased. On the basis of these root uptake experiments, it was suggested that the iodine concentrations in rice plants could be influenced by those in the soil solution.

### 4.2.3. *Technetium*

Technetium-99, a pure $\beta$ emitting radionuclide, is an artificial isotope which is accumulating in the environment due to its long physical half-life of $2.1 \times 10^5$ y. From the viewpoint of radioecology, $^{99}$Tc is another important radionuclide for long-term dose assessment because it is thought to be highly mobile in the soil environment. The behavior of $^{99}$Tc in paddy soils is interesting, because the chemical form of Tc is strongly affected by soil redox conditions: the most stable chemical form in aerobic surface soils is pertechnetate, $TcO_4^-$, but, under reducing conditions $TcO_4^-$ is transformed to less mobile forms such as $TcO_2$.

Table 2 lists activity concentrations of $^{99}$Tc and $^{137}$Cs in paddy soils (Tagami and Uchida, 1997). The $^{99}$Tc and $^{137}$Cs activity concentrations range from 0.022 to 0.11 Bq kg$^{-1}$ dry and 4.9 to 28 Bq kg$^{-1}$ dry, respectively. The activity ratios of $^{99}$Tc/$^{137}$Cs, shown in the last column of Table 2, range from $2.0 \times 10^{-3}$ to $5.8 \times 10^{-3}$. The activity ratio of $^{99}$Tc/$^{137}$Cs from fission is calculated to be $3.0 \times 10^{-4}$, with correction for physical decay, which is one order of magnitude lower than the measured ratios in the paddy soil samples.

The mechanisms of Tc accumulation in paddy fields can be explained by the change in the chemical form of Tc which is known to occur in flooded soils. Generally, during the planting period, the paddy field soils are flooded and subsequently the redox potentials decrease, so that relatively low redox conditions (sometimes less than 0 mV) are produced in the soils. Although $^{99}$Tc is expected to be in a soluble form, $TcO_4^-$, under aerobic conditions as in surface soil layers, this changes through a combination of factors such as the redox condition and microbial activity (Tagami and Uchida, 1996a, 1996b, 1999). Tc is probably transformed from its soluble $TcO_4^-$ form to less soluble forms such as $TcO_2$, $TcO(OH)_2$, $Tc_2S_7$ or $TcS_2$ under relatively low redox conditions (Lieser and Bausher, 1987; Brookins, 1988). The sulfides may be formed because $H_2S$ gas can be produced in water-

Table 2
Activity concentrations of $^{99}$Tc and $^{137}$Cs in Japanese paddy field soil samples and activity ratios of $^{99}$Tc to $^{137}$Cs (dry weight basis) (with permission from Elsevier)

| Sampling location | $^{137}$Cs (Bq/kg) | $^{99}$Tc (mBq/kg) | Activity ratio $^{99}$Tc/$^{137}$Cs |
|---|---|---|---|
| Ogata Village, Akita Pref. | $28.2 \pm 0.8$ | $110 \pm 30$ | $3.9 \times 10^{-3} \pm 1.1 \times 10^{-3}$ |
| Omagari City, Akita Pref. | $16.9 \pm 0.6$ | $34 \pm 5$ | $2.0 \times 10^{-3} \pm 3.0 \times 10^{-4}$ |
| Morioka City, Iwate Pref. | $10.1 \pm 0.6$ | $52 \pm 10$ | $5.1 \times 10^{-3} \pm 1.0 \times 10^{-3}$ |
| Koriyama City, Fukushima Pref. | $5.0 \pm 0.5$ | $29 \pm 3.6$ | $5.8 \times 10^{-3} \pm 9.5 \times 10^{-4}$ |
| Kawazoe Town, Saga Pref. | $4.9 \pm 0.5$ | $22 \pm 3$ | $4.5 \times 10^{-3} \pm 7.6 \times 10^{-4}$ |
| Imari City, Saga Pref. | $16.8 \pm 0.7$ | $88 \pm 15$ | $5.2 \times 10^{-3} \pm 9.2 \times 10^{-4}$ |

$\pm$: statistical errors in calculation (1 sigma).

logged soil during the growth of rice plants. These Tc forms have low mobility and can possibly be sorbed onto the soil solid phase much more readily than the anionic $TcO_4^-$.

### 4.3. *TFs of radionuclides from paddy field soil to rice*

In this section, TFs of several radionuclides and stable elements from paddy field soil to rice plants are discussed. In the case of rice plants, the TF may have been determined for unhulled rice, brown rice (hulled rice) or polished rice. When using a TF for rice grain it is preferable to state which of these forms applies as their constitutions differ: some reports do note which form has been measured.

#### 4.3.1. *$^{90}Sr$ and $^{137}Cs$*

There are several reports presenting TFs for Sr and Cs from soil to rice plants and these data are summarized in Table 3.

Komamura et al. (1996) reported the activity concentrations of global fallout $^{90}Sr$ and $^{137}Cs$ (measured using radiochemical methods) in paddy soil and brown rice samples collected from various parts of Japan. These authors also reported the concentrations of stable Sr ($Sr_{STAB}$, comprising 82% $^{88}Sr$, 10% $^{86}Sr$, 7% $^{87}Sr$ and 0.6% $^{84}Sr$) and Cs ($Cs_{STAB}$, comprising 100% $^{133}Cs$) in the same samples which were measured by inductively coupled plasma mass spectrometry (ICPMS). The average TF values were as follows: $^{90}Sr$, 0.0048; $Sr_{STAB}$, 0.00074; $^{137}Cs$, 0.0026; $Cs_{STAB}$, 0.00078. Tsukada et al. (2002a) also obtained similar TFs for both $^{137}Cs$ and stable Cs. It was found that the TFs of fallout $^{90}Sr$ and $^{137}Cs$ were 3–6 times higher than those of stable Sr and Cs. The results suggested that physico-chemical forms of global

Table 3
Reported TF values for Sr and Cs in rice

| Organization | Parts of crops | Nuclides | |
|---|---|---|---|
| | | Strontium | Cesium |
| Japan, NIAS (Komamura et al., 1996) | Polished rice (fallout nuclides) | $4.8 \times 10^{-3} \pm 3.4 \times 10^{-3}$ | $2.6 \times 10^{-3} \pm 2.8 \times 10^{-3}$ |
| | Polished rice (stable elements) | $7.4 \times 10^{-4} \pm 2.7 \times 10^{-4}$ | $7.8 \times 10^{-4} \pm 6.9 \times 10^{-4}$ |
| Japan, IES (Tsukada et al., 2002a) | Polished rice (fallout nuclides) | | $1.6 \times 10^{-3}$ ($2.1 \times 10^{-4}$–$1.2 \times 10^{-2}$) |
| | Polished rice (stable elements) | | $5.6 \times 10^{-4}$ ($1.1 \times 10^{-4}$–$2.8 \times 10^{-3}$) |
| Taiwan, RMC (Wang et al., 1998) | Polished rice (fallout nuclides) | $3.7 \times 10^{-2}$ (0.005–0.088) | $9.7 \times 10^{-2}$ (0.03–0.188) |
| | Brown rice (fallout nuclides) | $8.6 \times 10^{-2}$ (0.027–0.222) | $1.4 \times 10^{-1}$ (0.07–0.27) |
| Bangladesh, INST (Mollah et al., 1998) | Rice (pot experiment) | 0.78–0.89 | 0.25–0.33 |

fallout radionuclides and stable isotopes of both Cs and Sr are slightly different from each other.

As listed in Table 3, the TFs for both Sr and Cs are quite low, being less than 1.0 in several countries classified as (sub-)tropical regions. One of the reasons for these low TF values for rice grains is that both elements do not distribute uniformly in the rice plant. For Sr, only 0.7 and 1.1% of $^{90}Sr$ in the whole rice plant were found in polished rice and brown rice, respectively, whereas, 5.3% and 7.4% of the total $^{137}Cs$ were observed in polished rice and brown rice, respectively (Wang et al., 1998). Tsukada et al. (2002b) reported that 7 and 10% of stable Cs in a rice plant were distributed in polished rice and bran, respectively. These values were slightly higher than those obtained by Wang et al. (1998), but only marginally so.

From these studies, it can be concluded that: (a) distributions of both Cs and Sr in polished rice are smaller than those in other rice plant components and (b) the distribution ratio of Cs in polished rice is higher than that of Sr. These distribution characteristics of Cs and Sr in rice plants were derived from both the low and relatively close TF values of Cs and Sr for polished rice.

#### **4.3.2.** *Iodine*

In an experiment in which rice plants were grown on Andosol and Gray lowland soil samples, the TFs of radioiodine from the soil samples to brown rice were determined to be 0.007 and 0.002, respectively (Muramatsu et al., 1993). Since more $^{125}I$ was distributed in rice bran than in polished rice, it is reasonable to conclude that the amount of $^{125}I$ in polished rice will change with the degree of polishing. The weight ratio of brown rice to polished rice is usually about 1.0 : 0.9, which means that 10% of the grain surface layer, a part of the rice bran, is removed from brown rice during the polishing step. At this weight ratio, about 30% of the radioiodine remained in the polished rice (Muramatsu et al., 1989). Using this value, the TFs of radioiodine from the soils to polished rice were calculated to be about 0.002 for the Andosol and 0.0007 for the Gray lowland soil.

Takagi et al. (1985) analyzed stable iodine concentrations in paddy soil and rice samples collected in Japan to estimate the TFs for brown rice and polished rice. The TFs for polished rice ranged from 0.0001 to 0.008 and the average value was about 0.003, while the average TF for brown rice was about 0.007. Although the varieties of rice plant, soil types and experimental conditions were different from the radiotracer experiments carried out by Muramatsu et al. (1993), the TFs obtained in these studies were almost the same.

As described above, iodine sorbed on the soil solid phase under aerobic conditions is readily dissolved to the soil solution following flooding. Despite this, however, the TFs of iodine for brown rice and polished rice are extremely small. One explanation for this is that iodine is mainly distributed in the leaves of rice plants and translocation into the rice grains is limited.

#### **4.3.3.** *Technetium*

Yanagisawa et al. (1992) determined TFs for Tc uptake from soil to rice and wheat plants in laboratory radiotracer experiments. The TFs for rice plants, grown under typical paddy field conditions, were less than 0.005 for the hulled rice and about 1.1 for the lower leaf blade. For the wheat plants, which were grown under aerobic conditions, higher TFs of 0.027 for the hulled wheat and 230 for the lower leaf blade were found. The Tc activity concentrations in the grains of both plants were much lower than those in their leaves and TFs for Tc in the rice plants were lower than those for the wheat plants. The activity concentration of Tc in the soil solution collected from the flooded soil of rice pots decreased rapidly with time, while Tc in the soil solution of wheat pots decreased rather more slowly. It was suggested by Tagami and Uchida (1996b) that when a Tc tracer is added to a flooded paddy soil as $TcO_4^-$, Tc can easily be transformed to immobile forms due to the reducing conditions generated by microbial activity. Thus, the low TF observed for the rice plants could be explained by fixation of Tc in the soil in a chemically reduced form.

## 5. Radioecology of the Marshall Islands

### 5.1. *Introduction*

The Marshall Islands are located in the central Pacific, north of New Zealand (9°00N′, 168°00′E). Their soil is composed almost entirely of Ca-Mg-Sr carbonate since the soil parent material is coral. Due to the absence of clays, the cation exchange capacity of the soil is directly related to its organic matter content (Walker et al., 1997). Soil organic matter is derived largely from the steady contribution of litter from surrounding vegetation. In this unique ecosystem, Robison et al. (1997a) studied TFs for different plants as well as the spatial variation of TFs between atolls. The purpose of their work was to develop a fundamental understanding of $^{137}$Cs transfer from soil to tropical crops and fruits and of environmental processes controlling the mobility and availability of $^{137}$Cs in coral atoll soil.

Robison et al. also obtained TFs for $^{90}$Sr, Pu and $^{241}$Am. Interestingly, these TFs were not so high compared with those obtained by the CRP. This section focuses on the TFs for $^{137}$Cs in various tropical crops and fruits reported by Robison et al. (1997a). However, recently, it has been found that $^{99}$Tc still remains in the surface soils in the Marshall Islands and it appears that Tc, which was deposited on the surface soils more than 30 years ago, has sorbed strongly onto the soils in the tropical environment. This result is also described briefly below.

### 5.2. *TFs for $^{137}$Cs for various tropical crops*

TFs for $^{137}$Cs for various tropical crops and fruits from Bikini and Rongelap Islands are summarized in Table 4. Coconut palms (*Cocos nucifera*) are the most common tree on the Marshall Islands. Breadfruit (*Actocarpus altilis*), Pandanus (*Pandanus spp.*) and papaya (*Carica papaya*) are other important native food crop plants and tropical fruits. Sorghum and Chinese cabbage, which are not common crops on the islands, were also cultivated in a research field on the atolls to compare TFs obtained in other tropical areas as an activity of the IAEA/FAO CRP. The TFs range from a minimum of 0.3 for banana meat to a maximum of 144 for Chinese cabbage. Notably, the TFs for $^{137}$Cs are generally much higher than those reported for temperate areas where values typically range between 0.005 and 0.5 (IAEA, 1994).

As described above, the high TFs are attributed to the absence of clay minerals and the low concentration of available K in the soil. It is well known that the migration and availability of $^{137}$Cs for root uptake are strongly coupled with those of K (Anderson, 1988; Shaw and Bell, 1991). There was a possibility that salt sea spray may contribute to the source of available K in atoll soils and therefore influence $^{137}$Cs uptake into plants (Robison and

Table 4
TF values of $^{137}Cs$ in tropical crops and fruits in the Marshall Islands (Robison et al., 1997a; credit must be given to the University of California, Lawrence Livermore National Laboratory, and the Department of Energy under whose auspices the work was performed, when this information or a reproduction of it is used)

| Product | Location | TFs of $^{137}Cs$ | | |
|---|---|---|---|---|
| | | Median | Maximum | Minimum |
| Drinking coconut meat | Bikini Island | 13 | 64 | 2.8 |
| | Rongelap Island | 1.5 | 18 | 1.4 |
| *Pandanus* | Bikini Island | 35 | 82 | 11 |
| | Rongelap Island | 9.4 | 44 | 0.94 |
| Breadfruit meat | Bikini Island[a] | 2.9 | 7.4 | 0.44 |
| | Rongelap Island | 3.5 | 13 | 1.0 |
| Papaya meat | Bikini Island | 11 | 62 | 2.1 |
| Banana meat | Bikini Island | 0.78 | 1.2 | 0.27 |
| Coconut fronds | Bikini Island | 3.2 | 22 | 0.5 |
| Sorghum | Bikini Island | 8.1 | 24 | 0.81 |
| Chinese cabbage | Bikini Island | 36 | 144 | 2.9 |

Notes: TF values are expressed as $Bq\,kg^{-1}$ dry weight plant to $Bq\,kg^{-1}$ dry weight soil (see section 4).
[a]Breadfruit trees were transplants from Eneu Island.

Stone, 1992). Similarly, the availability of $^{137}Cs$ derived from fallout close to the detonation site might change over space and time, depending on the size, geochemical behavior, and aging of contaminated particles in the soil.

Cesium-137 uptake ratios in drinking coconut meat and Pandanus on Bikini Island were also compared with the ratios from Rongelap Island. TFs on Bikini Island were considerably higher that those on Rongelap Island (Table 4). Hence, there appeared to be a large coefficient of variability associated with TFs for $^{137}Cs$ in coral atoll soils of the Marshall Islands. This variability is present despite apparent similarities in the type, nature and development of the soil. Furthermore, the $^{137}Cs$ deposited on Bikini, Rongelap and Rongerik Atolls was derived largely from the BRAVO source term, and yet very significant differences in $^{137}Cs$ uptake in the same plant species were found.

### 5.3. *Technetium-99 concentrations in the surface soil in the Marshall Islands*

It is interesting to note that the $^{99}Tc$ activity concentrations in the three soil samples taken from surface soil on Bikini Island ranged from 0.73 to 1.11 $Bq\,kg^{-1}$, which are one or two orders of magnitude higher than in Japanese soils (Tagami et al., 2000). Using a reference date of 1954 the activity ratios of $^{99}Tc/^{137}Cs$ in Bikini surface soil ranged from $(0.7\text{–}1.1) \times 10^{-4}$ or around 50–70% of the theoretical ratio of $1.4 \times 10^{-4}$. Based on these results, it can be stated that persistent levels of $^{99}Tc$ are being retained in surface soil of the atoll, and that removal rates of $^{99}Tc$ and $^{137}Cs$ in surface soil are occurring on similar time-scales.

One soil sample was collected from Rongelap Atoll about 150 km east of Bikini. As was expected, the measured $^{99}Tc/^{137}Cs$ activity ratio in this sample was very similar to those values observed on Bikini Atoll, and could presumably be related back to the BRAVO source. Much higher $^{99}Tc/^{137}Cs$ activity ratios (range $(5.6–9.9) \times 10^{-4}$) were observed in soil samples collected from Enewetak Atoll. One plausible explanation may be related to the fact that some surface soil was removed during cleanup operations on Enewetak Atoll during the 1970's. $^{99}Tc$ was probably preferentially transported into deeper layers of soil and the surface soil samples collected during the 1990's are not truly representative of the relative total surface deposits of the two isotopes. Further work is expected to explain Tc cycling in the coral soil, rates of transport, and TFs into native plants.

## 6. Conclusions

The assessment of radionuclides for (sub-)tropical areas is probably different from that for temperate areas because agricultural products and dietary habits in the (sub-)tropical areas are different from those in temperate ones. For instance, livestock products including meat, eggs and milk make a major contribution to human diet in European and North American countries, whereas agricultural products, including cereals and vegetables are the main contributors to diet in Asian and South American countries. Consequently, radiological safety assessments for people who live in (sub-)tropical areas must take into account cereals and leafy vegetables as critical foodstuffs. In this chapter, therefore, soil to crop transfer factors (TFs) for radionuclides which were obtained by the CRP on 'transfer parameters in (sub-)tropical regions' by the IAEA and FAO in cooperation with the IUR, have been described.

The TFs for Cs and Sr in vegetables were the highest among the four crop groups reviewed and the TFs for Sr were generally higher than those for Cs. When CRP data in (sub-)tropical environments were compared with IUR data determined in temperate environments, the mean values agreed closely. Although there were no systematic differences in TFs between temperate and (sub-)tropical areas, the CRP results indicated that the deviation of TFs from average values was greater in the (sub-)tropical environments than in temperate zones.

Two unique cases within the broad category of (sub-)tropical environments, the paddy field ecosystem and the Marshall Island ecosystem, have been reviewed. Because rice plants grow under flooded soil conditions, the redox potential decreases in the paddy field ecosystem and it has been observed that the behavior of elements such as iodine and technetium are different under these conditions than in upland field systems. These differences appear to affect retention times in paddy fields and TFs from the soil to rice.

The observed TFs for $^{137}Cs$ in vegetation in the Marshall Islands were very high and up to 95% of the estimated radiation dose to humans via ingestion in the area is contributed by $^{137}Cs$ (Robison et al., 1997b). However, further study is clearly warranted if we are to understand fully the behavior of radionuclides in the Marshall Islands.

## Acknowledgements

The author would like to thank Dr. M.J. Frissel (the Netherlands) and Dr. K. Tagami (National Institute of Radiological Sciences, Japan) for their useful discussions and comments.

## References

Anderson, I. (1988). Potassium could cover up Bikini' s radioactivity. *New Scientist* **1642**, 17.

Brookins, D.G. (1988). In: *Eh-pH Diagrams for Geochemistry*. Springer-Verlag, Berlin, pp. 97–99.

Frissel, M.J. (1997). Soil to plant transfer of radionuclides in (sub-)tropical areas. Results of an IAEA-IUR co-ordinated research programme. In: *Proceedings of International Meeting on Influence of Climatic Characteristics upon Behavior of Radioactive Elements*, pp. 130–137. Rokkasho, Aomori, Japan, October 14–16.

Frissel, M.J., Deb, D.L., Fathony, M., Lin, Y.M., Mollah, A.S., Ngo, N.T., Othman, I., Robison, W.L., Skarlou-Alexiou, V., Topcuoglu, S., Twining, J.R., Uchida, S., Wasserman, W.A. (2002). Generic values for soil-to-plant transfer factors of radiocesium. *J. Environ. Radioact.* **58**, 113–128.

International Atomic Energy Agency (1994). Handbook of Parameters Values for The Prediction of Radionuclide Transfer in Temperate Environments. International Union of Radiologists, IAEA, Vienna. Technical Report Series No. 364.

Karube, J., Hasegawa, S., Toyomitsu, Y., Adachi, K., Tada, A. (1995). Basic aspects of paddy fields. In: Tabuchi, T., Hasegawa, S. (Eds.), *Paddy Fields in The World*. The Japanese society of irrigation, draining and reclamation engineering, Tokyo, pp. 246–268.

Komamura, M., Tsumura, A., Kodaira, K., Yuita, K., Yamasaki, S. (1996). Transfer factors of radionuclides from paddy soils to polished rice. In: Frissel, M.J., Brown, R.M., Uchida, S. (Eds.), *Proceedings of The Improvement of Environmental Transfer Models and Parameters*, pp. 155–164.

Lieser, K.H., Bausher, C. (1987). Technetium in the hydrosphere and in the geosphere 1. Chemistry of technetium and iron in natural waters and influence of redox potential on the sorption of technetium. *Radiochim. Act.* **42**, 205–213.

Ministry of Agriculture, Forestry and Fisheries; Division of Research, Secretariat of Minister (1988). *Shokuryo Jyukyu Hyo* (Tables of Food Supply and Demand). Nohrintokei Kyokai Press, Tokyo (in Japanese).

Mollah, A.S., Begum, A., Ullah, S.M. (1998). Determination of soil-to-plant transfer factors of $^{137}Cs$ and $^{90}Sr$ in the tropical environment of Bangladesh. *Radiat. Environ. Biophys.* **37**, 125–128.

Muramatsu, Y., Uchida, S., Sumiya, M., Ohmomo, Y., Obata, H. (1989). Tracer experiments on transfer of radio-iodine in the soil-rice plant system. *Water, Air, and Soil Pollution* **45**, 157–171.

Muramatsu, Y., Uchida, S., Yoshida, S. (1991). Radiotracer experiments on the desorption of iodine from paddy soil with and without rice plants. *Radioisotopes* **40**, 440–443.

Muramatsu, Y., Uchida, S., Ohmomo, Y. (1993). Root-uptake of radioiodine by rice plants. *J. Radiat. Res.* **34**, 214–220.

Robison, W.L., Stone, E.L. (1992). The effect of potassium on the uptake of $^{137}Cs$ in food crops grown on coral soils: coconut at Bikini Atoll. *Health Physics* **62** (6), 496–511.

Robison, W., Conrado, C., Hamilton, T. (1997a). A comparative study on $^{137}Cs$ transfer from soil to vegetation in the Marshall Islands. In: *Proceedings of International Meeting on Influence of Climatic Characteristics upon Behavior of Radioactive Elements*, pp. 122–129. Rokkasho, Aomori, Japan, October 14–16, 1997.

Robison, W., Bogen, K., Conrado, C. (1997b). An updated dose assessment for resettlement options at Bikini Atoll-U.S. nuclear test site. *Health Phys.* **73**, 100–114.

Shaw, G., Bell, J.N.B. (1991). Competitive effects of potassium and ammonium on caesium uptake kinetics in wheat. *J. Environ. Radioactivity* **13**, 283–296.

Simon, S., Graham, J. (1997). Findings of the first comprehensive radiological monitoring program of the Republic of the Marshall Islands. *Health Phys.* **73**, 66–85.

Tagami, K., Uchida, S. (1996a). Microbial role in immobilization of technetium in soil under waterlogged conditions. *Chemosphere* **33**, 217–225.

Tagami, K., Uchida, S. (1996b). Aging effect on technetium behaviour in soil under aerobic and anaerobic conditions. *Toxicol. Environ. Chem.* **56**, 235–247.

Tagami, K., Uchida, S. (1997). Concentration of global fallout Tc-99 in rice paddy soils collected in Japan. *Environ. Pollut.* **95**, 151–154.

Tagami, K., Uchida, S. (1999). Chemical transformation of technetium in soil during the change of soil water conditions. *Chemosphere* **38**, 963–971.

Tagami, K., Uchida, S., Hamilton, T., Robison, W. (2000). Measurement of technetium-99 in Marshall Islands soil samples by ICP-MS. *Applid Radiat. Isotop.* **53**, 75–79.

Takagi, H., Kimura, T., Kobayashi, H., Iwashima, K., Yamagata, N. (1985). Transfer of iodine from paddy soil to rice grain. *Hoken Butsuri* **20**, 251–257 (in Japanese).

Tsukada, H., Hasegawa, H., Hisamatsu, S., Yamasaki, S. (2002a). Transfer of $^{137}Cs$ and Stable Cs from paddy soil to polished rice in Aomori, Japan. *J. Environ. Radioact.* **59**, 351–363.

Tsukada, H., Hasegawa, H., Hisamatsu, S., Yamasaki, S. (2002b). Rice uptake and distribution of radioactive $^{137}Cs$, stable $^{133}Cs$ and K from soil. *Environ. Poll.* **117**, 403–409.

Uchida, S., Okabayashi, H. (1988). Transfer Factors. In: Saiki, M., Ohmomo, Y. (Eds.), *Transfer Factors of Radionuclides from Soils to Agricultural Products. Environmental Parameters Series 1 (RWMC-88-P-11).* Radioactive Waste Management Center, Tokyo (in Japanese).

Uchida, S., Tsukada, H. (2004). Radioecology of Rice. Task Group of the International Union of Radioecology. http://www.iur-uir.org/taskgroups.cgi?id=14.

Ueda, T., Suzuki, Y., Nakamura, R. (1974). Transfer of Cs-137 and Sr-90 from the environment to the Japanese population via marine organisms. In: *Proceedings of the Seminar on Radiological Safety Evaluation of Population Doses and Application of Radiological Safety Standards to Man and the Environment.* IAEA, Vienna, pp. 501–511. IAEA-SM-184/7.

UNSCEAR (United Nations Scientific Committee on the Effects of Atomic Radiation) (1982). *Source and Effects of Ionizing Radiation.* United Nations, New York. United Nations Publication.

Walker, R., Gessel, S., Held, E. (1997). The ecosystem study on Rongelap Atoll. *Health Phys.* **73**, 223–233.

Wang, J.J., Wang, C.J., Huang, C.C., Lin, Y.M. (1998). Transfer factors of $^{90}Sr$ and $^{137}Cs$ from paddy soil to the rice plant in Taiwan. *J. Environ. Radioact.* **39**, 23–34.

Watanabe, T. (1992). Water budget in paddy field lots. In: Murty, V.V.N., Koga, K. (Eds.), *Proceedings of The International Workshop on Soil and Water Engineering for Paddy Field Management, Bangkok, Thailand*, pp. 1–12.

Yamamoto, M., Kofuji, H., Tsumura, A., Yamasaki, S., Yuita, K., Komamura, M., Komura, K., Ueno, K. (1994). Temporal feature of global fallout $^{237}Np$ deposition in paddy field through the measurement of low-level $^{237}Np$ by high resolution ICP-MS. *Radiochim. Acta* **64**, 217–224.

Yanagisawa, K., Muramatsu, Y., Kamata, H. (1992). Tracer experiments on the transfer of technetium from soil to rice and wheat plants. *Radioisotopes* **41**, 397–402.

# Naturally occurring radionuclides from industrial sources: characteristics and fate in the environment

Edward R. Landa

*US Geological Survey, 430 National Center, Reston, Virginia 20192, USA*

## 1. Introduction

Technologically enhanced naturally occurring radioactive material (TENORM) refers to earth materials (including ores, their processing products and waste by-products; water; natural gas) in which human activities have increased, in comparison to the unaltered state, the concentration of naturally occurring radionuclides, and/or enhanced the potential for radiation exposure to humans or biota. The human activities may include extraction (e.g., mining of ores, drilling of oil, gas and water wells), physical processing (e.g., grinding, particle size separation, aeration, temperature and pressure changes) and chemical processing (e.g., acid treatment) of the materials. In this chapter, we will focus on radionuclides of the uranium (U) and thorium (Th) decay series (Figure 1).

TENORM is typically associated with non-nuclear industries. Uranium milling tailings (UMT) are a component of the nuclear fuel cycle, and have generally been considered distinct from TENORM. This is solely a regulatory distinction. There is a large body of knowledge on this form of TENORM that has recently been reviewed by the author (Landa, 2004), and issues germane to other forms of TENORM will be discussed here. This chapter will focus on geochemical factors that concentrate radionuclides in certain media; industrial processes that partition these nuclides to various fractions; waste disposal and management; and environmental processes that influence radionuclide mobility. It will not cover areas such as regulatory approaches, sampling and radioanalytical methods, and pathway modeling to evaluate radiation dose and potential health effects in occupationally exposed individuals or the general population. Minimally processed materials, such as building stones and mineral water, and indoor radon (Rn) will not be covered. For additional information on specific forms of TENORM, including their geographic distributions and typical radionuclide concentrations, and coverage of industrial activities not reported in detail here (e.g., clay and ceramics production, sulfuric acid production from pyrites, production of specific metals such as tin, aluminum, iron, copper, gold, titanium, niobium and rare earth elements), the reader is referred to the recent overview by the International Atomic Energy Agency (2003).

RADIOACTIVITY IN THE ENVIRONMENT
VOLUME 10 ISSN 1569-4860/DOI: 10.1016/S1569-4860(06)10010-8

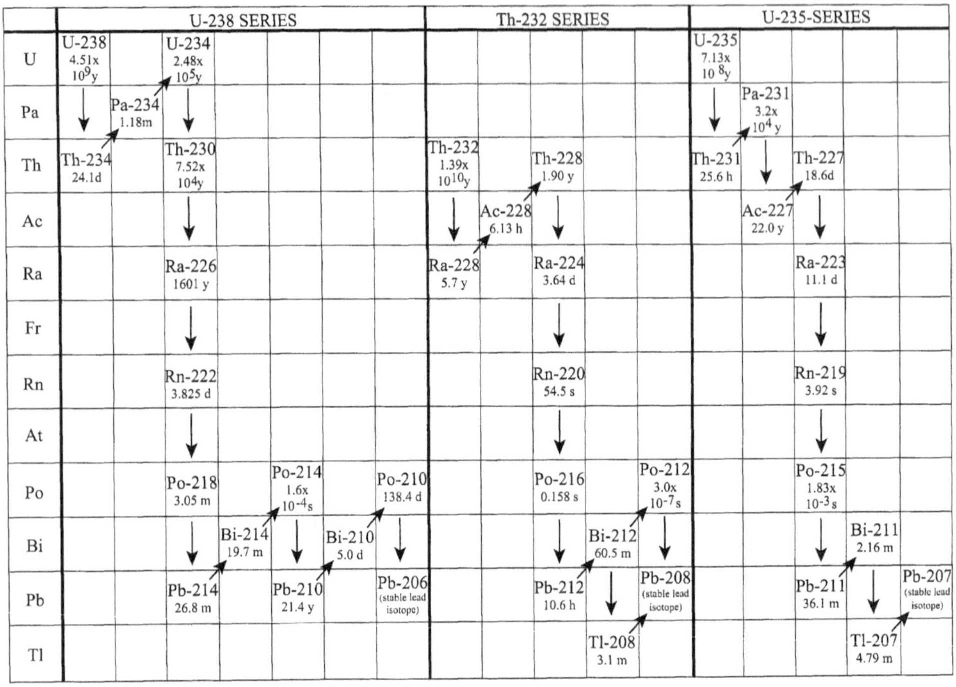


Fig. 1. The uranium ($^{238}U$), thorium ($^{232}Th$), and actinium ($^{235}U$) decay series. The arrows show the mode of decay (vertical arrow = alpha particle emission; 45°-angle arrow = beta particle emission). Half-lives are given for each nuclide.

TENORM is coming under increasing regulatory scrutiny by national and international bodies (e.g., Scholten and van der Steen, 1998; National Academy of Sciences [U.S.A.], 1999). In this chapter, we will explore similarities and differences across various classes of TENORM. These differences are critical in knowing when caution must be exercised in extrapolating regulatory guidance from one TENORM class to another. The breadth of materials identified by investigators to-date as having elevated concentrations of U and Th series activity is impressive. Some of these, such as $^{210}Pb$ deposits in natural gas refinery equipment and $^{226}Ra$ scales in paper pulp processing equipment and domestic water supply lines, represent unexpected occurrences to many in the radiation protection community. With increasing awareness of the issue and greater screening and surveillance, other novel occurrences are likely to be uncovered in the future.

## 2. Geochemical behavior of naturally occurring radionuclides in the lithosphere

The geochemical concepts of partitioning of elements during rock formation, and of cycling of elements between rock, water and air in near-surface environments are critical to understanding the occurrence in the lithosphere, and movement of U, Th, and their daughter products in the biosphere. Fractional crystallization of magmas results in the formation of igneous rocks of differing mineralogies and compositions. U and Th atoms, due to size and charge, tend to be excluded from the structures of minerals formed in the early stages of magma cooling and differentiation. The concentration of these excluded elements tends to increase in the remaining magma. Late-crystallizing minerals thus tend to be high in U and Th. U and Th can also be concentrated by sedimentary processes, such as groundwater mobilization and redeposition, and accumulation of organic materials (Bliss, 1978). Where U and Th concentrations are high enough, these rocks represent potentially exploitable ores to industrial societies.

In the western United States, there has been extensive mining and milling of U for the production of nuclear fuels. The need to regulate health and safety aspects of U mining and of UMT disposal was recognized during the 1950s–1970s, and addressed in federal legislation such as the Atomic Energy Act of 1954 and the Uranium Mill Tailings Radiation Control Act of 1978. The extraction and processing of other rocks and geo-fluids with U and Th (and/or decay product) enrichments below ore-grade, but above ambient-concentrations, is the source of the other group of TENORM—a group with less well-defined regulatory guidelines and boundaries. Industrial activities may, in some cases, concentrate certain radionuclides to levels above those of the raw materials, and above those of already regulated waste materials.

The geochemical behavior of a radionuclide in a given decay-series can be expected to vary with the atomic number (i.e., the element) of the parent and daughter products. For example, in the U decay series, nuclides of U, Th, radium (Ra), Rn, and other elements occur. The chemical properties of these radioelements range from an inert gas (Rn) to a particle-reactive tetravalent cation (Th). These properties determine the fate of the radionuclides in fuel and mineral-processing flowsheets, their subsequent transport in surficial disposal environments and, ultimately, their biological availability and uptake. Thus, knowledge of these properties and their impact on radionuclide partitioning in industrial and environmental processes is essential to defining source terms and assessing dose.

## 3. Form, fate and transport of radionuclides in the biosphere

Regulations for controlling exposure of the public to radionuclides are generally dose-based. As these doses result from interaction of humans with radionuclides contained in environmental media—air, water, soil, biota—a knowledge of the behavior of naturally occurring radionuclides in these media is needed. It is important to know:

- the concentrations and physical/chemical forms of the radionuclides in a specific class of TENORM,
- the differential mobilities of the various radionuclides in a decay chain in the TENORM environment,
- how technological processes have changed the chemical form of nuclides, release rates of radionuclides to air, water and biota, and the impact of these changes on potential human radiation exposure,
- how the TENORM will evolve with time (e.g., weathering reactions, biological community changes, radioactive decay).

### 3.1. *Uranium*

Oxidation/reduction processes play a major role in the occurrence and behavior of U. The dominant U valence states which are stable in geologic environments are the uranous ($U^{4+}$) and uranyl ($U^{6+}$) states, the former being far less soluble. U transport generally occurs in oxidizing surface and ground waters as the uranyl ion, $UO_2^{2+}$, or as uranyl fluoride-, phosphate-, or carbonate-complexes. The uranyl ion and uranyl fluoride complexes dominate in oxidizing, acidic waters, while the phosphate- and carbonate-complexes dominate in near-neutral and alkaline oxidizing waters, respectively. Hydroxyl-, silicate-, organic-, and sulfate-complexes may also be of importance, the latter especially in mining and milling operations using sulfuric acid as a leaching agent. Maximal sorption of uranyl ions on natural materials (organic matter, iron-, manganese- and titanium-oxyhydroxides, zeolites, and clays) occurs at pH 5.0 to 8.5. The sorption of uranyl ions by such natural media appears to be reversible, and for U to be "fixed" and thereby accumulate, it requires reduction to $U^{4+}$ by the substrate or by a mobile phase such as $H_2S$ (Landa, 1980).

### 3.2. *Thorium*

In aqueous systems, only the $Th^{4+}$ oxidation state is known to exist. The tetravalent ion undergoes hydrolysis in aqueous solutions above pH 2–3, and is subject to extensive sorption by clay minerals and humic acid at near neutral pH. In near-neutral pH and alkaline soils, precipitation of Th as a highly insoluble hydrated oxide phase and coprecipitation with hydrated ferric oxides may, in addition to adsorption reactions, be important mechanisms for the removal of Th from solution. Because of sorption and precipitation reactions, and the slow solution rate of Th-bearing minerals, Th levels in natural waters are generally low.

At low pH, such as in an acid-leach U mill, Th becomes more soluble. Acid-leach milling may dissolve from 30 to 90% of the Th in the ore. Acidic effluents (pH 2.5) from U mills in the Grants Mineral Belt of New Mexico contain from 5.6 to 6.3 $\mathrm{kBq\,l^{-1}}$ of $^{230}Th$. The solubilized Th can be subsequently precipitated if the acidic effluent is neutralized, either by contact with natural media or by process additions of limestone to the waste solutions. The high inventory of soluble $^{230}Th$ in such an effluent made it the radionuclide of greatest mobility when, in 1979, a dam at a UMT impoundment in New Mexico failed, sending effluent down an arroyo (Weimer et al., 1981). Similarly, under acidic conditions at some U mills, $^{230}Th$ has been shown to have migrated considerably deeper into the subsoil than $^{226}Ra$ (U.S. Department of Energy, 1993).

As organic complexes of Th have been shown to be of importance in natural water systems, the presence of organic complexing agents in U mill effluents has been offered as a partial explanation for the low distribution coefficients observed for $^{230}Th$ on subsoil materials equilibrated with tailings pond water from a Wyoming U mill (Landa, 1980).

### 3.3. *Radium*

Ra exhibits only the ($2^+$) oxidation state in solution, and its chemistry resembles that of barium. Ra forms water-soluble chloride, bromide, and nitrate salts. The phosphate, carbonate, selenate, fluoride, and oxalate salts of Ra are slightly soluble in water, while Ra sulfate is relatively insoluble in water [$K_{sp} = 4.25 \times 10^{-11}$ at 20°C]. Ra in U ore is only slightly soluble in $H_2SO_4$, but highly soluble in HCl and $HNO_3$, presumably because of the greater solubility of $RaCl_2$ and $Ra(NO_3)_2$, as compared to $RaSO_4$.

The hydrated ion of Ra is the smallest in the alkaline earth series, and hence it would tend to be preferentially retained by ion exchange. In alkaline solutions, anionic complexes of Ra with organic ligands, such as EDTA (ethylenediamine tetraacetic acid) and citric acid, are known to occur. Means et al. (1978) suggest that EDTA-mobilization may be responsible for apparently elevated levels of Ra seen in water and soil sampled around a radioactive waste disposal trench at the Oak Ridge National Laboratory burial ground.

Ra does not form discrete minerals, but can coprecipitate with many minerals including calcium carbonate, hydrous ferric oxides, and barite ($BaSO_4$). Ra may be sorbed by clay minerals, colloidal silicic acid, manganese oxides and organic matter. While, unlike U, only a single valence state exists for Ra, the dissolution or precipitation of sorbing phases such as barite and ferric hydrous oxides under changing oxidation/reduction conditions can influence the mobility of Ra. Ground waters low in sulfate, with high ionic strengths and high contents of calcium and barium, are conducive to the transport of Ra.

The water leaching data suggest that UMT in the environment may constitute a long-term source for Ra contamination of contacting surface and ground waters. The same is probably true of other TENORM wastes in which the $^{226}Ra$ is associated with sparingly soluble minerals such as $BaSO_4$. Indeed some tailings in Canada that have been exposed to weathering forces for several years apparently do show evidence of $^{226}Ra$ depletion (Landa, 1980).

## 4. Bioavailability

Much in the same way that nutrients locked up in unweathered rock fragments may not be available for plant uptake, the chemical forms of radionuclides in TENORM may greatly influence their biological availabilities. Total radionuclide concentrations in TENORM may not be reflective of the biologically labile pool, and some sub-segment of that total may be of greater value in assessing biological uptake and resultant dose. For example, in assessing the hazard due to inhalation of Rn decay products, Rn which does not escape the Ra-bearing mineral matrix is not a concern. The Rn emanation coefficient (REC) is the fraction of the Rn formed in a Ra bearing solid which escapes to the atmosphere. This can vary widely in natural and industrial materials; for example, while $^{226}Ra$ coprecipitated with barium sulfate generally emanates less than 1% of the $^{222}Rn$ generated, that coprecipitated with or sorbed on Fe hydrous oxide can emanate nearly 100% (Hahn, 1936). For TENORM with high emanation coefficients, Rn release and Rn decay product inhalation may be the pathway of concern. For TENORM with low emanation coefficients, whole body, gamma exposure may provide the limiting dose.

For the ingestion pathway, the solubility of the TENORM particles in bodily fluids (i.e., lung serum for inhaled particles, stomach acid) can exert a major influence on the uptake of radionuclides by humans. In-vivo solubility can be affected by particle size, position of nuclides in less accessible, particle interiors vs. those on particle surfaces (i.e., heat-volatilized $^{210}Pb$ and $^{210}Po$, condensed onto cooling particle surfaces in flue dust of thermal process phosphate plants, and in coal combustion fly ash), and dissolution kinetics. In soil and aquatic environments, radionuclide solubility may vary with pH, presence of complexing or precipitating ions in contacting solutions and redox conditions. Where biological uptake data are not available, leaching tests using dilute acids to assess readily dissolvable fractions, or salt solutions to assess ion-exchangeable fractions have been used to provide an index of biological availability. Madruga et al. (2001) have investigated the bioavailability of $^{226}Ra$ to plants growing on UMT. 1 M ammonium acetate or calcium chloride extracted about 15% of the total $^{226}Ra$ of the UMT. The use of exchangeable $^{226}Ra$ concentrations in the determination of soil-to-plant transfer factors will better enable plant availability comparisons to be made between different TENORM substrates.

Winde and van der Walt (2004) have noted that salt crusts on gold-uranium mill tailings in South Africa may be consumed by grazing cattle. These materials have U concentrations that are an order of magnitude higher than those in the underlying tailings. The crusts consist of moderately- and highly-soluble salts such as gypsum ($CaSO_4 \cdot 2H_2O$), epsomite ($MgSO_4 \cdot 7H_2O$) and goslarite ($ZnSO_4 \cdot 7H_2O$). Radionuclides and trace elements that have co-migrated upward to the surface with the bulk constituents of these minerals can be assumed to be soluble and bioavailable. In the United States, molybdenum poisoning in cattle grazing in the vicinity of a uraniferous-lignite ashing facility in North Dakota, a uranium ore storage area in Texas, and a uranium mill in Colorado have been reported. In the Texas case, these poisonings were attributed to "moly blooms", efflorescing salts enriched in molybdenum that occur on exposed faces of U-mine overburden (Landa, 1984).

## 5. Oil and natural gas production and processing

Oil and natural gas reservoirs commonly contain large quantities of accompanying water; this water is typically saline and tends to be high in chlorides and low in sulfates (due to sulfate reducing conditions in the subsurface environment). These production waters come to the surface during pumping operations and, after separation of the water from the oil and gas, require disposal. These waters tend to be low in U and Th (Table 1), but may contain elevated levels of $^{226,228}Ra$. The high Ra content of such production waters was first recognized around 1927

Table 1

Reported concentration ranges of uranium- and thorium-decay series radionuclides in crude oil, natural gas, production waters, pipe scales and sludges. Units as shown. "–" indicates not reported. Data from Jonkers et al. (1997) and written communication from G. Jonkers (2005); used with the permission of the author

| Radionuclide | Crude oil ($mBq\,g^{-1}$) | Natural gas ($Bq\,m^{-3}$) | Production water ($Bq\,l^{-1}$) | Pipe scale ($Bq\,g^{-1}$) | Sludges ($Bq\,g^{-1}$) |
|---|---|---|---|---|---|
| $^{238}U$ | 0.0001–10 | – | 0.0003–0.1 | 0.001–0.5 | 0.005–0.01 |
| $^{226}Ra$ | 0.1–40 | – | 0.002–1200 | 0.1–15,000 | 0.05–800 |
| $^{222}Rn$ | – | 5–200,000 | – | – | – |
| $^{210}Pb$ | – | 0.005–0.02 | 0.05–180 | 0.02–2000 | 0.1–1300 |
| $^{210}Po$ | 0–10 | 0.002–0.08 | – | 0.02–1.5 | 0.004–160 |
| $^{232}Th$ | 0.03–2 | – | 0.0003–0.001 | 0.001–0.07 | 0.002–0.01 |
| $^{228}Ra$ | – | – | 0.3–180 | 0.05–2800 | 0.5–50 |

during investigations in the Ukhta River basin of the Komi region of northwestern Russia by famed biogeochemist Vladimir Ivanovich Vernadsky and colleagues. Since the 1950s, soils in the same vicinity, with elevated Ra concentrations from long-term contact with similar waters (without accompanying oil), have been the focus of multifaceted radioecological studies by the Russian Academy of Sciences (Taskaev et al., 2003).

In the U.S., more than 90% of these production waters are disposed of by injection underground, sometimes in enhanced oil recovery wells, or sometimes solely with waste disposal as a goal. The remainder is disposed by surface discharge to earthen evaporation/seepage pits, or to streams or wetlands (Smith, 1992). Some production waters have been applied to roads for dust control on dirt roads (Rittiger and Yusko, 1996). At offshore wells on the Outer Continental Shelf of the Gulf of Mexico, overboard disposal of produced well solids (i.e., formation sand, TENORM-scale) is banned, but overboard disposal of production water (treated to remove such solids) is allowed (Minerals Management Service [U.S.], 1996a). Discharge of production water into the sea is typical practice in Brazil (Jerez Vegueria et al., 2002), Europe (Betti et al., 2004) and elsewhere. In the North Sea, pipe scale and sludge from offshore oil production platforms is discharged overboard (van Weers, 2002, Betti et al., 2004).

At some facilities in Texas, Ra has been removed from production waters by treatment with activated charcoal prepared from walnut hulls. This spent charcoal is thus rendered a solid TENORM waste (Ruth McBurney, Texas Department of Health, oral communication, 1993). In Pennsylvania, a brine treatment facility using pH adjustment and flocculation techniques to remove metals, yielded a sludge which was dewatered and sent to a landfill. The sludge, which contained about 925 $Bq\,kg^{-1}$ of $^{226}Ra$ and 850 $Bq\,kg^{-1}$ $^{228}Ra$, triggered portal radiation detectors at the landfill, and this initiated an investigation by the Pennsylvania Department of Environmental Protection (Rittiger and Yusko, 1996).

As the produced water flows through pipes at the oil field, temperatures tend to drop and solutes tend to precipitate, forming a scale consisting of barium sulfates (barite), strontium sulfate (celestite), calcium sulfate (gypsum), calcium carbonate, iron oxides, silica, and other compounds along the interior walls. Ra tends to coprecipitate with these solids, resulting in a radioactive scale. $^{226}Ra$ concentrations as high as 15,000 $Bq\,g^{-1}$ have been reported (Table 1), but typical concentrations are in the 4 to 370 $Bq\,g^{-1}$ range. Scale production tends to increase in older well fields, as the amount of water produced relative to oil tends to decline with time. Practices such as injection of water to enhance oil production can also result in increased scale, especially if such waters are high in soluble sulfates (Hartog et al., 1998).

Exposure scenarios associated with these scales include gamma-exposure of workers at oil production platforms and soil contamination at pipe-reaming facilities. Operations such as the latter are conducted to maintain flow at the oil wells, or to decontaminate pipes prior to recycling for other applications. Scale can be removed from pipes and other production equipment by mechanical methods including cutting and shearing, and high-pressure water-, sand- and cryogenic-carbon dioxide pellet-blasting (Lancée et al., 1997). Chemical decontamination methods using salts of amino carboxylic acids and proprietary reagents are available for the dissolution of scale and other surficial TENORM materials; Ra can be precipitated from these spent solutions and the solid concentrate disposed (Coll, 1997, Lancée et al., 1997). Mechanical cleaning can generate a considerable amount of dust. Work in progress at Texas A&M University is measuring airborne dust levels, particle size distributions, and areal distribution of ejected scale deposited at the pad site of a simulated "pipe rattling" operation (Berry and Cezeaux, 2004). Zielinski et al. (2001) have demonstrated how $^{228}Ra/^{226}Ra$ activity ratios can possibly be used to date and identify sources of the scale, an issue of forensic interest.

Assessing the potential release of Rn from pipe scale at soil contamination sites and at planned disposal sites requires a knowledge of the REC of the material. Pure barium sulfate (barite) has a very low REC, reported to be about 0.2 to 0.3% (Hahn, 1936, Wahl and Bonner, 1951). While often predominantly barite, oilfield pipe scale can have other components such as calcium carbonate and silica. Measured RECs of pipe scales and of soils contaminated by oil field sludge (see below) tend to be higher than the pure barite value, ranging from about 2 to 14% (Herbert et al., 1995, Rood et al., 1998, White and Rood, 2001).

Sludges are related deposits, typically found settled out on the bottoms of equipment and storage tanks at various points in the oil/gas/water separation processing stream (Table 1). The sludges are often oily, and disposal in burn pits was a common practice. Past disposal practices such as landspreading (to promote biodegradation of oil), and dilution with clean soil to radionuclide concentrations below regulatory control are becoming less common due to radiological and non-radiological concerns. Mohamad Puad and Muhd Noor (2004) have recently looked at radionuclide behavior during combustion of sludge at temperatures from 100 to 800°C, and for time periods from 1 to 2.5 hours. In laboratory tests with sludge in open crucibles that were heated in an oven, up to about 70% of the $^{238}U$, and about 20–25% of the $^{226,228}Ra$ and $^{232}Th$ were volatilized by 800°C combustion for 2.5 hours. In actual commercial incinerators, the volatile radionuclides are likely to condense onto particulates in the stack that are either emitted as fly ash or trapped by bag filters or other pollution control devices. The model developed by the authors here would appear to be applicable to other TENORM combustion scenarios, such as peat and coal-fired power plants.

$^{222}Rn$ is carried from the reservoir with the natural gas, and its decay products tend to plate out as films on the interior surfaces of pipes, valves and equipment in the gas plant (Gesell, 1975). While short-lived, gamma-emitting Rn decay products such as $^{214}Bi$ may pose an exposure hazard to plant workers, the longer-lived decay products $^{210}Pb$ and $^{210}Po$ are of concern in terms of environmental fate following disposal of scrap metal from such operations, as well as for occupational exposure during maintenance and repair on disassembled equipment (Summerlin and Prichard, 1985). Recent reports have shown that $^{210}Pb$ can be present in much higher activities than that derived from the $^{222}Rn$ present in the gas ("unsupported $^{210}Pb$"); the transport and deposition mechanisms for the lead have been discussed by Hartog et al. (1996) and Schmidt (1998). The deposition of metallic lead is thought to be an electrochemical process related to the corrosion of iron; $Pb^{2+}$ is reduced to $Pb^0$ as $Fe^0$ is oxidized to $Fe^{2+}$. In support of this proposed mechanism, it is of interest to note that Bland and Chiu (1996) found unsupported $^{210}Pb$ (32 $Bq\,g^{-1}$; with $^{210}Po$ in approximate equilibrium) to be associated with fine-grained rust particles that accumulated in the interior of liquified natural gas transport tanks.

In some cases where gas–water mixtures are present, several millimeters of lead-bearing minerals (oxides, carbonates, chlorides and sulfides; metallic lead; lead-bearing calcite) can build up on piping and equipment at natural gas production facilities (Schmidt, 1998). Lead deposits (up to 1600 Bq $^{210}Pb\,g^{-1}$) with or without accompanying Ra-bearing barites occur as scales at both oil and gas production facilities, as well as geothermal fields (Hartog et al., 1996,

Jonkers et al., 1997, Hartog et al., 1998, Schmidt, 1998). Mercury (Hg) may also be present in some of these Pb scales (Kolb and Wojcik, 1985, Schmidt, 1998); the occurrence of mercury has received particular attention in gas fields in Europe, and disposal of the produced waters, scales and sludges, and decontamination of equipment and pipes for recycling has received special attention in light of this co-mingling of radioactive and non-radioactive contaminants (Hamm, 1998, Müssig and Free, 1998).

Landfill disposal of scale and sludge is one of the more common disposal practices (Smith et al., 1999). Insights into the fate of TENORM-bearing phases such as barite and metallic lead may be gleaned from investigations of the fate of related materials in similar settings. For example, Ulrich et al. (2003) have shown the dissolution of detrital barite grains in alluvium at a landfill associated with sulfate-reducing conditions in leachate impacted areas. Studies of lead bullets and shot residing in soils (e.g., Cao et al., 2003) and water bodies at outdoor shooting ranges may serve as good analogs for the environmental fate of disposed Pb-rich scales and the transport to groundwater of $^{210}$Pb.

A hydrocarbon seep zone in sediments of the Gulf of Mexico (Aharon et al., 2001) offers a natural analog to barite-rich pipe scale dumped overboard into the sea at offshore oil production platforms. Chimneys and crusts of barite grow rapidly (4–9 cm in thickness $y^{-1}$) where Ba and Ra rich brines react with sulfate in seawater. These deposits form in sediment zones of active microbial sulfate reduction and pyrite occurs in the deposits. In areas where the venting of fluids and the growth of barite has ceased, it would be of interest to examine barite at the anoxic benthic mud interface for evidence of secondary dissolution. The ready supply of soluble sulfate from seawater may mitigate against microbial metabolism of a barite substrate. Laboratory studies by Phillips et al. (2001) and Pardue and Guo (1998) on a barite-rich pipe scale and coastal sediments contaminated by oilfield production waters suggest that mobilization of Ra under non-marine, sulfate-reducing conditions may occur.

While the discussion above has emphasized the potentially deleterious effects of sulfate reduction on Ba and Ra mobility from TENORM, waste management scenarios are often complex, and will require trade-offs of risks and benefits. For example, anaerobic conditions may be needed to suppress pyrite oxidation, and the resultant acid mine drainage, in TENORM-bearing wastes (see for example, Fernandes et al., 1996). At the Pantex nuclear weapons plant in Texas, sulfate has been added to soils to precipitate and immobilize soluble Ba, a contaminant associated with the production of explosives. These soils also contain elevated concentrations of nitrosamine explosives; the biodegradation of these hazardous compounds in the soil is enhanced under sulfate reducing conditions, and for this reason the site managers have chosen to add wood chips and manure to a composting system in order to promote highly reducing conditions (S.M. Stoller Corp., 2000). At explosives production sites at Los Alamos National Laboratory, where barium sulfate precipitates from effluent treatment have been buried, subsequent mobilization of Ba has been noted (Jackson et al., 1990).

Pipe scale and sludges can be disposed of by deep subsurface encapsulation in abandoned well bores, or by hydraulic fracturing and injection into permeable formations. Geologic and engineering criteria for such disposal on the Outer Continental Shelf of the Gulf of Mexico have been issued by the Minerals Management Service [U.S.] (1996b). The feasibility of disposal of oil field pipe scale and sludges in salt caverns (mined out cavities in salt domes) has been recently looked at by Argonne National Laboratory (Veil et al., 1998). Like formation waters in oil and gas reservoirs, geothermal reservoirs may produce waters with elevated radium concentrations; these can result in enhanced radium concentrations in pipe scales and in residues ("filter cake") from steam production at geothermal power plants (Roberts et al., 1998).

## 6. Phosphate fertilizer and elemental phosphorus production

Up to about 0.02% U can substitute for positions typically occupied by atoms of calcium in the structure of the mineral carbonate fluorapatite (Durrance, 1986). This mineral commonly occurs in phosphate rock, the ore for the production of phosphoric acid and elemental phosphorus. Commercial extraction of U has occurred at several phosphoric acid plants in Florida and Louisiana (U.S. Department of Energy, 1996) and at sites in other nations.

Two basic manufacturing processes ("wet" and "thermal") are involved. The wet process treats the ore with sulfuric acid to yield phosphoric acid and hydrated calcium sulfate (gypsum). This "phosphogypsum" (PG) waste contains trace amounts of $^{226}RaSO_4$ coprecipitated with $CaSO_4 \cdot 2H_2O$. About 70–90% of the $^{226}Ra$ in the ore entering the extraction circuit partitions to the PG (Osmond and Cowart, 1985, Bolivar et al., 2002); typical $^{226}Ra$ concentrations for PG in Florida and Spain are about 800–1200 $Bq\,kg^{-1}$, and 400–1000 $Bq\,kg^{-1}$ respectively (Birky et al., 1998, Rechcigl et al., 1998, El-Mrabet et al., 2003, Aguado et al., 2004). The volumes of this waste are large, as the process yields about 5 tons of PG for every ton of phosphoric acid produced. The PG waste has been disposed of on both land and in rivers. Land disposal generally results in large piles (called "gyp stacks"). Hull and Burnett (1996) found process waters contained in these stacks to have high ionic strength, low pH (1.4–2.5), low concentrations of $^{226}Ra$ (0.08–0.30 $Bq\,l^{-1}$; due to high sulfate concentrations), but high concentrations of U and $^{210}Pb$ (up to 65 $Bq\,l^{-1}$).

Studies on the Mississippi River and the Rhine estuary have shown that where PG is discharged to such fresh- or brackish waters, the gypsum dissolves rapidly, releasing the Ra which either remains in solution or sorbs onto suspended sediment (Kraemer and Curwick, 1991, Pennders et al., 1992). $^{210}Pb$ and $^{210}Po$ in the PG occur as an insoluble residue that settles in the estuarine discharge zone, whereas the gypsum dissolves. The mineralogical occurrence of the $^{210}Pb$ and $^{210}Po$ in the PG is probably similar to that observed by Landa et al. (1994) in acid-leach UMT effluent—lead sulfate (anglesite) crystallizes in a gypsum matrix. Such "poikiolitic anglesite" appears to be the major host of $^{210}Pb$ in PG (written communication, Carter D. Hull, Department of Nuclear and Radiological Engineering, University of Florida; Nov. 15, 1994).

Assuming the deposition of PG, in an anaerobic, sulfate-reducing environment, microbially enhanced dissolution of the gypsum is likely to occur. However, release of the $^{210}Pb$ to contacting water in this anoxic environment is unlikely, due to the highly insoluble nature of PbS ($K_{sp,\,18^{\circ}C} = 3 \times 10^{-28}$). Investigations on PG carried out at Florida State University suggest that this may not be true for the daughter product $^{210}Po$, whose chemistry is analogous to that of sulfur. These studies indicate that bulk dissolution of the gypsum may be accompanied by the solubilization and release of $^{210}Po$ to contacting ground water (Harada et al., 1989, LaRock et al., 1996). Thus, it is critical to consider the changes in chemical properties as one moves through a decay series, the geochemical hosts of the radionuclides, and the ambient conditions when considering the aqueous mobility of radionuclides associated with TENORM.

Table 2
Concentrations of uranium-decay series radionuclides in raw materials (phosphate rock), products (dicalcium phosphate (DCP); calcium chloride (82%)), and waste sludges at phosphate plants using HCl-extraction process. Mean concentration (dry weight basis) in Bq $kg^{-1}$. Modified from data in Gafvert et al. (2001) and used here with permission from Elsevier

| | $^{238}U$ | $^{230}Th$ | $^{226}Ra$ | $^{210}Pb$ |
|---|---|---|---|---|
| Phosphate rock | 838 | 825 | 855 | 843 |
| DCP | 954 | 10 | 15 | 169 |
| Sludge 1 ($CaF_2$) | 1483 | 3408 | 1183 | 833 |
| Sludge 2 ($Mg(OH)_2$) | 22 | 32 | 2111 | 9405 |
| $CaCl_2$ | <0.01 | <0.01 | 825 | 70 |

In estuarine sediments in Spain with long term histories of discharge of phosphate fertilizer manufacturing effluents (liquid and PG slurries), sequential selective extraction studies showed the largest fraction of the $^{226}Ra$ to reside in the iron (Fe) and manganese (Mn) oxide/hydrous oxide phase extractable by hydroxylamine hydrochloride (Aguado et al., 2004). This suggests that considerable dissolution of the PG, and release of its coprecipitated Ra occurs in the riverine/estuarine waters of the discharge zone. Subsequent sorption and/or coprecipitation of the Ra by Fe-Mn oxides/hydrous oxides results in the observed sediment pattern. The Ra should remain fixed in these phases while oxidizing conditions persist in the sediment, but may be remobilized under Fe-Mn reducing conditions; for analogs in UMT see Landa et al. (1991).

While the sulfuric acid extraction is the most widely used wet-process variant, some phosphate plants use a hydrochloric acid (HCl) extraction of the ore to produce either fertilizers or feed supplements (such as dicalcium phosphate, DCP) for domestic animals. Because of the much higher solubility in HCl, most of the $^{226}Ra$ is found in a by-product $CaCl_2$ solution that is either discharged to local rivers (Paridaens and Vanmarcke, 2001), or sold for use in the oil industry, the food industry, or as a road salt (Gafvert et al., 2001). In the case of river discharge, locally elevated $^{226}Ra$ concentrations were noted in bank sediments. A variety of by-product sludges associated with fluorine and magnesium present in the ore may remove radionuclides from the chloride effluents (Table 2). A $CaF_2$ sludge (designated as sludge 1 in Table 2) tends to carry U that was present in solution. Approximately 0.5 ton of $CaF_2$ is typically produced per ton ($P_2O_5$-basis) of ore treated, and this is disposed in piles on land (International Atomic Energy Agency, 2003). (Note: Naturally occurring $CaF_2$ (the mineral fluorite; as an ore often termed "fluorspar") can also be uraniferous (Landa and Councell, 2000).) U-bearing $CaF_2$ solids are also a disposal issue at $UF_6$-handling nuclear fuel cycle facilities, such as enrichment plants. $^{230}Th$ is also removed from solution with the $CaF_2$ by either sorption or coprecipitation (Landa et al., 1995). Sludge 2 is predominantly $Mg(OH)_2$; it carries a remarkable amount of $^{210}Pb$ and $^{210}Pb$ out of solution. The 82% $CaCl_2$ solution noted in Table 2 results from evaporation (done to concentrate the salt for commercial use) of the process stream solution after separation of sludge 2; it is high in $^{226}Ra$. If no commercial use is available, the dilute $CaCl_2$ solution is disposed to surface water bodies. Pretreatment may include $BaSO_4$ precipitation for Ra removal (International Atomic Energy Agency, 2003); this step was added in 1991 to the waste management plan for the phosphate plant in Belgium studied by Paridaens and Vanmarcke (2001), and reduced $^{226}Ra$ concentrations in the discharged effluent from 20 $Bq\,l^{-1}$ to 3 $Bq\,l^{-1}$.

Another alternative extraction process for phosphate ore uses nitric acid ($HNO_3$). Ra here is expected to partition to either waste calcium carbonate ($CaCO_3$) or calcium oxide (CaO) that is used primarily in cement production. $CaCO_3$ generated from the neutralization of the initial acidic waste stream in the HCl-process noted above (i.e., where HCl is used in lieu of $HNO_3$) have $^{226}Ra$ concentrations of about 2000–4000 $Bq\,kg^{-1}$ (International Atomic Energy Agency, 2003). This $HNO_3$ process yields ammonium phosphate and ammonium nitrate as its fertilizer products. The technology is new, and radionuclide mass balances are not available. Some guidance may be gleaned from Burnett et al. (1996) who showed how PG can be converted to fertilizer ammonium sulfate; the waste product here is also $CaCO_3$, and it, rather than the ammonium sulfate, contains most of the inventory of $^{238}U$, $^{226}Ra$, $^{210}Pb$, and $^{210}Po$.

Given the potential for high solubility of thorium isotopes in sulfuric acid, the partitioning of $^{230}Th$ between phosphoric acid and PG is of interest. Limited mass balance data were available in early studies of wet-process phosphate fertilizer manufacturing with which to characterize the partitioning, and conflicting predictions were evident (Mustonen, 1985, Osmond and Cowart, 1985, Papastefanou, 2001). Garcia-León et al. (1998) suggests that about 40% of the Th in feed ore will be present in the PG fraction. Landa et al. (1995) showed limited removal of Th by coprecipitation with gypsum, but high removal by other sulfate minerals such as anglesite and barite; the presence of poikiolitic anglesite in PG has been noted above, and this phase may play a role in $^{230}Th$ retention. Garcia-León et al. (1998) showed a clear $^{230}Th$ signature in the upper portions of cores of estuarine sediments receiving PG slurry and other plant effluents.

Pipe scales containing up to 3700 $kBq\,kg^{-1}$ of $^{226}Ra$ are seen at some wet-process plants. X-ray diffraction reveals the presence of barite, gypsum and other minerals. Typical radon emanation coefficients of about 5% are reported (Lardinoye et al., 1982).

The thermal process involves heating the phosphate rock to about 1400°C to yield an elemental phosphorus product and a $CaSiO_3$ vitreous slag waste with a $^{226}Ra$ content of about 1300–1500 $Bq\,kg^{-1}$ (International Atomic Energy Agency, 2003). A low Rn emanation coefficient from this glassy material limits the Rn exposure pathway. Another atmospheric pathway involves the volatilization of $^{210}Pb$ and $^{210}Po$ associated with the heating of the ore. Boothe (1977) showed that a major portion of the $^{210}Po$ in the ore is volatilized in the electric furnace used. Elemental phosphorus, the commercial product here, condenses out from the stack. Some of the $^{210}Po$ escapes the stack, and some condenses on dust particles in the stack that are collected by emission control devices. At a thermal plant in the Netherlands, where such stack off-gases are vented through a wet-scrubber for emissions control, the water from that system is discharged to an estuary. In sharp contrast to the case of the wet-process PG effluent noted above, only about 30% of the $^{210}Pb$ and 10% of the $^{210}Po$ are dissolved in the thermal plant effluent; with about 200-fold dilution with seawater, the $^{210}Po$ figure increases to 100% (Pennders et al., 1992). As the bioavailability of a soluble radionuclide can be expected to be much higher than that of its insoluble form, the importance of understanding TENORM processing and geochemical forms of radionuclides when doing dose assessments is clear.

The gaseous loss of $^{210}Po$ and $^{210}Pb$ and its condensation on fine particles is a common phenomenon in a host of TENORM-generating industries. The boiling points of Po and Pb are

962°C and 1740°C respectively. Consequently, significant quantities of $^{210}Po$, unsupported by $^{210}Pb$, were observed in the dust collected by an electrostatic precipitator at the stack of a tin refinery studied by Baxter et al. (1996). The Olympic Dam mill in South Australia treats copper-uranium ores in a furnace; volatilization of $^{210}Po$ in the furnace and its deposition in the stack environs is thought to be the reason for high concentrations of this radionuclide in tissues of herbivorous lizards sampled in the tailings disposal area (Read and Pickering, 1999). Baddeleyite ($ZrO_2$) and zircon ($ZrSiO_4$) are minerals used in the production of fused zirconia for use at high temperatures. This material is produced by heating zircon and/or baddeleyite to about 2800°C. At these high temperatures both Pb and Po are volatile. At a plant treating baddeleyite containing 7 $kBq\,kg^{-1}$ of $^{210}Po$ and $^{210}Pb$, activity concentrations of 600 $kBq\,kg^{-1}$ of $^{210}Po$, and 200 $kBq\,kg^{-1}$ of $^{210}Pb$ were observed in the flue dust (Hipkin et al., 1998). Volatilization of Pb compounds may occur at lower temperatures than that of metallic Pb. For example the boiling point of $PbCl_2$ is 971°C. This may explain the stack losses of $^{210}Pb$ seen at temperatures around 1300–1400 °C in scrap metal melting operations handling pipes and equipment contaminated by oil field scales (Stadge and Kreh, 1998). It is of interest to note that in non-TENORM situations such as forest fires, where temperatures typically do not exceed about 800°C, volatilization of $^{210}Po$ and $^{210}Pb$, as well as physical resuspension associated with burning, is thought to contribute to the elevated atmospheric concentrations of these radionuclides observed (Los Alamos National Laboratory, 2003).

An interesting observation by Boothe (1977) was the possible volatile loss of part of the $^{230}Th$ inventory during furnace heating of the phosphate ore. Similarly, Tadmor (1986) showed that about 10% of the $^{230}Th$ in combusted coal may be volatile and released from stacks.

Almost all of the U in the ore entering the wet-process extraction circuit partitions to the phosphoric acid (Birky et al., 1998), which is further processed to produce a variety of fertilizers. For example, triple super phosphate fertilizer is manufactured by reacting phosphoric acid with phosphate rock; thus both Ra and U are elevated in this product, and the long-term application of phosphate fertilizers to soils may increase their contents of these radioelements. Over a 50–80 year application period, the concentrations of $^{238}U$ and $^{226}Ra$ in the plough layer could be increased from a few percent to several times background levels (Pfister et al., 1976, National Council on Radiation Protection and Measurements, 1987). Gypsum is often applied to soils as a fertilizer source of calcium and sulfur. PG represents an inexpensive and readily available by-product source of gypsum to agricultural users. To limit radiation exposure (principally by direct gamma radiation and indoor Rn inhalation exposure), the U.S. Environmental Protection Agency (1992) issued a ruling that bans the agricultural usage of PG containing more than 370 $Bq\,kg^{-1}$ (10 pCi/g) of $^{226}Ra$. By extrapolation from a three-year field trial, Rechcigl et al. (1998) showed that annual application for 100 years of PG with about 800 $Bq\,kg^{-1}$ $^{226}Ra$ (at a rate typically recommended for crop usage of Ca and S) would increase the soil $^{222}Rn$ flux from the central Florida field site investigated about four-fold.

## 7. Water and wastewater treatment residuals

Conventional water treatment processes designed to remove suspended solids and dissolved chemical contaminants from drinking water supplies will also remove radionuclides. During the period of atmospheric nuclear weapons testing, the U.S. Public Health Service and others did much work on removal of fission products such as $^{89,90}Sr$ from water supplies (Straub, 1971). A variety of treatments, including lime-soda ash softening and phosphate coagulation, which were shown to be effective for radiostrontium removal, can also remove significant quantities of other alkaline earth metals, including Ra (Menetrez and Watson, 1983). Gäfvert et al. (2002) showed coagulation and flocculation with either aluminum sulfate or ferric chloride, to form $Al(OH)_3$ or $Fe(OH)_3$, was effective at removing more than about 85% of U, Th, and Po at a municipal water treatment plant in Sweden but Ra removal was poor (30% or less). Barium sulfate precipitation has long been used to treat effluents at U mines and mills for removal of $^{226}Ra$. It has also been employed to treat water supplies with elevated Ra concentrations (U.S. Environmental Protection Agency, 1982).

The radionuclide concentrations in the sludges generated by these treatments will be a function of the raw water radionuclide concentrations and the radionuclide removal efficiencies. Lime softening sludges from water supplies in Illinois and Wisconsin, with raw water $^{226}Ra$ concentrations of 37–185 $mBq\,l^{-1}$, have $^{226}Ra$ concentrations of 220–1100 $Bq\,kg^{-1}$ (U.S. Environmental Protection Agency, 1993). The sludges are most often disposed of in on-site lagoons, or at municipal landfills, with little regulatory control. At some municipal wastewater treatment plants, elevated levels of Ra in sewage sludge have been attributed to residuals discharged to the sewer systems by drinking water treatment plants (Interagency Steering Committee on Radiation Standards [U.S.], 2003). Reject waters from reverse osmosis systems may have high Ra concentrations. Present disposal methods for such saline waters include discharge to ditches, storm sewers, surface water bodies, subsurface drainfields (Sorg et al., 1980), and deep-well injection. Sarkar et al. (1998) have investigated land application of such saline waters (with and without dilution by sewage effluent) as an alternative disposal option, and modeled the sorption and leaching to ground water of $^{226}Ra$ with irrigation of a sandy soil. Sorption decreased with decreasing soil pH, suggesting that caution be used in application of such waters to acidic soils.

As noted above, hydrous ferric oxides alone do not appear to be good Ra sorbers under typical water treatment conditions. In contrast, hydrous Mn oxides (HMO) are excellent sorbents. For example, HMO-impregnated acrylic fiber has a long history of use for quantitative removal of Ra isotopes for analysis in oceanographic and hydrological investigations (Moore and Reid, 1973). This same ability can be harnessed in water treatment plants using mixing of raw water with preformed flocs of HMO, followed by sand bed filtration, to efficiently remove Ra (Valentine et al., 1990). The Ra-bearing HMO removed from the sand bed by periodic back-washing may be released to the sewage treatment plant or otherwise disposed (e.g., to lagoon or river, or applied to land). The fate of this Ra will depend, in part, upon redox conditions in the disposal environment. Under anaerobic conditions where microbial reduction of manganese oxides might occur, Ra release to contacting waters may occur.

Research at the University of Iowa begun in the mid-1990s has uncovered the occurrence of $^{226}Ra$-rich scales in water supply pipes. $^{226}Ra$ concentrations of up to 15,000 $Bq\,kg^{-1}$ have been observed (Fisher et al., 2000). In contrast to the predominantly $BaSO_4$ oil field pipe scales, these are composed largely of Fe-Mn oxides/hydrous oxides. Such Fe-Mn minerals tend to have much higher Rn emanation coefficients than barite. Thus, the presence of these materials in water supply lines can create $^{222}Rn$ sources in close proximity to the user, with minimal time for radioactive decay prior to consumption. Increased, and probably unrecognized, radiation exposure by ingestion and inhalation exists in this case. In one town studied, where the water leaving the treatment plant had a $^{222}$Rn concentration of 15 Bq $l^{-1}$, these scales in the distribution system resulted in water reaching a consumer with a remarkable 759 Bq $l^{-1}$ $^{222}$Rn concentration (Field et al., 1995). The incremental increase in $^{222}$Rn concentration in delivered water associated with such pipe scale deposits has been termed the "Delta factor" to reflect both the change in concentration and the town (Delta, Iowa) where the phenomenon was first noted (Field et al., 1995). The site of the discovery was fortuitous in terms of scientific nomenclature, especially when one notes that subsequent studies were performed in What Cheer, Iowa (Fisher et al., 1998).

Fe-rich scales (up to 2% Mn) with $^{226}$Ra concentrations of 1284 to 3613 Bq $kg^{-1}$ (RECs of 28 to 38%) have recently been reported in water well pipes in Saudi Arabia (Askoy et al., 2002). Determination of the geographic extent of these occurrences, and their impact on degradation of water quality is an area deserving of additional study. Also of note is the observation by the University of Iowa group that scale-like deposits on filter media can result in dramatic increases in airborne $^{222}$Rn in water plants during back-washing of the filter beds; these findings have significant implication for worker exposure (Fisher et al., 1996).

Where water supplies with elevated levels of U or Ra are used, and disposed to a sewage treatment plant or septic system, the radionuclides will partition between the biosolids and the wastewater effluent. In septic systems, NaCl-rich effluents from the recharging of water softening units may potentially influence this partitioning by ion exchange displacement of radionuclides from the biosolids. Sewage and septic systems are geochemically complex. The concentration and chemical form of radionuclides in residual solids can be influenced by:

- sorption to organic materials or microbial biomass [see, for example, Tsezos et al. (1987) on use of return sludge (consisting mostly of microbial biomass) from a municipal wastewater treatment plant, for removal of $^{226}$Ra from U mine and mill wastewater],
- microbial transformations such as U reduction,
- coprecipitation with mineral phases such as calcium-, iron- and aluminum phosphates.

## 8. Coal mining and combustion

The reducing conditions under which coals form are conducive to the accumulation of U. Typical mucks, peats, lignites and coals contain about 0.05 to 3 ppm U. Th is strongly adsorbed by peats, and the typical coal contains from 1 to 10 ppm Th (Boyle, 1982). Upon combustion of coals, most of the U, Th and decay products remain within the ash. Radionuclide concentrations are enhanced within the ash fraction, both in bulk by virtue of the destruction of the large amount of organic matter, and selectively by redeposition of volatile species on fine ash fractions. TENORM issues related to coal usage include consideration of the coal and ash themselves, and of discharge waters from coal mines.

In the Upper Silesian coal district of southern Poland many of the mines discharge saline waters (high in Na and Cl; low in $SO_4$) with elevated Ra ($^{226}$Ra and $^{228}$Ra) concentrations. Prior to reaching the surface discharge point, part of the Ra inventory initially present may have coprecipitated as $BaSO_4$ deposits within the mines; these materials can have $^{226}$Ra

Table 3
Chemical composition of water flowing through the old mine working in the Silesia Coal Mine, Poland (from Pluta, 2001; used with the permission of the author and publisher (Springer))

| Chemical components | Water flowing into old mine working | Water flowing out of old mine working |
|---|---|---|
| pH | 7.4 | 6.0 |
| $Cl^-$ ($g\,dm^{-3}$) | 48.6 | 33.3 |
| $SO_4^{2-}$ ($g\,dm^{-3}$) | <0.01 | 0.27 |
| $Ba^{2+}$ ($g\,dm^{-3}$) | 0.12 | <0.01 |
| Fe-total ($g\,dm^{-3}$) | 0.024 | 0.034 |
| $^{226}Ra$ ($Bq\,dm^{-3}$) | 23.3 | 0.3 |

Note the drop in pH associated with the oxidation of pyrite and marcasite in the coal-bearing strata. This is accompanied by an increase in soluble $SO_4^{2-}$ concentrations; a portion of this soluble sulfate precipitates as $Ba(Ra)SO_4$ in the mine workings.

concentrations as high as 400,000 $Bq\,kg^{-1}$. Such a high degree of removal of Ba and Ra from solution is seen in zones of the mine workings where sulfate is produced by the oxidation of the iron sulfide ($FeS_2$) minerals pyrite and marcasite (Table 3). This underground, Ra-coprecipitation process can be enhanced for protection of surface water quality by:

- mixing high sulfate brines with the Ra-rich low sulfate brines, or
- placing sulfate-bearing materials such as gypsum and anhydrite (these materials are routinely emplaced in the old mine working to serve as fire-protection walls), or industrial wastes such as fly ash and slag from coal combustion within the old mine workings through which the water flows (Pluta and Zuber, 1995, Chalupnik et al., 2001, Pluta, 2001).

While most coals will have decay products in secular equilibrium with the parent, a young, post-glacial peat deposit in northeastern Washington State with about 0.1% U shows less than 10% in-growth of decay product activity. This deposit has been exploited as a U source, and the lack of decay product activity rendered the tailings here more benign than those at a typical U mill, a factor that was considered in the licensing decision (Stohr and Erickson, 1984).

Disequilibrium may also be seen in combustion products. Data on fly ash presented by Baxter et al. (1996) suggest strong enrichment (with respect to other U-series nuclides) in $^{210}Pb$ and $^{210}Po$, due to volatilization and subsequent condensation on the fly ash particles. The deposition of fly ash in the environs surrounding a coal-fired power plant has resulted in the halo of elevated concentrations of $^{210}Pb$ and $^{210}Po$ seen in surface (0–10 cm) soil samples (Glöbel and Andres, 1985) (Figure 2). Enrichment in $^{210}Pb$ and $^{210}Po$ in fine particles in furnace stacks, noted earlier for thermal phosphate production, is also seen in other industrial processes involving high temperature. Waste products here, collected by emission control devices and sent for disposal in landfills, include thermal pretreatment off-gas cleaning dust (from sintering of iron ore) at steel making plants, and powder waste at fused zirconia refractory plants (International Atomic Energy Agency, 2003).

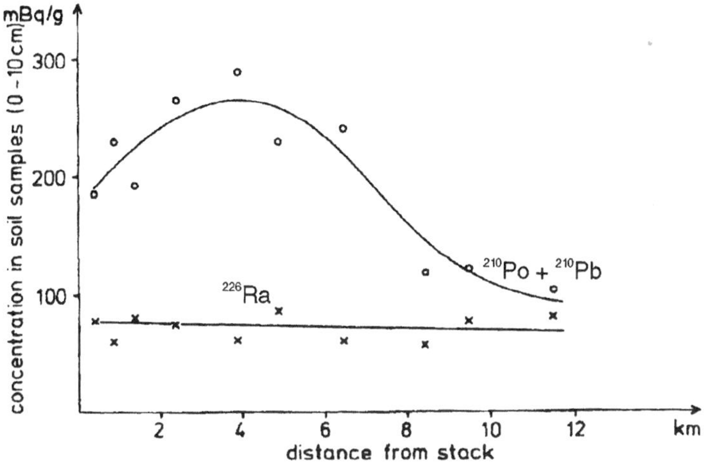


Fig. 2. Concentration of $^{210}$Po, $^{210}$Pb and $^{226}$Ra in the upper 10 cm of the soil as a function of the distance from the stack of a coal-fired power plant (from Glöbel and Andres, 1985; used here with permission from Elsevier).

Rn emanation from fly ash is a possible exposure pathway from both ash disposal piles, and from use of fly ash as a concrete aggregate. As glassy materials such as fly ash tend to have low Rn emanation coefficients (<1%; van der Lugt and Scholten, 1985), this mitigates exposures. Bem et al. (1998) have presented data suggesting that in some cases, the REC of fly ash may be high enough to produce a $^{210}$Pb anomaly in overlying soil at recently reclaimed sites; weathering periods were minimal and this appears unlikely to be the cause of increased Rn release. Disposed ash from a lignite-fired power plant in the former Yugoslavia was shown to leach U to the shallow groundwater (concentrations of up to 40 mBq $l^{-1}$ $^{238}$U; Vukovic et al., 1996).

## 9. Mineral sands

Mineral sands refer to deposits of rutile, ilmenite, zircon, monazite and other high specific gravity minerals that contain elevated levels of U and Th. The sands may be subject to digestion and processing for recovery of metals such as zirconium, titanium and rare earth elements, or used with minimal chemical processing for foundry applications.

Chlorination is a process in which ores are treated with chlorine gas and then water, to recover soluble metallic-chloride salts. Chlorination of zircon-bearing sands was used at a plant in Oregon to extract zirconium, niobium (columbium[1]), tantalum and hafnium. The process rendered Ra, as well as these economic-metals, water soluble. The finely-ground process tailings contained about 18,500 Bq $kg^{-1}$ of $^{226}$Ra, much occurring presumably as soluble

[1] Columbium and niobium are synonymous names for element 41; columbium was the name given in 1801, and niobium (Nb) was the name officially designated by the International Union of Pure and Applied Chemistry in 1950. While "niobium" is general used in the chemical literature, "columbium" is still commonly discussed today in mining and metallurgy.

$RaCl_2$. Seepage water at this tailings disposal site contained up to 1665 $Bq\,l^{-1}$ (Bliss, 1978, Boothe et al., 1980).

Foundries use these sands to create molds for casting steel alloy parts. The casting molds are eventually disposed of in landfills. In the U.S., state regulations restricting the disposal of radioactive material, and the use of portal radiation monitors at landfills has, in some cases, made it difficult for the industry to dispose of discarded casting molds (Anonymous, 1995). The Rn emanation coefficients of these accessory minerals tend to be low, ~0.1 to 5% for zircon and monazite as compared to ~10 to 40% for soils and UMT (Barretto et al., 1972, Landa, 1987). Such differential environmental mobility factors should be considered for the atmospheric pathway in radiological dose assessments for these wastes.

## 10. Site reclamation

Where pyrites ($FeS_2$) are present in coal deposits, metallic ores, uranium ores and mill tailings, their oxidation can create sulfuric acid and acid mine drainage problems. For example, $FeS_2$ oxidation has been implicated in the profound and rapid leaching losses in Th (about 97% in 20 years) observed in the upper 15 cm of UMT disposed at Elliot Lake, Ontario (Moffett and Tellier, 1978). Fernandes et al. (1996) have shown that due to the potential contamination of water resources, $FeS_2$ oxidation needs to be considered in dose assessments and decommissioning plans at uranium mill tailings facilities. Cover materials have been used in the reclamation of abandoned mine lands and mill tailings to help reestablish vegetation and to provide barriers to the influx of oxygen (in the case of pyritic waste rock and tailings) and the emission of Rn (a major remedial action goal in the case of UMT and other TENORM). Organic-rich cover materials and amendments such as peat, sewage sludge, manure, and paper-mill waste offer the potential to provide soil fertility and moisture-capacity benefits, and enhanced oxygen consumption by microbial activity. While such usage is not presently widespread at UMT sites in the United States, wood chips have been used as a cover at a UMT site in eastern Washington (U.S. Department of Energy, 1996).

While such materials would appear to be of value in limiting oxygen penetration to underlying materials, and hence limiting $FeS_2$ oxidation, the effects of surface-applied organic materials on subsurface chemical processes and ground-water chemistry are complex and uncertain. Cravotta (1998) showed that although a surface-applied mixture of sewage sludge and wood chips promoted revegetation of pyritic coal mine spoil (the overburden or non-ore grade material removed in mining), the material:

- was not an effective barrier to $O_2$ influx to underlying zones,
- contributed trace metals and nutrients to ground water, and
- promoted the formation of acid mine drainage.

The sludge/wood mixture, applied at a rate of 135 metric tonnes per hectare, did not slow the rate of $FeS_2$ oxidation and acid formation; indeed nitrate added with the sludge/wood may have enhanced $FeS_2$ oxidation. Concentrations of acidity, sulfate, metals (Fe, Mn, Al, Cd, Cu, Cr, Ni, Zn), and nitrate ($NO_3^-$), and abundances of iron-oxidizing bacteria (*Thiobacillus ferrooxidans*) were elevated in the ground water from sludge/wood-treated spoil relative to untreated spoil having a similar mineral composition. Gaseous and dissolved $O_2$ concentrations, however, did not differ between the treatments. Concentrations of $NO_3^-$ were elevated in vadose-zone water samples from sludge/wood-treated spoil, frequently exceeding 10 mg $l^{-1}$. Downgradient decreases in $NO_3^-$ to less than 3 mg $l^{-1}$ and increases in $SO_4^{2-}$ concentrations in underlying ground water were consistent with $NO_3^-$ removal by oxidation of $FeS_2$. $NO_3^-$ leached from the sludge/wood could have increased $FeS_2$ oxidation because of (1) chemical oxidation of $FeS_2$ by $NO_3^-$ and (2) microbial utilization of $O_2$ or $NO_3^-$ to oxidize reduced S or Fe. Where such processes are active, the leaching of U and Th from TENORM might occur. At these disposal sites, $NO_3^-$ can originate from either organic material amendments, or from process reagents. Even when $NO_3^-$-containing process reagents are not used in TENORM-generating industries, the possibility of in-situ $NO_3^-$ production by the action of ammonium-oxidizing bacteria (*Nitrosomonas* spp.) needs to be considered (Landa, 2004).

## 11. Assessing radionuclide partitioning in TENORM disposal environments

The discussion of TENORM in this chapter has been structured around two distinct approaches: (1) materials flow analysis, and (2) consideration of biogeochemical cycling of radionuclides. The sequential application of these approaches should be considered a key part of the conceptual framework for radiological assessments of TENORM.

Materials flow analysis (MFA) assesses the partitioning of substances during industrial processing to products and wastes; it provides a cradle-to-grave view that considers sources as well as reuse of waste materials (e.g., agricultural use of PG) and final disposition of products (e.g., soil enrichment in U by phosphate fertilizer application). MFA is central to the field of "industrial ecology," an emerging area of interest in environmental policy and research (Thomas et al., 2003), and is the approach of choice in the recent International Atomic Energy Agency (2003) guidance document on TENORM.

Baxter et al. (1996) have noted that for some TENORM industries (in their specific case, a tin refinery), radionuclide budgets and pathways are not well understood. However, not only pathways, but sources also await discovery. The pulp and paper industry is certainly not one where most people would have expected the occurrence of TENORM. However, in the late 1990s, piping and tanks from paper mills in Pennsylvania and Ontario triggered radiation detectors at scrap metal facilities (Yusko, 1996, Gray, 2000). Scales, thought to be $CaSO_4$ and $BaSO_4$, with very high concentrations of $^{226,228}Ra$ have since been reported in other pulp and paper mills. $^{228}Ra$ concentrations of 11,000 to 250,000 Bq $kg^{-1}$, and $^{226}Ra$ concentrations of 15,000 to 120,000 Bq $kg^{-1}$ have been reported in scales. The uses of sulfate-containing process reagents (e.g., alum, sulfuric acid), and recycling of process water seem to be factors in scale formation (Gray, 2000, Fisher and Easty, 2003). At least one report noted a $^{230}Th$ concentration in a scale sample of 4000 Bq $kg^{-1}$ (Gray, 2000). With Chernobyl (rather than TENORM) concerns as the impetus, MFA-type studies in Europe on fission products in pulp mills have provided good data on $^{90}Sr$ (Krosshavn et al., 1996, Vaca et al., 1999) and its chemical analog Ra (Vaca et al., 2001). Clearly, many opportunities for cross-cutting research exist. Other TENORM sources will probably be uncovered in the future by a combination of hypothesis testing, routine surveillance, and chance.

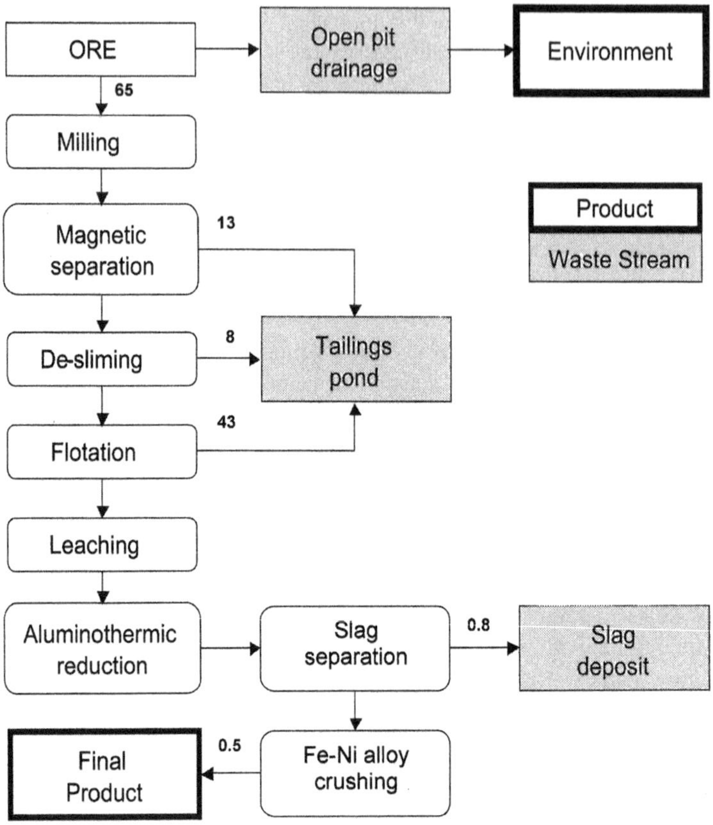


Fig. 3. Flowsheet of operational process at a Brazilian niobium processing facility. Numbers associated with flowsheet arrows are mass fluxes (ton $h^{-1}$) for given steps. Ore, solid wastes and final products are shown. Adapted from Pires do Rio et al. (2002), and International Atomic Energy Agency (2003); used here with the permission from the IAEA and from Elsevier.

MFA offers a way to track radionuclides for monitoring and regulatory control, and to suggest ways to mitigate TENORM problems. Knowing mass fluxes at various steps in an industrial process and radionuclide concentrations for wastes and products, one can calculate radionuclide fluxes at each step (as an example from the niobium industry, see Figure 3 and Tables 4 and 5). These data can then be used in dose estimates for various pathways for the different waste materials (Pires do Rio et al., 2002).

Another use of MFA is for radiation source avoidance; for example:

- Mine operators may use selective techniques to avoid strata with elevated radionuclide concentrations.
- The use of scale inhibitors or in-situ removal of radionuclides from oil-field production fluids by the introduction of sorbents down-hole can limit the buildup of TENORM in piping and equipment (Lancée et al., 1997).

Table 4
Concentrations of U and Th series decay-series radionuclides ($Bq\,kg^{-1}$) in ores, solid wastes and final products at a Brazilian niobium processing facility. Adapted from Pires do Rio et al. (2002), and International Atomic Energy Agency (2003); used here with the permission from the IAEA and from Elsevier

| Material | Radionuclide concentration ($Bq\,kg^{-1}$) | | | | |
|---|---|---|---|---|---|
| | $^{228}Ra$ | $^{238}U$ | $^{226}Ra$ | $^{210}Pb$ | $^{232}Th$ |
| Ore | 4550 | 3390 | 7890 | 904 | 2040 |
| Magnetic separation waste | 843 | 941 | 1730 | 293 | 315 |
| De-sliming waste | 6700 | 6900 | 6180 | 1754 | 3080 |
| Flotation waste | 4985 | 3400 | 7530 | 1040 | 1840 |
| Slag | 34,189 | 5160 | 435 | 16,700 | 6450 |
| Final product | 64 | 41 | – | 23 | 258 |

Table 5
Fluxes of U and Th series decay-series radionuclides ($Bq\,y^{-1}$) in ores, solid wastes and final products at a Brazilian niobium processing facility. Adapted from International Atomic Energy Agency (2003); used here with the permission from the IAEA and from Elsevier

| Material | Radionuclide fluxes ($10^{12}\,Bq\,y^{-1}$) | | | | |
|---|---|---|---|---|---|
| | $^{228}Ra$ | $^{238}U$ | $^{226}Ra$ | $^{210}Pb$ | $^{232}Th$ |
| Ore | 2.5 | 1.9 | 4.3 | 0.49 | 1.1 |
| Magnetic Separation waste | 0.092 | 0.10 | 0.19 | 0.032 | 0.034 |
| De-sliming waste | 0.45 | 0.46 | 0.42 | 0.12 | 0.21 |
| Flotation waste | 1.8 | 1.2 | 2.7 | 0.38 | 0.66 |
| Slag | 2.4 | 0.036 | 0.0030 | 0.12 | 0.44 |
| Final product | 0.00025 | 0.00016 | – | 0.000088 | 0.0015 |

- U may be recovered (for nuclear fuel cycle use) from phosphoric acid production lines prior to fertilizer production (Nazari et al., 2004), and from copper leaching operations (U.S. Environmental Protection Agency, 1999) prior to site reclamation. It has been estimated that about 13,000 metric tons of $U_3O_8$ could be recovered worldwide annually from phosphoric acid (Nazari et al., 2004).
- Gold ores in the Witwatersrand basin of South Africa contain elevated concentrations of U ($\sim$100 ppm average). Depending on market conditions, this U may or may not be extracted from the ore. With a typical U extraction efficiency of about 90%, this means that tailings here can vary by an order of magnitude in U content (Winde and van der Walt, 2004). The sulfuric acid leaching used for U removal can affect both the total U inventory in the tailings and its leaching potential during tailings storage (Landa, 1982). The Witwatersrand tailings have been associated with U contamination of adjacent streams (Winde and van der Walt, 2004).

Other strategies using MFA focus on selection of appropriate reuse/recycle options. For example, residues with high Rn emanation coefficients may be better suited for use in roadbed-construction rather than for manufacture of residential-building materials. Additional processing of residues may allow for reuse; e.g., U can be extracted from $CaF_2$ residues by HCl extraction to levels below regulatory concern to allow for free release and industrial use. MFA that includes U- and Th-series radionuclides can provide guidance when new uses of by-products are proposed; for example, the use of gypsum produced by neutralization of sulfuric acid effluents from mineral sands processing facilities as a sulfur source to crops (Summers et al., 2003).

Where considerations of MFA come to an end—in disposal or use of radionuclide-bearing materials—biogeochemical considerations come to the forefront, and can help predict the environmental partitioning of the radionuclides to soil and sediments, water, and air as a result of in-situ physical, chemical and biological processes. It is critical to consider the changes in chemical properties as one moves through a decay series, the geochemical hosts of the radionuclides, and the ambient conditions when considering the aqueous mobility of radionuclides associated with TENORM. Recent work at the U.S. Nuclear Regulatory Commission (Veblen et al., 2004) has used microprobe analysis of weathering rinds in buried slags from tin and niobium/tantalum extraction to quantify fluxes of U and Th from these surfaces. Investigations such as these can aid in the development of leaching and dose assessment models addressing the long-term fate of TENORM in the environment.

## References


Aguado, J.L., Bolívar, J.P., García-Tenorio, R. (2004). Sequential extraction of $^{226}$Ra in sediments from an estuary affected by anthropogenic inputs of natural radionuclides. *Journal of Environmental Radioactivity* **74**, 117–126.

Aharon, P., Van Gent, D., Fu, B., Scott, L.M. (2001). Fate and effects of barium and radium-rich fluid emissions from hydrocarbon seeps on the benthic habitats of the Gulf of Mexico offshore Louisiana. Minerals Management Service Report OCS Study, MMS 2001–004. Available at http://www.gomr.mms.gov/homepg/regulate/environ/techsumm/2001/2001-004.html.

Anonymous (1995). New rules could bring ceramics industry under LLW standards. Nuclear Waste News, business Publishers Inc., Silver Spring, MD. Nov. 9, 1995, p. 435.

Askoy, A., Al-Jarallah, M., Al-Haddad, M.N. (2002). Natural radioactivity in the scale of water well pipes. *Journal of Environmental Radioactivity* **61**, 33–40.

Barretto, P.M.C., Clark, R.B., Adams, J.A.S. (1972). Physical characteristics of radon-222 emanation from rocks, soils, and minerals: its relation to temperature and alpha dose. In: Adams, J.A.S., Lowder, W.M., Gesell, T.F. (Eds.), *Proceedings of The Second International Symposium on The Natural Radiation Environment, August 7–11, Houston*, pp. 731–740. U.S. Energy Research and Development Administration report CONF-720805-P1.

Baxter, M.S., MacKenzie, A.B., East, B.W. (1996). Natural decay series radionuclides in and around a large metal refinery. *Journal of Environmental Radioactivity* **32**, 115–133.

Bem, E.M., Bem, H., Wieczorkowski, P. (1998). Studies of radionuclide concentrations in surface soil and around fly ash disposal sites. *Science of the Total Environment* **220**, 215–222.

Berry, R.O., Cezeaux, J.R. (2004). Methods employed in measurement of pipe scale released air by a dry rattling process. WM'04 Conference, February 29–March 4, Tucson, AZ, 5 p.

Betti, M., de las Heras, L.A., Janssens, A., Henrich, E., Hunter, G., Gerchikov, M., Dutton, M., van Weers, A.W., Nielsen, S., Simmonds, J., Bexon, A., Sazykina, T. (2004). Results of the European Commission Marina II Study Part II—effects of discharges of naturally occurring radioactive material. *Journal of Environmental Radioactivity* **74**, 243–254.

Birky, B.K., Tolaymat, T., Warren, B.C., Bolch, W.E., Ammons, R., McNally, T., Nall, J.W. (1998). Evaluation of exposure to technologically enhanced naturally occurring radioactive materials (TENORM) in the phosphate industry. Florida Institute of Phosphate Research, FIPR Publication #05-046-155, 273 p., available at http://www1.fipr.state.fl.us/Publications.

Bland, C.J., Chiu, N.W. (1996). Accumulation of $^{210}Pb$ activity on particulate matter in LPG rail cars. *Applied Radiation and Isotopes* **47**, 925–926.

Bliss, J.D. (1978). Radioactivity in selected mineral extraction industries: a literature review. U.S. Environmental Protection Agency, Technical Note ORP/LV-79-1, 53 p.

Bolivar, J.P., Garcia-Tenorio, R., Mas, J.L., Vaca, F. (2002). Radioactive impact in sediments from an estuarine system affected by industrial wastes releases. *Environment International* **27**, 639–645.

Boothe, G.F. (1977). The need for radiation controls in the phosphate and related industries. *Health Physics* **32**, 285–290.

Boothe, G.F., Stewart-Smith, D., Wagstaff, D., Dibblee, M. (1980). The radiological aspects of zircon sand use. *Health Physics* **38**, 393–398.

Boyle, R.W. (1982). *Geochemical Prospecting for Thorium and Uranium Deposits*. Elsevier, New York, 498 p.

Burnett, W.C., Schultz, M.K., Hull, C.D. (1996). Radionuclide flow during the conversion of phosphogypsum to ammonium sulfate. *Journal of Environmental Radioactivity* **32**, 33–51.

Cao, X., Ma, L.Q., Chen, M., Hardison Jr., D.W., Harris, W.G. (2003). Weathering of lead bullets and their environmental effects at outdoor shooting ranges. *Journal of Environmental Quality* **32**, 526–534.

Chalupnik, S., Michalik, B., Wysocka, M., Skubacz, K., Mielnikow, A. (2001). Contamination of settling ponds and rivers as a result of discharge of radium-bearing waters from Polish coal mines. *Journal of Environmental Radioactivity* **54**, 85–98.

Coll, L. (1997). *Chemical Decontamination of NORM Scales*. CORPEX Technologies Inc., Research Triangle Park NC, brochure, 6 p.

Cravotta, C.A. (1998). III Effect of sewage sludge on formation of acidic ground water at a reclaimed coal mine. *Ground Water* **36**, 9–19.

Durrance, E.M. (1986). *Radioactivity in Geology: Principles and Applications*. Wiley, New York, 441 p.

El-Mrabet, R., Abril, J.M., Periáñez, R., Manjón, G., Garcia-Tenorio, R., Delgado, A., Andreu, L. (2003). Phosphogypsum amendment effect on radionuclide content in drainage water and marsh soils from southwestern Spain. *Journal of Environmental Quality* **32**, 1262–1268.

Fernandes, H.M., Franklin, M.R., Veiga, L.H.S., Freitas, P., Gomiero, L.A. (1996). Management of uranium mill tailing: geochemical processes and radiological risk assessment. *Journal of Environmental Radioactivity* **30**, 69–95.

Field, R.W., Fisher, E.L., Valentine, R.L., Kross, B.C. (1995). Radium-bearing pipe scale deposits: implications for national waterborne radon sampling methods. *American Journal of Public Health* **85**, 567–570.

Fisher, E.L., Fuortes, L.J., Field, R.W. (1996). Occupational exposure of water-plant operators to high concentrations of radon-222 gas. *Journal of Occupational and Environmental Medicine* **38**, 759–764.

Fisher, E.L., Fuortes, L.J., Ledolter, J., Steck, D.J., Field, R.W. (1998). Temporal and spatial variation of waterborne point-of-use $^{222}Rn$ in three water distribution systems. *Health Physics* **74**, 242–248.

Fisher, E.L., Fuortes, L.J., Valentine, R.L., Mehrhoff, M., Field, R.W. (2000). Dissolution of $^{226}$Radium from pipe-scale deposits in a public water supply. *Environment International* **26**, 69–73.

Fisher, R.P., Easty, D.B. (2003). Results of surveys at United States pulp and paper mill for the presence of scales and precipitates containing naturally occurring radioactive material (NORM). *Health Physics* **84**, 518–525.

Gafvert, T., Holm, E., Roos, P. (2001). Radionuclide fluxes at a plant manufacturing dicalcium phosphate for domestic animals. *Journal of Environmental Radioactivity* **54**, 61–73.

Gäfvert, T., Ellmark, C., Holm, E. (2002). Removal of radionuclides at a waterworks. *Journal of Environmental Radioactivity* **63**, 105–115.

Garcia-León, M., Garcia-Tenorio, R., Abril, J.M., Garcia-Orellana, I., Gutierrez, E., Laissaoui, A., Martinez-Aguirre, A., Periañez, R. (1998). The phosphoric acid production as a source of environmental radioactivity: the Spanish case. Second International Symposium on the Treatment of Naturally Occurring Radioactive Materials Proceedings, November 10–13, Krefeld, Germany, pp. 90–94.

Gesell, T.F. (1975). Occupational radiation exposure due to $^{222}Rn$ in natural gas and natural gas products. *Health Physics* **29**, 681–687.

Glöbel, B., Andres, C. (1985). Investigations of fly-ash and soil samples in the environment of a coal-fired power plant. *Science of the Total Environment* **45**, 63–67.

Gray, P. (2000). NORM deposition in the pulp and paper industry. The NORM Report v. 7, p. 33.

Hahn, O. (1936). *Applied Radiochemistry*. Cornell University Press, Ithaca, New York, 278 p.

Hamm, M. (1998). A new recycling-plant for metals, contaminated by naturally occurring radionuclides. Second international symposium on the treatment of naturally occurring radioactive materials Proceedings, November 10–13, Krefeld, Germany, pp. 132–136.

Harada, K., Burnett, W.C., LaRock, P.A. (1989). Polonium in Florida groundwater and its possible relationship to the sulfur cycle and bacteria. *Geochimica et Cosmochimica Acta* **53**, 143–150.

Hartog, F.A., Knaepen, W.A.I., Jonkers, G. (1996). Radioactive lead: an underestimated NORM issue? Proceedings of the 1995 American Petroleum Institute and Gas Research Institute naturally occurring radioactive material conference, Houston, Texas, October 17–18, pp. 59–72.

Hartog, F.A., Knaepen, W.A.I., Jonkers, G. (1998). Origin and encounter of $^{210}$Pb in E&P Facilities. Second international symposium on the treatment of naturally occurring radioactive materials Proceedings, November 10–13, Krefeld, Germany, pp. 53–57.

Herbert, M.B., Scott, L.M., Zrake, S.J. (1995). A radiological characterization of remediated tank battery sites. *Health Physics* **68**, 406–410.

Hipkin, J., Paynter, R.A., Shaw, P.V. (1998). Exposures to work to ionising radiation due to the use of naturally occurring radioactive materials in industrial processes. *Applied Radiation and Isotopes* **49**, 205–209.

Hull, C.D., Burnett, W.C. (1996). $^{238}$U decay-series nuclides in fluids within a Florida phosphogypsum storage stack. In: Proceedings of the 29th midyear topical meeting of the Health Physics Society. NORM/NARM: Regulation and Risk Assessment, pp. 111–120.

International Atomic Energy Agency (2003). Extent of environmental contamination by naturally occurring radioactive material (NORM) and technological options for mitigation. Technical Report Series no. 419, Vienna. Available at http://www-pub.iaea.org/MTCD/publications/PubDetails.asp?pubId=6789.

Interagency Steering Committee on Radiation Standards (2003). ISCORS Assessment of Radioactivity in sewage sludge: recommendations on management of radioactive materials in sewage sludge and ash at publicly owned treatment works. ISCORS Technical Report 2003-4 U.S. Department of Energy-EH-0668 & U.S. Environmental Protection Agency 832-R-03-002B. Available at www.iscors.org/library.html.

Jackson, P.J., Torres, A.P., Delhaize, E., Pack, E., Bolender, S.L. (1990). The removal of barium ion from solution using *datura innoxia* mill. Suspension culture cells. *Journal of Environmental Quality* **19**, 644–648.

Jerez Vegueria, S.F., Godoy, J.M., Miekeley, N. (2002). Environmental impact studies of barium and radium discharges by produced waters from the "Bacia de Campos" oil field offshore platforms, Brazil. *Journal of Environmental Radioactivity* **62**, 29–38.

Jonkers, G. (2005). Personal communication.

Jonkers, G., Hartog, F.A., Knaepen, W.A.I., Lancee, P.F.J. (1997). Characterisation of NORM in the oil and gas production (E&P) Industry. International symposium on radiological problems with natural radioactivity in the non-nuclear industry, September 8–10, Amsterdam.

Kolb, W.A., Wojcik, M. (1985). Enhance Radioactivity due to natural oil and gas production and related radiological problems. *Science of the Total Environment* **45**, 77–84.

Kraemer, T.F., Curwick, P.B. (1991). Radium isotopes in the lower Mississippi River. *Journal of Geophysical Research* **96**, 2797–2806.

Krosshavn, M., Bjornstad, H.E., Engh, L., Lien, H.N., Salbu, B. (1996). Speciation of $^{137}$Cs and $^{90}$Sr in discharges from the pulp industry and the marine recipient. *Journal of Environmental Radioactivity* **32**, 157–164.

Lancée, P., Eylander, J.G.R., Hartog, F.A., Jonkers, G. (1997). Review on E&P NORM volume reduction techniques and waste disposal routes. In: International symposium on radiological problems with natural radioactivity in the non-nuclear industry, September 8–10, Amsterdam, section 6.2.

Landa, E.R. (1980). Isolation of uranium mill tailings and their component radionuclides from the biosphere—some earth science perspectives. U.S. Geological Survey Circular 814, 32 p.

Landa, E.R. (1982). Leaching of radionuclides from uranium ore and mill tailings. *Uranium* **1**, 53–64.

Landa, E.R. (1984). Leaching of molybdenum and arsenic from uranium ore and mill tailings. *Hydrometallurgy* **13**, 203–211.

Landa, E.R. (1987). Radium-226 contents and Rn emanation coefficients of particle-size fraction of alkaline, acid and mixed U mill tailings. *Health Physics* **52**, 303–310.

Landa, E.R. (2004). Uranium mill tailings: nuclear waste and natural laboratory for geochemical and radioecological investigations. *Journal of Environmental Radioactivity* **77**, 1–27.

Landa, E.R., Councell, T.B. (2000). Uranium content and leachable fraction of fluorspars. *Health Physics* **79**, 286–290.

Landa, E.R., Phillips, E.J.P., Lovley, D.R. (1991). Release of $^{226}$Ra from uranium mill tailings by microbial Fe(III) reduction. *Applied Geochemistry* **6**, 647–652.

Landa, E.R., Stieff, L.R., Germani, M.S., Tanner, A.B., Evans, J.R. (1994). Intense alpha-particle emitting crystallites in uranium mill wastes. *Nuclear Geophysics* **8** (5), 443–454.

Landa, E.R., Le, A.H., Luck, R.L., Yeich, P.J. (1995). Sorption and coprecipitation of trace concentrations of thorium with various minerals under conditions simulating an acid uranium mill effluent environment. *Inorganica Chimica Acta* **229**, 247–252.

Lardinoye, M.H., Weterings, K., van de Berg, W.B. (1982). Unexpected $^{226}$Ra build-up in wet-process phosphoric-acid plants. *Health Physics* **42**, 503–514.

LaRock, P., Hyun, J.H., Boutelle, S., Burnett, W.C., Hull, C.D. (1996). Bacterial mobilization of polonium. *Geochimica et Cosmochimica Acta* **60**, 4321–4328.

Los Alamos National Laboratory (2003). AIRNET Data Evaluation during the Cerro Grande fire: interpretation of preliminary air monitoring data gathered during the Cerro Grande fire by the LANL's air quality group. Available at http://www.lanl.gov/orgs/rres/maq/AirConcCGF.htm#Introduction.

Madruga, M.J., Brogueira, A., Alberto, G., Cardoso, F. (2001). $^{226}$Ra bioavailablity to plants at the Urgeirica uranium mill tailings site. *Journal of Environmental Radioactivity* **54**, 175–188.

Means, J.L., Crerar, D.A., Duguid, J.O. (1978). Migration of radioactive waste: radionuclide mobilization by complexing agents. *Science* **200** (4349), 1477–1481.

Menetrez, M.Y., Watson, J.E. (1983). Natural radioactivity in North Carolina ground water supplies. University of North Carolina, Water Resources Research Institute, report WRRI-UNC-83-208, 30 p.

Minerals Management Service (U.S.) (1996a). Environmental assessment for issuance of notice to lessees and operators of Federal oil and gas leases on the Outer Continental Shelf, Gulf of Mexico Region: Guidelines for the offshore storage and sub-seabed disposal of waste resulting from the development and production of oil and gas on the Outer Continental Shelf; May 8, 1996.

Minerals Management Service (U.S.) (1996b). Notice to lessees and operators of Federal oil and gas leases on the Outer Continental Shelf, Gulf of Mexico Region: Guidelines for the offshore storage and sub-seabed disposal of waste resulting from the development and production of oil and gas on the Outer Continental Shelf (5-8-96) NTL no. 96-03.

Moffett, D., Tellier, M. (1978). Radiological investigations of an abandoned uranium tailings area. *Journal of Environmental Quality* **7**, 310–314.

Mohamad Puad, H.A., Muhd Noor, M.Y. (2004). Behaviors of $^{323}$Th, $^{238}$U, $^{228}$Ra and $^{226}$Ra on combustion of crude oil terminal sludge. *Journal of Environmental Radioactivity* **73**, 289–305.

Moore, W.S., Reid, D.F. (1973). Extraction of radium from natural waters using manganese-impregnated acrylic fibers. *Journal of Geophysical Research* **78**, 8880–8886.

Müssig, S., Free, M. (1998). Environmentally friendly disposal of mercury and NORM-contaminated waste in the exploration and production industry. Second international symposium on the treatment of naturally occurring radioactive materials, Proceedings, November 10–13, Krefeld, Germany, pp. 142–146.

Mustonen, R. (1985). Radioactivity of fertilizers in Finland. *Science of the Total Environment* **45**, 127–134.

National Academy of Sciences (U.S.), National Research Council (1999). Evaluation of Guidelines for the exposures to technologically enhanced naturally occurring radioactive materials. National Academy Press, Washington, DC, 294 p. Available at http://books.nap.edu/catalog/6360.html.

National Council on Radiation Protection and Measurements (1987). Radiation exposure of the U.S. population from consumer products and miscellaneous sources, NCRP report no. 95.

Nazari, K., Maragheh, M.G., Rad, A.J. (2004). Studies on extraction of uranium from phosphoric acid using PN-1200 extractant. *Hydrometallurgy* **71**, 371–377.

Osmond, J.K., Cowart, J.B. (1985). Radioelement migration in natural and mined phosphate terrains. Florida Institute of Phosphate Research, FIPR Publication #05-002-027, 142 p. Available at http://www1.fipr.state.fl.us/Publications.

Papastefanou, C. (2001). Radiological impact from atmospheric releases of $^{238}$U and $^{226}$Ra from phosphate rock processing plants. *Journal of Environmental Radioactivity* **54**, 75–83.

Pardue, J.H., Guo, T.Z. (1998). Biogeochemistry of $^{226}$Ra in contaminated bottom sediments of oilfield waste pits. *Journal of Environmental Radioactivity* **39**, 239–253.

Paridaens, J., Vanmarcke, H. (2001). Radium contamination of the banks of the river Laak as a consequence of the phosphate industry in Belgium. *Journal of Environmental Radioactivity* **54**, 53–60.

Pennders, R.M.J., Koster, H.W., Lembrechts, J.F. (1992). Characteristics of $^{210}$Po and $^{210}$Pb in effluents from phosphate-producing industries: a first orientation. *Radiation Protection Dosimetry* **45**, 737–740.

Pfister, H., Philipp, G., Pauly, H. (1976). Population dose from natural radionuclides in phosphate fertilizers. *Radiation and Environmental Biophysics* **13**, 247–261.

Phillips, E.J.P., Landa, E.R., Kraemer, T., Zielinkski, R. (2001). Sulfate-reducing bacteria release barium and radium from naturally occurring radioactive material in oil-field barite. *Geomicrobiology Journal* **18**, 167–182.

Pires do Rio, M.A., Amaral, E.C.C., Fernandes, H.M., Rochedo, E.R.R. (2002). Environmental radiological impact associated with non-uranium industries: A proposal for screening criteria. *Journal of Environmental Radioactivity* **59**, 1–17.

Pluta, I. (2001). Barium and radium discharged from coal mines in the Upper Silesia, Poland. *Environmental Geology* **40**, 345–348.

Pluta, I., Zuber, A. (1995). Origin of brines in the Upper Silesian Coal Basin (Poland) inferred from stable isotope and chemical data. *Applied Geochemistry* **10**, 447–460.

Read, J., Pickering, R. (1999). Ecological and toxicological effects of exposure to an acidic, radioactive tailings storage. *Environmental Monitoring and Assessment* **54**, 69–85.

Rechcigl, J.E., Alcordo, I.S., Roessler, C.E., Littell, R.C., Stieff, L.R. (1998). Radiological impact of applying phosphogypsum to tilled land cropped to annual ryegrass. *Soil Science* **163**, 834–840.

Rittiger, C.L., Yusko, J.G. (1996). An overview of NORM data collected in Pennsylvania specific to oil and gas production. In: Proceedings of the 29th midyear topical meeting of the Health Physics Society. NORM/NARM: Regulation and Risk Assessment, pp. 129–132.

Roberts, C.J., Quinby, J.B., Duggan, W.P., Yuan, Y. (1998). Disposal options and case-study pathway analyses. *Applied Radiation and Isotopes* **49**, 241–258.

Rood, A.S., White, G.J., Kendrick, D.T. (1998). Measurement of $^{222}$Rn flux, $^{222}$Rn emanation, and $^{226,228}$Ra concentration from injection well pipe scale. *Health Physics* **75**, 187–192.

Sarkar, D., O'Connor, G.A., Ruple, G.J., Sartain, J.B. (1998). Reuse of Carlton reject water: II. Fate and transport of $^{226}$Ra. *Soil and Crop Sciences Society of Florida Proceedings* **58**, 38–44.

Schmidt, A.P. (1998). Lead precipitates from natural gas production installations. *Journal of Geochemical Exploration* **62**, 193–200.

Scholten, L.C., van der Steen, J. (1998). International symposium on radiological problems with natural radioactivity in the non-nuclear industry. *Applied Radiation and Isotopes* **49**, 273–274.

Smith, K.P. (1992). An overview of naturally occurring radioactive materials (NORM) in the petroleum industry. Argonne National Laboratory report ANL/EAIS-7, 37 p.

Smith, K.P., Blunt, D.L., Williams, G.P., Arnish, J.J., Pfingston, M., Herbert, J., Haffenden, R.A. (1999). An assessment of the disposal of petroleum in nonhazardous landfills. *Environmental Science and Technology* **37**, 2060–2066.

S.M. Stoller Corporation (2000). Final compost deployment implementation report (Pantex Plant).

Sorg, T.J., Forbes, R.W., Chambers, D.S. (1980). Removal of radium-226 from Sarasota County, Fla., drinking water by reverse osmosis. *Journal of the American Water Works Association* **72**, 230–237.

Stadge, G., Kreh, R. (1998). Problems occurring with activity determination in naturally occurring radioactive materials. Second International Symposium on the Treatment of Naturally Occurring Radioactive Materials Proceedings, November 10–13, Krefeld, Germany, pp. 38–42.

Stohr, J.S., Erickson, J.L. (1984). Regulation of a post-glacial uranium deposit in the state of Washington. In: Sixth symposium on uranium mill tailings management. Colorado State University, Geotechnical Engineering Program, pp. 15–23.

Straub, C.P. (1971). Radioactivity. In: *AWWA: Water Quality and Treatment—A Handbook of Public Water Supplies*. McGraw Hill, New York, pp. 441–462.

Summerlin, J. Jr., Prichard, H.M. (1985). Radiological health implications of lead-210 and polonium-210 accumulations in LPG refineries. *American Industrial Hygiene Association Journal* **46**, 202–205.

Summers, R., Clarke, M., McCafferty, P., Pope, T. (2003). A slowly soluble, sulfur fertiliser from a by-product of mineral sands processing. *Australian Journal of Soil Research* **41**, 1213–1227.

Tadmor, J. (1986). Atmospheric release of volatilized species of radioelements from coal-fired plants. *Health Physics* **50**, 270–273.

Taskaev, A.I., Landa, E.R., Guryev, D.V., Golovko-Butler, N., Kraemer, T.F. (2003). Vodnyi: a long-term, low-level radiation exposure field site in Russia. *Japanese Journal of Health Physics* **38**, 332–343.

Thomas, V., Theis, T., Lifset, R., Grasso, D., Kim, B., Koshland, C., Pfahl, R. (2003). Industrial ecology: policy potential and research needs. *Environmental Engineering Science* **20**, 1–9.

Tsezos, M., Baird, M.H.J., Shemilt, L.W. (1987). The use of immobilised biomass to remove and recover radium from Elliot Lake uranium tailing streams. *Hydrometallurgy* **17**, 357–368.

U.S. Department of Energy (1993). Generic protocol for thorium-230 cleanup/verification at UMTRA project sites. Uranium mill tailings remedial action project document. DOE-Albuquerque.

U.S. Department of Energy (1996). Integrated data base report—1995: U.S. spent nuclear fuel and radioactive waste inventories, projections, and characteristics; DOE report DOE/RW-0006, Rev. 12, 271 p.

U.S. Environmental Protection Agency (1982). Disposal of radium-barium sulfate sludge from a water treatment plant in Midland, South Dakota. Technical Assistance Program Report PB-83-181875.

U.S. Environmental Protection Agency (1992). 40 CFR Part 61, National emissions standards for hazardous air pollutants, National emissions standards for radon emissions from phosphogypsum stacks. Final rule. Federal Register, 57, no. 107, June 3, 1992, pp. 23305–23320.

U.S. Environmental Protection Agency (1993). Diffuse NORM waste—waste characterization and preliminary risk assessment. U.S. Environmental Protection Agency, Office of Radiation and Indoor Air report RAE-9232/1-2, vol. 1 (draft; may 1993).

U.S. Environmental Protection Agency (1999). Technologically enhanced naturally occurring radioactive materials in the southwestern copper belt of Arizona. EPA Report 402-R-99-002. Available at http://www.epa.gov/radiation/pubs.htm#tenorm.

Ulrich, G.A., Breit, G.N., Cozzarelli, I.M., Suflita, J.M. (2003). Sources of sulfate supporting anaerobic metabolism in a contaminated aquifer. *Environmental Science and Technology* **37**, 1093–1099.

Vaca, F., Manjón, G., García-León, M. (1999). $^{90}$Sr in an alkaline pulp mill located in the South of Spain. *Journal of Environmental Radioactivity* **46**, 327–344.

Vaca, F., Manjón, G., García-León, M. (2001). Radioactive discharges from an alkaline pulp mill located in the South of Spain. *Journal of Environmental Radioactivity* **52**, 91–97.

Valentine, R.L., Spangler, K.M., Meyer, J. (1990). Removing radium by adding preformed hydrous manganese oxides. *American Water Works Association Journal* **82** (2), 66–71.

van der Lugt, G., Scholten, L.C. (1985). Radon emanation from concrete and the influence of using flyash in cement. *Science of the Total Environment* **45**, 143–150.

van Weers A.W. (2002). Inputs of natural radionuclides from non-nuclear industries into the north east European waters, NORM sources from Marina II. International Union of Radiocology Conference September 2–5, Monaco.

Veblen, L.A., Farthing, D.J., O'Donnell, E., Randall, J.D. (2004). Characterization of radioactive slags. NUREG 1703, U.S. Nuclear Regulatory Commission. Available at http://www.nrc.gov/reading-rm/doc-collections/nuregs/staff/.

Veil, J.A., Smith, K.P., Tomasko, D., Elcock, D., Blunt, D.L., Williams, G.P. (1998). Disposal of NORM—contaminated oil field wastes in salt caverns. U.S. Department of Energy. Available at www.ead.anl.gov/pub/dsp_detail.cfm?PubID=983.

Vukovic, Z., Mandic, M., Vukovic, D. (1996). Natural radioactivity of ground waters and soil in the vicinity of the ash repository of the coal-fired power plant "Nikola Tesla" A—Obrenovac (Yugoslavia). *Journal of Environmental Radioactivity* **33**, 41–48.

Wahl, A.C., Bonner, N.A. (Eds.) (1951). *Radioactivity Applied to Chemistry*. John Wiley & Sons, New York, 604 p.

Weimer, W.C., Kinnison, R.R., Reeves, J.H. (1981). Survey of radionuclide distributions resulting from the Church Rock, New Mexico, uranium mill tailings pond dam failure. U.S. Nuclear Regulatory Commission report NUREG/CR-2449.

White, G.J., Rood, A.S. (2001). Radon emanation from NORM-contaminated pipe scale and soil at petroleum industry sites. *Journal of Environmental Radioactivity* **54**, 401–413.

Winde, F., van der Walt, I.J. (2004). The significance of groundwater-stream interactions and fluctuating stream chemistry on waterborne uranium contamination of streams—a case study from a gold mining site in South Africa. *Journal of Hydrology* **287**, 178–196.

Yusko, J.G. (1996). An overview of Pennsylvania's experience with NORM. In: Proceedings of the 4th Annual Beneficial Reuse Conference (University of Tennessee, October 22–24, Knoxville, TN).

Zielinski, R.A., Otton, J.K., Budahn, J.R. (2001). Use of radium isotopes to determine the age and origin of radioactive barite at oil-field production sites. *Environmental Pollution* **113**, 299–309.

# Protection of the environment from exposure to ionising radiation

David Copplestone[a], Steve Jones[a], Rob Allott[a], Peter Merrill[a], Jordi Vives[b]

[a]*Environment Agency, PO Box 12, Richard Fairclough House, Knutsford Road, Warrington, WA4 1HG*
[b]*Westlakes Scientific Consulting, Westlakes Science and Technology Park, Moor Row, Cumbria, CA24 3LN*

## 1. Introduction

The current recommendations of the International Commission on Radiological Protection (ICRP) (ICRP, 1991) state that:

> *"The Commission believes that the standard of environmental control needed to protect man to the degree currently thought desirable will ensure that other species are not put at risk. Occasionally, individual members of non-human species might be harmed, but not to the extent of endangering whole species or creating imbalance between species."*

Whilst this statement may well be true in circumstances where pathways to man are present in the environment in which other species are exposed (IAEA, 1992; UNSCEAR, 1996a, 1996b), there has been a growing acceptance that it is not a sufficiently sound basis to demonstrate positively that the natural environment is protected from the impacts of anthropogenic radioactivity. Furthermore there is now a pressing need in many countries, which are implementing conservation-based legislation, to be able to demonstrate protection of the environment in its own right.

For example, the Rio Declaration on Environment and Development (United Nations, 1992) set out important philosophical principles for the protection of the environment and established the concept that the environment should be protected in its own right, in addition to the requirement to protect it as a resource for humans to inhabit and exploit. During the 1990s, and indirectly as a result of the Rio Declaration, national regulatory bodies in Canada and Sweden came under pressure to consider the impact of anthropogenic radioactivity on the natural environment because their regulatory framework for radioactive emissions and waste disposals explicitly required protection of man and the environment to be considered. Largely as a consequence of this, a series of symposia on the protection of the environment from ionising radiation (the SPEIR conferences) were convened in Stockholm (1996), Ottawa (1999) and Darwin (2002). In parallel, several regulatory agencies began to develop methodologies to assess the impact of anthropogenic radioactivity on the natural environment because of the

RADIOACTIVITY IN THE ENVIRONMENT
VOLUME 10 ISSN 1569-4860/DOI: 10.1016/S1569-4860(06)10011-X

need to demonstrate that there was no impact of authorised releases of radioactivity under specific articles of legislation (e.g. in the UK the need to comply with the EU Birds and Habitats Directives (Council Directives 79/409/EEC on the conservation of wild birds and 92/43/EEC on the conservation of natural habitats, and wild flora and fauna)). They began to apply these methodologies as part of their normal regulatory activities.

This chapter summarises recent developments in this area, sets out likely future developments and trends, and identifies potential implications for both regulators and operators and provides, as an example, an impact assessment for an estuarine site in England as determined using the Environment Agency's current assessment methodology. This example highlights the overall tiered assessment approach being adopted in several assessment methodologies and indicates the knowledge gaps and limitations inherent in the assessment process.

## 2. The current conceptual basis for protection of the environment from ionising radiation

Before summarising specific developments, it may be helpful to discuss the general problems faced by any framework or methodology which attempts to demonstrate protection of the natural environment from adverse impacts attributable to anthropogenic radioactivity.

The first major problem is, of course, the complexity and variety of the natural environment. Protecting man from the impacts of anthropogenic radioactivity released into the environment also involves a great deal of complexity, but this can be managed because only pathways back to a single species *(Homo sapiens)* need to be considered. Anatomical and biokinetic models for 'reference man', together with coefficients of radiation risk from human epidemiological studies, can be used as the basis of a scientifically based assessment of risk.

In the case of the natural environment, it would clearly be impossible to consider every possible species of organism which may be affected and simplification is essential if any approach is to be workable. The currently accepted approach is to simplify matters by considering 'reference organisms' which inhabit 'reference ecosystems'. Reference organisms are intended to be representative of the range of habitats occupied, radionuclide uptake behaviours, and sizes and shapes (affecting dosimetric calculation) which would characterise a real ecosystem.

The second major problem is deciding what the objectives and criteria for protection should be. Clearly the aim should be to ensure 'no deleterious effect', but there are areas for debate on whether populations or individual organisms should be the object of protection. Here there appears to be some consensus that the main focus should be on protection of populations, but at the current state of knowledge this can only be assessed by considering likely effects on individual organisms and using generalisation to assess how any likely effects at the individual level may become manifest at the population level. There is general agreement that the effects, or endpoints, at individual level which will need to be considered are:

- Enhancement of mortality.
- Enhancement of morbidity.
- Reduction of reproductive success.
- Enhancement of scorable cytogenetic effects.

The third major problem is how to assess the likelihood of these effects in individual organisms within an exposed population. The generally accepted approach, following on from the radiological protection framework for humans, is to:

- Determine the distribution of radionuclides in the ecosystem by measurement or modelling.
- Calculate the absorbed radiation doses to organisms.
- Assess the likelihood of effects from dose-response relationships derived from laboratory experiments or field studies of heavily exposed populations.

The principal variation in approach here is whether the approach is to evaluate explicitly the likelihood of effects and consider whether the result is acceptable (as, for example, the FASSET approach—see below) or whether to evaluate the results of the assessment against pre-set criteria in terms of dose (as, for example, the Environment Agency in England and Wales and US-DoE approaches—see below).

Although the concept of basing protection on the assessment of radiation dose is broadly accepted, gaps in the data required to perform a comprehensive assessment of dose to exposed organisms has prompted debate as to whether simpler approaches based on toxicity testing and environmental concentration standards would be preferable. However, closer examination suggests the possible alternatives offer no real simplification.

## 3. State of the art at the end of the 20th century

Coincidentally, the turn of the century has marked a point at which practical developments in this area have accelerated substantially. The state of the art in 2000 may be summarised as follows:

- *Reference organisms and reference ecosystems:* the concept of reference organisms had been promulgated in a number of publications (e.g. Pentreath, 1999).
- *Radionuclide transfer models and parameters:* Few radionuclide transfer models, or compilations of associated transfer parameters, intended for the assessment of radionuclide uptake by organisms other than those forming part of the human food-chain are available. Derivation of appropriate transfer factors required basic searches of the radioecological literature.
- *Dosimetric models:* Basic methods for the calculation of radiation doses to organisms of various sizes, represented as ellipsoidal shapes with uniform internal distribution of radionuclides, were documented (e.g. NCRP, 1991). However, result tabulation was limited to a series of energy dependency curves for a limited number of organism sizes and dimensions.
- *Dose-effect relationships and dose rate criteria:* Effects of ionising radiation on organisms of many kinds had been considered in reviews by IAEA and UNSCEAR, and broad conclusion reached that exposure of populations of terrestrial organisms to dose rates of 40 $\mu$Gy h$^{-1}$, and exposure of populations of aquatic organisms to dose rates of 400 $\mu$Gy h$^{-1}$, would be unlikely to lead to observable effects in populations (IAEA, 1992; UNSCEAR, 1996a, 1996b). However, no general guidance was available on dose rates at which particular effects (e.g. enhancements of mortality and morbidity, reduction in repro-

ductive success, enhancement of the incidence of scorable cytogenetic effects) might be observed in different taxa.

Subsequent advances in the state of the art, both in terms of the framework for protection and the technical and scientific basis for carrying out assessments, are described in the sections which follow.

## 4. Developments at the national level

### 4.1. *UK*

During 2000, the Environment Agency (EA) in England and Wales faced questions from English Nature, in their role as a statutory consultee on the determination of radioactive waste discharge and disposal authorisations under the Radioactive Substances Act 1993. English Nature asked, not unreasonably, how they could sensibly take a view on the acceptability of proposed discharges when no assessments were being made of the potential impact on wildlife. The need to implement the EU Birds and Habitats Directives (see below) in the UK raised the same issue. As a result, the Environment Agency and English Nature jointly developed methods for assessing the impact of ionising radiation on wildlife. The resulting report (Copplestone et al., 2001) comprised:

- A review of pathways for the exposure of plants and animals to ionising radiation;
- A review of the effects of ionising radiation on biota;
- A review of existing regulatory frameworks for environmental protection;
- A review of methods for dose calculation, including review of the relative biological effectiveness of different radiation types and its treatment in dose calculations;
- Description of a methodology for calculation of radiation doses in 'reference' terrestrial, freshwater, and coastal marine ecosystems, including the specification of 'reference organisms', default values for radionuclide transfer factors, and dose coefficients for radionuclides uniformly distributed within the organism (internal dose coefficients) and uniformly distributed within the surrounding environmental medium (external dose coefficients);
- Recommendation of dose rate criteria which could be used to judge the acceptability of the assessed doses;
- Incorporation of the methodology into an Excel workbook based dose calculator.

The dose calculation methods, and some aspects of the overall approach, were based on further development of techniques established during an earlier work on a pilot assessment of the impact of a generic deep repository for low and intermediate level radioactive waste in the UK (Jones et al., 2003).

The review of effects essentially supported the IAEA/UNSCEAR dose rate criteria; however, it also recommended that, as uncertainties in dose assessment can be substantial, situations in which assessed doses exceed 5% of the IAEA/UNSCEAR criteria (i.e. 2 $\mu$Gy h$^{-1}$ for terrestrial ecosystems, 20 $\mu$Gy h$^{-1}$ for aquatic ecosystems) should be subject to more detailed examination in order to reduce uncertainty.

Subsequent to the report by Copplestone et al. (2001), the Environment Agency assessment method was extended to incorporate additional radionuclides and now comprises the following.

- The *marine coastal ecosystem*. This includes as reference organisms benthic bacteria, phytoplankton, zooplankton, macrophytes, fish eggs, benthic molluscs, benthic crustacea, pelagic fish, benthic fish, seabirds, seals and whales. Concentration factor data and dose coefficients are provided for these organisms for the radionuclides $^{3}$H, $^{14}$C, $^{32}$P, $^{60}$Co, $^{90}$Sr, $^{99}$Tc, $^{106}$Ru, $^{125}$I, $^{129}$I, $^{131}$I, $^{137}$Cs, $^{210}$Po, $^{234}$Th, $^{238}$U, $^{239}$Pu and $^{241}$Am. However, there are significant gaps in the concentration factor data for a number of these organisms.
- The *freshwater ecosystem*. This includes as reference organisms benthic bacteria, phytoplankton, zooplankton, macrophytes, benthic molluscs, benthic crustacea, pelagic fish, benthic fish, aquatic mammals, and waterbirds. Concentration factor data and dose coefficients are provided for the same radionuclides as the marine ecosystem. Here also there are significant gaps in the concentration factor data for a number of organisms.
- The *terrestrial ecosystem*. This includes as reference organisms soil bacteria, lichens, trees, shrubs, herbs, seeds, fungi, caterpillars, ants, bees, woodlice, earthworms, herbivorous and carnivorous mammals, rodents, birds, bird eggs, and reptiles. Concentration factor data and dose coefficients are provided for $^{3}$H, $^{14}$C, $^{32}$P, $^{35}$S, $^{41}$Ar, $^{60}$Co, $^{85}$Kr, $^{90}$Sr, $^{106}$Ru, $^{129}$I, $^{131}$I, $^{137}$Cs, $^{226}$Ra, $^{234}$Th, $^{238}$U, $^{239}$Pu and $^{241}$Am. As for the aquatic ecosystems, there are significant gaps in the concentration factor data for a number of organisms; for the terrestrial system the lack of concentration factor data is particularly marked.

Having developed a methodology, the Environment Agency is now routinely asking site operators to consider the effects of proposed discharges on wildlife during the process of determining radioactive discharge authorisations.

Implementation of the EU Birds and Habitats Directives in the UK, through the Conservation (Natural Habitats) Regulations 2000, has provided the Environment Agency with a further impetus to extend the methodology. These Regulations require the Agency to ensure that no authorised emissions (radioactive or otherwise) have an adverse impact on conservation sites designated under the Regulations (collectively referred to as the Natura 2000 sites. In order to achieve this the Agency has further extended the methodology of Copplestone et al. (2001) by:

- Identifying the organisms which are of prime conservation significance at each of the Natura 2000 sites ('feature organisms');
- Determining which of the reference organisms from the Copplestone et al. (2001) methodology provides the best surrogate for each 'feature organism' in terms of dose coefficients;
- Identifying from the literature, where available, suitable concentration or transfer factor values for radionuclide uptake by 'feature organisms';
- Where (as is often the case) reliable concentration or transfer factors cannot be derived, assigning pessimistic concentration or transfer factor values for use in a 'screening' assessment.

This staged approach has allowed the Agencies to fold in new science as it has developed in this field. Further details are given in the example assessment provided below. In addition, the Environment Agency has produced a compilation of basic radionuclide data relevant to the assessment of impacts on wildlife (Kelly and Thorne, 2003). Finally, the Environment Agency recognises that basic data on the effects of irradiation are scarce or non-existent for some taxa, particularly in the case of chronic irradiation and in the case of high LET irradiation from alpha emitters. They have produced guidance on protocols for experimental studies of this type (Wood et al., 2003), as a precursor for such studies in the future (e.g. Hingston et al., 2004).

### 4.2. *USA*

The US Department of Energy (USDoE) has set internal criteria and procedures for the protection of wildlife both on and around the extensive facilities for which it has responsibility (Figure 1). Criteria are based directly on dose standards; a limit of 10 mGy $d^{-1}$ has been set for aquatic biota (USDoE, 1993) and consultation documents propose extension to include limits of 10 mGy $d^{-1}$ for terrestrial plants and 1 mGy $d^{-1}$ for terrestrial animals (USDoE, 1996).

An extensive assessment manual, with accompanying software (USDoE, 2002), has been developed. The methodology is based on a 'screening' approach in which site conditions are first tested against derived soil and water concentrations (Biota Concentration Guides, BCGs) before (if necessary) proceeding to more sophisticated analyses:

Biota Concentration Guides are derived using pessimistic organism–media concentration factors, and conservative dose coefficients which assume the organism is simultaneously infinitely small (for calculation of external doses), and infinitely large (for calculation of internal doses).

If the assessment proceeds to more detailed stages, radionuclide uptake by the site specific organisms are assessed by kinetic-allometric models. These models make assumptions about the variation of food (and hence radionuclide) intake and clearance rates, based on power

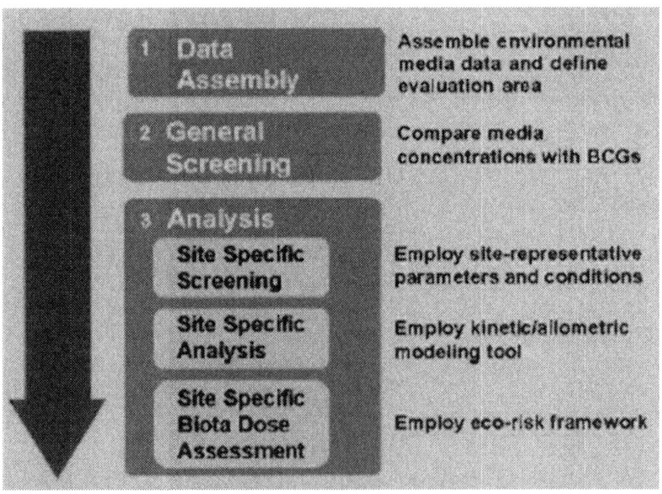


Fig. 1. The US DoE graded approach for assessing the impact of ionising radiation on biota (USDoE, 2002).

functions of body mass. Even at the more detailed stages of analysis, organism geometry does not appear to be taken into account when assessing doses.

The methodology is very well documented and clear guidance is provided for assessors. However there are some technical weaknesses, principally:

- The kinetic-allometric models lack detailed justification and are unlikely to provide accurate predictions of radionuclide uptake by biota, unless validated at a site specific level.
- The assessments appear to focus attention at organisms from the higher trophic levels—e.g. fish, terrestrial mammals—without considering whether effects on lower trophic levels—e.g. phytoplankton, soil invertebrates—may compromise ecosystem function.
- The treatment of dosimetry remains simplistic and conservative even at the more detailed levels of assessment.

The methodology has now been incorporated into the RESRAD-BIOTA code, which addresses the weakness in dosimetry by introducing size dependent dose conversion factors (USDoE, 2004).

These methods are being applied systematically at all USDoE sites.

### 4.3. *Canada*

The Canadian Environmental Protection Act (1999) and the Canadian Environmental Assessment Act (2003) require the identification, assessment and management of toxic substances (including radionuclides and ionising radiation) and require ecological risk assessments to be carried out for major new 'projects' and major existing facilities. These requirements apply to most major nuclear activities in Canada.

The responsibility for ensuring that assessments are carried out falls to the Canadian Nuclear Safety Commission, which in turn requires assessments to be carried out and submitted by the site operators.

A radionuclide-specific methodology is to be published in 2004. However, at present there is no prescriptive methodology; the approach taken is to undertake a generic human health and ecological risk assessment, considering both radioactive and non-radioactive stressors (see, for example, Swami et al., 1999).

A number of significant issues have arisen from experience in carrying out these assessments (Mihok, 2004). For example:

- Industry does not accept the ecological risk assessment approach as a mature science because it not infrequently identifies risk where no ecosystem stress is apparent. This, presumably, results from use of conservatism in the assessment as a response to lack of firm data.
- There are major difficulties in translating risk quotients based on individual effects to Canadian criteria in terms of 'magnitude, extent, duration, frequency and permanence' of effects at the population level.
- Public concern focuses on particular organisms at individual sites ('Valued Ecosystem Components') and use of reference or surrogate organisms, or data from another site, does not create confidence.

### 4.4. *Australia*

The Supervising Scientist Division of the Australian Department of the Environment and Heritage has carried out an ecological risk assessment of impacted aquatic ecosystems for the Ranger uranium mine in the Northern Territories (Johnson et al., 2005). Their stated objective (derived from Australian criteria for regulation of chemical stressors) was to ensure that 99% of potentially affected species would be protected.

The assessment included both radioactive and non-radioactive stressors. For non-radioactive stressors the results of extensive site-specific toxicity research were used as the benchmarks of harm; for ionising radiation, in the absence of clearer criteria, the IAEA/UNSCEAR value of 400 $\mu Gy\,h^{-1}$ was used.

Extensive site-specific measurement data were available for concentrations of both radioactive and non-radioactive stressors. For radionuclides, dose rate calculations were made using the spreadsheets of Copplestone et al. (2001).

The main findings were:

- Chemical toxicity of mine effluents was more limiting than radiological effects for protection of both humans and wildlife.
- For radiological effects, protection of wildlife was more limiting than protection of humans, largely because human exposure pathways were more remote from the source than were wildlife exposure pathways.

However, early work by the Australian Nuclear Science and Technology Organisation (Twining et al., 2003) in applying the Australian criteria for protection from chemical stressors to available effects data for radionuclides suggests that dose rate criteria on this basis may be substantially lower than 400 $\mu Gy\,h^{-1}$.

In addition, the Australian Radiation Protection and Nuclear Safety Agency (ARPANSA) has published a brief review of the issues surrounding radiation protection of the environment (Cooper, 2003). This concludes that, in developing the Australian National Radioactive Waste Repository:

> *"... it is proposed to rely on the ICRP system of radiation protection and the framework of dose limitation for members of the public to ensure that non-human biota are provided an adequate degree of protection. It is recognised that adherence solely to established dose limits does not ensure total protection of non-human species, and other strategies are in place to provide the required level of protection."*

However, it is not clear from the report what these additional strategies comprise, and how their efficacy or otherwise would be assessed.

### 4.5. *France*

There are currently no regulatory requirements for assessing impact of radionuclides on wildlife. However, COGEMA has carried out two significant case studies. These consider impacts on the Rhone due to discharges from the Marcoule site (St-Pierre et al., 1999), and impacts on the coastal waters around the Cotentin peninsula due to discharges from the Cap de la Hague reprocessing plant (COGEMA, 2003).

In both cases the reports were based on measurements from the routine site monitoring programmes together with any available data from special investigations or other studies. As a result, the assessments are focused on those organisms for which monitoring data exist; although this gives quite comprehensive coverage the studies do not attempt an *a priori* categorisation of organisms which are likely to be present in the ecosystems studied and be important from the viewpoint of ecosystem function and/or conservation. The dosimetric approach was similar to that of Copplestone et al. (2001), and cross-checks in the course of peer review have confirmed close agreement in the dose coefficients used.

The reports conclude in both cases that radiation dose rates to the organisms studied from anthropogenic radioactivity are only a small fraction of the dose rates due to background sources and are well below the IAEA/UNSCEAR guidance values. As a result no adverse effects could be expected.

## 5. Developments at the international level

### 5.1. *EU Frameworks 5 and 6—the FASSET and ERICA projects*

The FASSET project commenced in November 2000 under the EC 5th Framework programme, and concluded in October 2003. Its objective was to develop a framework for the assessment of environmental impact of ionising radiation in European ecosystems. It has produced six scientific reports, all of which are available from the FASSET web site (www.fasset.org).

The FASSET framework is based on a conventional stepwise environmental assessment approach, comprising:

1. Planning the assessment: regulatory requirements, stakeholder views, aims and objectives.
2. Problem formulation: describing practice or activity and potentially impacted ecosystems, initial hazard analysis, decision on complexity of assessment required.
3. Exposure and effects analysis: quantifying exposure of relevant organisms to stressors, determination of prevalence and severity of any effects which may be expected.
4. Risk characterisation: identifying, evaluation and prioritising resulting risks to the environment.
5. Decision and management: determining appropriate management action—permissions and consents, remediation, amelioration.

The FASSET project has focused on steps 2 and 3. It has provided a basis for describing ecosystems in terms of the ‘reference’ systems of forests, semi-natural pastures and heathlands, agricultural land, wetlands, freshwaters, marine waters, and brackish waters. These systems are ‘populated’ by reference organisms which are broadly similar in nature and range to those used in Copplestone et al. (2001). Data are provided on available radionuclide transfer factors; dose coefficients are provided as a basis for calculation of dose rates; and example models of radionuclide transfer are provided. The FASSET Radiation Effects Database (FRED) provides a basis for assessing the likely prevalence and severity of any effects which may be expected on the basis of calculated dose rates.

It should be noted that the FASSET framework explicitly avoids setting 'dose limits' for biota, instead aiming to identify whether any effects may be expected so that their acceptability may be considered.

The FASSET project involved no new experimental work and radionuclide transfer factors were extracted from the available literature. Thus, the conclusions of earlier work, described above, that there are significant data gaps for the terrestrial environment and, to a somewhat lesser extent, the freshwater aquatic environments, have been confirmed. Report 5 within the FASSET framework (above) contains some guidance on how allometric relationships may be used to fill gaps in certain limited circumstances, but this is still an area which requires attention.

The ERICA project is an EC 6th Framework project which aims to extend the work of FASSET into the risk characterisation and management steps, and also to fill gaps revealed by the FASSET project where this is feasible within the project scope. ERICA does make some provision for additional experimental work and field studies on radionuclide transfers to biota, and radiation effects; however the scope has been limited by available funding.

An important part of the ERICA project is a series of case studies which will initially evaluate the utility of the FASSET approach and data at a number of sites, feeding back significant issues into the ongoing ERICA work, and subsequently evaluate the improvement achieved by the results of ERICA. These case studies comprise:

- Conservation sites on saltmarsh and dunes impacted by emissions from Sellafield, UK.
- Agricultural land impacted by emissions from Sellafield, UK.
- Marine ecosystems impacted by emissions from oil and gas platforms, Norway.
- Freshwater ecosystems (river Loire) impacted by emissions from nuclear power plants, France.
- Terrestrial ecosystems impacted by deposition from the Chernobyl accident, Ukraine.
- Terrestrial ecosystems impacted by emissions from uranium extraction industry, Komi, Russia.

### 5.2. *The EC MARINA II project*

The original EC MARINA project considered the radiological impact on humans of radioactive discharges to the marine environment from all EU member states. The Marina II project, completed in 2002, updated this original study but also included an assessment of the impacts on marine biota (Sazykina and Kryshev, 2002). This extension was driven in part by the interest of the OSPAR Radioactive Substances Committee in 'environmental quality' as related to radionuclides.

The report concludes:

- An adequate methodology has been developed for the assessment of doses to marine organisms.
- Doses to marine organisms varied within a very wide range up to $10^{-4}$ Gy day$^{-1}$ in the most impacted zones.
- The highest doses were found for biota close to the Sellafield plant and also for biota close to industries processing naturally occurring radioactive materials.
- For these areas of highest dose, deterministic effects on organisms would not be expected.

- Doses to organisms due to discharges from the offshore oil industry require further investigation.

## 5.3. *Activities of the International Union of Radioecology*

The International Union of Radioecology (IUR) is an international professional organisation for those involved in the field of radioecology. As a result, its membership is heavily biased towards researchers from academic and government research institutions. It has limited funds and does not sponsor a significant amount of research from its own resources; rather, it convenes meetings and workshops and provides limited funds for travel to facilitate the co-ordination of research activities.

Since 1997, the IUR has been quite active in the area of radiation impacts on ecosystems, not least because the issue is likely to drive radioecology research requirements over the next decade.

Significant publications from the IUR over this period are:

- a compilation of data on calculation of doses to, and effects on, biota (IUR, 2000);
- output of a consensus conference on the general principles for protection of the environment from ionising radiation (Strand and Oughton, 2002);
- a report on the current status and future work in this area as of 2002 (IUR, 2002).
- a Statement of the IUR: Protection of the Environment in the 21st Century: Radiation Protection of the Biosphere including Humankind (IUR, 2003).

These IUR publications are useful as a summary of the issues, and the consensus conference makes a useful contribution to overall principles, but contribute relatively little substance to the process of developing an assessment compared to the national level and FASSET work described above. The 2003 Statement draws attention to the knowledge gaps that exist, and the consequent need for continuing research.

IUR is moving to establish a network of research organisations to promote collaboration and resource-effective research that will address these knowledge gaps. The network is expected to be ready to operate in 2005. However, IUR will not be funding the work; the purpose of the network is to utilise available sources of funding at, largely, national level.

## 5.4. *Activities of the Nuclear Energy Agency*

In parallel with the developing work of the ICRP (see below), the Nuclear Energy Agency of the Organisation for Economic Co-operation and Development has determined that it should provide a series of fora for stakeholders to debate the issues relating to radiological protection of the environment and provide input to ICRP's development of recommendations.

Three fora were planned—at the beginning of ICRP's development, following the issue of draft considerations by ICRP, and following the issue of recommendations.

The first forum was held in Taormina, Sicily in February 2002 and brought together operators, regulators, scientists, politicians and NGOs. A summary report and workshop proceedings are available from the NEA (NEA, 2002, 2003). Conclusions from the forum were as follows.

- The environment should be protected. In general, it is felt that the environment has been protected by current human radiological protection principles and the regulation which has flowed from them but the level of protection has not been demonstrated.
- There is sufficient knowledge to proceed with the development of recommendations on which regulation could be based. Scientific studies, such as the FASSET programme, will be needed; their role should be to identify gaps in knowledge and fill them in a prioritised way in support of policy and regulatory needs.
- Environmental stresses from radiation should not be considered in isolation from other pollutants and stressors in the environment. The resources devoted to protection of the environment from ionising radiation should be proportional to the scale of the hazard.
- ICRP is the organisation best placed to address the development of recommendations. ICRP's development should include wide circulation of draft materials, conducting of 'feasibility testing' and accommodation of comments from stakeholders.

The principal strategic output from this first forum was the 'legitimisation' of the involvement of ICRP in the issue and, effectively, positioning ICRP as the lead organisation internationally.

### 5.5. *Activities of the International Commission on Radiological Protection*

In May 2000 the Main Commission of the ICRP established a Task Group, reporting directly to it, in order to produce a report on the protection of the environment. The Group, chaired by Lars-Eric Holm of SSI, Sweden, was tasked with the aim of developing both a protection policy and suggesting a framework for environmental protection based on scientific and ethical-philosophical principles.

The report of this Task Group is now available as ICRP Publication 91 (ICRP, 2003). The main recommendation of the Task Group is that:

> "*the Commission therefore needs to revise its current system of protection, and particularly:*
>
> - *develop a comprehensive approach to the study of the effects on, and protection of, all living matter with respect to the effects of ionising radiation*;
> - *develop a system of radiological protection that includes protection of non-human species with a clear set of objectives and principles, and an agreed set of quantities and units applicable to all living things*;
> - *interpret basic knowledge of radiation effects in species other than humans so that they can be used in an environmental context, for example, in setting criteria or benchmarks of protection at the appropriate level of hierarchy (individuals or populations)*;
> - *develop a small set of primary reference fauna and flora, plus their relevant data bases so that others can develop more area and situation specific numerical approaches to assessment and management of risks to non-human species*;
> - *show its commitment to protection of non-human species and let this be reflected in the organisation of work and in the composition of experts*;
> - *plan regular reviews and revisions of this new system as new knowledge develops*."

ICRP's thinking as set out in the Task Group report is clearly aligned to the reference organism—reference ecosystem approach although indications are that ICRP will consider a more limited range of reference organisms[1] than have been used in the FASSET and UK

[1] The reference organisms being considered by ICRP are rodent, duck, frog, freshwater fish, marine flatfish, bee, crab, marine snail, earthworm, pine tree, grass, seaweed (ICRP, 2004).

| Derived consideration level | Relative dose level (incremental annual dose) | Likely effect on individuals | Aspects of concern |
|---|---|---|---|
| Level 5 | >1000 normal | Early mortality | Possible remedial action considered |
| Level 4 | >100 normal | Reduced reproductive success | Concern dependent on what fauna and flora, and their numbers, likely to be affected |
| Level 3 | >10 normal | Scorable DNA damage | Concern dependent upon size and nature of area affected |
| Level 2 | Normal background range | | No action considered |
| Level 1 | <Normal background | Low | No action considered |

Fig. 2. The proposed ICRP system of "derived consideration levels".

Environment Agency approaches. There is a tension here between simplicity of the system of protection and making it sufficiently comprehensive to be convincing. If the range of reference organisms is too restricted, it may not be possible to adequately 'map' the actual organisms of interest at a particular site onto the set of reference organisms (see, for example, Copplestone et al. 2003).

The indications are also that ICRP may not recommend specific dose limits. Their thinking appears aligned towards a system of 'derived consideration levels' (Figure 2) expressed as multiples of natural background exposures, for example:

This is, perhaps, in anticipation of the difficulty in interpreting radiation effects data for a wide range of organism types. However, it will demand that levels of natural background exposure are well understood; this is not the case at present, and it is clear that natural background exposures cover a wide range and vary substantially between taxa (e.g. Brown et al., 2004).

Recently ICRP have announced that they are to set up a new Committee to further develop their recommendations in this area. Establishing a new Committee is a major step for ICRP and clearly signals the importance they attach to this topic.

### **5.6.** *Activities of the United Nations Scientific Committee on the Effects of Atomic Radiation*

The United Nations Scientific Committee on the Effects of Atomic Radiation (UNSCEAR) has published a substantial review of the effects of ionising radiation on biota together with data on radiation doses to which biota have been exposed, both from anthropogenic and (to a limited extent) from natural sources (UNSCEAR, 1996a, 1996b). UNSCEAR's current work programme includes identification of dose response relationships for effects on biota and case studies at contaminated sites. A further report is planned for 2006.

## 5.7. *Activities of the International Atomic Energy Agency*

The IAEA has published some of the earliest substantial material on the topic of protection of biota from ionising radiation (IAEA, 1988, 1992) and so has a long history of engagement with the topic. More recently, IAEA has convened experts meetings in 2000 and 2001, and has used a group of consultants to produce a number of technical documents.

TECDOC-1091 (IAEA, 1999) set out to examine the need for development of criteria and approaches for protection of the environment from the effects of ionising radiation, and to promote discussion which might eventually lead to the development of an IAEA Safety Standard. It reviewed developments occurring in IAEA member states, concluding that development of guidance and criteria is desirable, and that the state of knowledge is sufficient to move forward although improved knowledge will be required in certain areas.

TECDOC-1270 (IAEA, 2002) discussed ethical considerations involved in protecting the environment from ionising radiation. From this report a number of general principles have been drawn:

- *any radiation exposure should not affect the capability of the environment to support future generations of humans and biota (principle of sustainability)*;
- *any radiation exposure should not have any deleterious effect on any species, habitat, or geographic feature that is endangered or is under ecological stress or is deemed to be of particular societal value (principle of conservation)*;
- *any radiation exposure should not affect the maintenance of diversity within each species, amongst different species, and amongst different types of habitats and ecosystems (principle of maintaining biodiversity)*;
- *the management of any source of radiation exposure of the environment should aim to achieve an equitable distribution of the benefits of the source of the radiation exposure and the harm to the environment resulting from the radiation exposure, or to compensate for any inequitable damage (principle of environmental justice)*; *and*
- *in decisions on the acceptability and appropriate management of any source of radiation exposure of the environment, the different ethical and cultural views held by those humans affected by the decisions should be taken into account (principle of respect for human dignity).*

IAEA sponsored an international conference on the protection of the environment from ionising radiation in Stockholm in October 2003 (IAEA, 2003). This conference reached the conclusion that:

> "*While accepting that there remain significant gaps in knowledge and that there needs to be continuing research... there is an adequate knowledge base to proceed and (the conference) strongly supported the development of a framework for environmental radiation protection*"

and:

> "*the time is ripe for launching a number of international initiatives to consolidate the present approach to controlling radioactive discharges to the environment by taking explicit account of the protection of species other than humans*".

Most recently, IAEA has drafted an Action Plan (IAEA, 2005) on protection of the environment from ionising radiation which was discussed at an expert's meeting in the first week of June 2004. The draft Action Plan proposes:

- IAEA should establish a Steering Committee to review and co-ordinate the actions of UNSCEAR, ICRP, IUR and IAEA together with nominated senior experts;

- IAEA should continue to foster information exchange through workshops, conferences, training courses and other fora;
- IAEA should develop Safety Requirements that set out the principles for environmental protection. This should proceed in parallel with UNSCEAR and ICRP's work but should not be finalised until UNSCEAR and ICRP have published;
- IAEA will assist member states to develop capabilities to conduct assessments through case studies and model inter-comparisons;
- Following publication of ICRP's recommendations, IAEA will review its existing standards and guidance documentation to include consideration of the protection of non-human species.

## 6. Example of the Environment Agency assessment approach in England and Wales

The Environment Agency has a duty to comply with the EU Birds and Habitats Directives (Council Directives 79/409/EEC on the conservation of wild birds and 92/43/EEC on the conservation of natural habitats, and wild flora and fauna) when planning and undertaking all of its regulatory and operational activities within England and Wales. These European Directives were introduced into UK legislation by the Conservation (Natural Habitats) Regulations 1994, as amended in 2000. Under these Regulations the Agency has obligations to review relevant existing authorisations, consents, licences and permissions (hereafter collectively referred to as permissions) to ensure that no Agency authorised permission results in an adverse effect, either directly or indirectly, on the integrity of identified European Sites within England and Wales. In this context the sites consist of classified, or potential, Special Protection Areas (SPA) created under the Birds Directive and candidate, or designated, Special Areas of Conservation (SAC) under the Habitats Directive. These sites are collectively referred to as Natura 2000 sites.

One of the stressors requiring assessment is ionising radiation. However, it should be noted that the need for this is being driven by conservation pressure and not from a real or perceived radiological hazard in the environment. As mentioned previously, a staged assessment methodology has been derived and the steps taken are outlined below.

### 6.1. *Overview of the assessment approach*

#### 6.1.1. *Stages 1 & 2 Assessments*

Stages 1 and 2 assessments have been used to determine the number of potentially impacted Natura 2000 sites following the approach described in EA, 2002. The Stage 1 assessment filtered out applications and activities authorised by the Agency that, by virtue of their nature or location, could not conceivably have an effect on the features of interest of given European sites. This was conducted using simple rules to determine the likelihood of an authorised discharge reaching a Natura 2000 site so, for example atmospheric releases are only considered further if they occur within 1 km of the Natura 2000.

Having identified a number of potentially impacted sites in Stage 1 (mainly intertidal areas of estuaries), the Stage 2 assessment then reviewed the maximum permissible radioactive discharge levels from authorised sites and compared these to defined screening levels (Allott

Table 1
Terrestrial reference organisms as listed in Copplestone et al. (2001)

| Reference organism (and example species) | Reference dimensions (mm) | Mass (kg) fresh weight (FW) |
|---|---|---|
| Lichen | 100 × 5 × 5 | $1.31 \times 10^{-3}$ |
| Moss | 100 × 10 × 5 | $2.62 \times 10^{-3}$ |
| Tree (root) | 100 × 2 × 2 | $2.10 \times 10^{-4}$ |
| Shrub (root) | 100 × 2 × 2 | $2.10 \times 10^{-4}$ |
| Herb (root) | 100 × 2 × 2 | $2.10 \times 10^{-4}$ |
| Germinating seed | 6 × 1 × 1 | $1.80 \times 10^{-6}$ |
| Fungal fruiting body | 30 × 15 × 10 | $2.63 \times 10^{-3}$ |
| Caterpillar | 30 × 7 × 7 | $7.70 \times 10^{-4}$ |
| Social Insect—ants | 5 × 3 × 3 | $2.00 \times 10^{-5}$ |
| Social Insect—bee | 20 × 15 × 10 | $2.00 \times 10^{-3}$ |
| Wood louse | 15 × 6 × 3 | $1.00 \times 10^{-3}$ |
| Earthworm | 100 × 5 × 5 | $3.50 \times 10^{-3}$ |
| Herbivorous mammal (rabbit) | 300 × 150 × 100 | $8.00 \times 10^{-1}$ |
| Carnivorous mammal (fox) | 670 × 350 × 180 | 5.50 |
| Small burrowing rodent (mouse) | 100 × 20 × 20 | $2.00 \times 10^{-2}$ |
| Woodland bird (grouse) | 350 × 150 × 150 | 1.50 |
| Bird egg | 40 × 25 × 25 | $1.30 \times 10^{-3}$ |
| Reptile (grass snake) | 1200 × 60 × 60 | 2.26 |

and Dunn, 2001)). Research conducted by Copplestone et al. (2001) was used to underpin the derivation of the discharge screening levels. This approach used the concept of reference organisms, which have been defined as "*a series of entities that provide a basis for the estimation of radiation dose rate to a range of organisms which are typical, or representative, of a contaminated environment. These estimates, in turn, would provide a basis for assessing the likelihood and degree of radiation effects. It is important that they are not a direct representation of any identifiable animal or plant*" (Strand and Larsson, 2001). The reference organism concept therefore provides a series of organism types, which can be considered representative of different trophic levels (see Table 1).

It is possible to determine relevant ecological parameters for a species representative of the reference organism (for example, a herbivorous mammal might be represented by a rabbit). The ecological parameters provide information on prey, predators and the time spent in different compartments of the ecosystem, for example for a duck, how much time is spent on sediment, surface of the water, flying etc. It is also possible to define a simplified geometry (usually as an ellipsoid) for the purposes of dosimetric calculations (full details are provided in Copplestone et al., 2001). The equations used are given below. For each reference organism a concentration factor has been defined relative to soil, water or air depending upon the type of assessment (freshwater, estuarine/marine or terrestrial) and radionuclide. Figure 3 provides an overview of the assessment methodology process.

The assessment approach was developed for the following radionuclides for different ecosystems, as follows. Terrestrial only: $^{35}S$, $^{41}Ar$, $^{85}Kr$, $^{226}Ra$. Aquatic only: $^{99}Tc$, $^{125}I$,

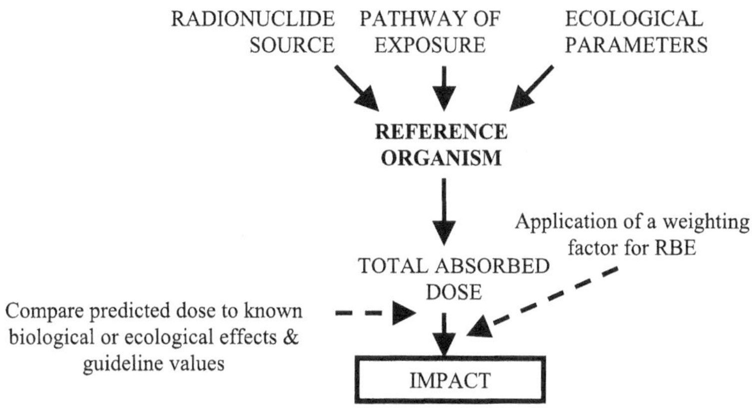


Fig. 3. Overview of the assessment process given in Copplestone et al. (2001).

$^{210}$Po. Both terrestrial and aquatic: $^{3}$H, $^{14}$C, $^{32}$P, $^{60}$Co, $^{90}$Sr, $^{106}$Ru, $^{129}$I, $^{131}$I, $^{137}$Cs, $^{234}$Th, $^{238}$U, $^{239}$Pu and $^{241}$Am.

The dosimetry equations used are as follows:

Terrestrial ecosystems (soil)

$$\dot{H} = \sum_i \left(C_i^{\mathrm{soil}} \times \mathrm{CF}_i^{\mathrm{soil}} \times \mathrm{DPUC}_{\mathrm{total},i}^{\mathrm{int}}\right) + \left(\left(f_{\mathrm{soil}} + 0.5 f_{\mathrm{surface}}\right) C_i^{\mathrm{soil}} \times \mathrm{DPUC}_{\mathrm{total},i}^{\mathrm{ext}}\right), \quad (1)$$

where:

$\sum_i$ represents summation over all nuclides;
$C^{\mathrm{soil}}$ is the activity concentration of the radionuclide in surface soil;
$\mathrm{CF}^{\mathrm{soil}}$ is the concentration factor for the organism referenced to soil;
$f_{\mathrm{soil}}$ is the fraction of time the organism spends under the soil surface; and
$f_{\mathrm{surface}}$ is the fraction of time the organism spends on the ground surface.

Aquatic ecosystems

$$\dot{H} = \sum_i \left(C_i^{\mathrm{water}} \times \mathrm{CF}_i^{\mathrm{water}} \times \mathrm{DPUC}_{\mathrm{total},i}^{\mathrm{int}}\right) + \left(\left(f_{\mathrm{sediment}} + 0.5 f_{\mathrm{surface}}\right) C_i^{\mathrm{sediment}} + \left(f_{\mathrm{water}} \times C_i^{\mathrm{water}}\right)\right) \times \mathrm{DPUC}_{\mathrm{total},i}^{\mathrm{ext}}\right), \quad (2)$$

where:

$C^{\mathrm{water}}$ and $C^{\mathrm{sediment}}$ are the radionuclide activity concentrations in water and sediment respectively;
$\mathrm{CF}^{\mathrm{water}}$ is the concentration factor for non-human species referenced to water;
$f_{\mathrm{sediment}}$ is the fraction of time spent buried in sediment;
$f_{\mathrm{surface}}$ the fraction of time spent on the sediment surface; and

$f_{\text{water}}$ the fraction of time spent free swimming in the water column or on the water surface.

Terrestrial ecosystems (air)

$$\dot{H} = \sum_i \left(C_i^{\text{air}} \times \text{CF}_i^{\text{air}} \times \text{DPUC}_{\text{total},i}^{\text{int}}\right) + \left(\left(f_{\text{soil}} + 0.5 f_{\text{surface}}\right) C_i^{\text{soil}} \times \text{DPUC}_{\text{total},i}^{\text{ext}}\right), \quad (3)$$

where:

$\sum_i$ represents summation over all nuclides;
$C^{\text{air}}$ is the activity concentration of the radionuclide in air;
$C^{\text{soil}}$ is the concentration of the radionuclide in surface soil, calculated from the air concentration and the relevant concentration ratio;
$\text{CF}^{\text{air}}$ is the concentration factor for the organism referenced to air;
$f_{\text{soil}}$ is the fraction of time the organism spends under the soil surface; and
$f_{\text{surface}}$ is the fraction of time the organism spends on the ground surface.

Similar equations have been defined for the aquatic situation.

As a result of the Stage 2 assessments and the use of a very conservative screening level of 2 $\mu\text{Gy h}^{-1}$, a number of authorisations were highlighted as potentially impacting on one or more Natura 2000 sites. These require further assessment using a tiered approach. Most of the assessments that exceeded the Stage 2 screening level were for Natura 2000 sites located in estuarine situations which received the radionuclides principally by aquatic transport. It should be noted that one of the main reasons identified as to why an authorisation failed the Stage two screening level was because of the selection of a suitable radionuclide analogue for use in the assessment. This is because not all radionuclides authorised for release are included in the list of radionuclides used in Copplestone et al. (2001). For example $^{99m}$Tc is released from medical institutions in relatively large quantities but it has a short (6 hour) half-life. It is usually released into the sewer system, which means that there will be a time delay in the $^{99m}$Tc reaching any Natura 2000 site and hence radioactive decay will reduce the activity concentration arriving at the site. Furthermore, the short half-life means a rapid reduction in activity concentration of any $^{99m}$Tc that reaches the Natura 2000 site. However $^{99m}$Tc was not included in the original list of radionuclides used in Copplestone et al. (2001) and so has to be represented by an analogue. In the case of the Stages 2 and 3 assessments this has been done using another gamma emitting radionuclide, $^{137}$Cs, which has a 30-year half-life. As a consequence the dose rate from $^{99m}$Tc to the organisms are over estimated.

**6.1.2.** *Stage 3 assessment*

Stage 3 assessments are ongoing to determine the significance of any effect resulting from a single permission or combination of permissions on the Natura 2000 site(s) identified by Stages 1 and 2. There is a particular need to assess the potential impact on listed species present in the legislation. The legislation requires that these habitat assessments are completed by 2008.

In order to achieve this all the listed species that may be present at the Natura 2000 sites potentially impacted by the authorisations identified in Stage 2 needed to be stylised and represented by one of the reference organism geometries described in Copplestone et al. (2001).

Table 2
Feature species (only common names given) identified at the Natura 2000 sites potentially impacted by the Stage 2 screened authorisations. The number of species is given in brackets. Those species marked with * are found in the Tees Natura 2000 sites

| | |
|---|---|
| Bird species (55) | Avocet, Bar-tailed Godwit, Bewicks Swan, Bittern, Black-tailed Godwit, Brent goose, Chough, Common Scoter, Common Tern, Cormorant, Curlew, Dartford Warbler, Dunlin, Gadwall, Gannet, Golden Plover, Great Crested Grebe, Grey plover, Guillemot, Hen Harrier, Honey Buzzard, Kittewake, Knot*, Lapwing, Lesser Black-backed Gull, Little Tern*, Manx Shearwater, Marsh Harrier, Mediterranean Gull, Nightjar, Oystercatcher, Peregrine, Pink-footed Goose, Pintail, Puffin, Razorbill, Redshank*, Ringed Plover*, Ruff, Sanderling, Sandwich Tern*, Scaup, Shelduck, Short-Eared Owl, Shoveler, Snipe, Stone Curlew, Storm Petrel, Teal, Tufted Duck, Turnstone, White-fronted Goose, Whooper Swan, Wigeon, Woodlark |
| Plant species (4) | Early Gentian, Fen Orchid, Petal Wort, Shore Dock |
| Terrestrial invertebrates (2) | Southern Damselfly, Stag Beetle |
| Amphibians (2) | Great Crested Newt, Natterjack Toad |
| Terrestrial mammals (4) | Bechsteins Bat, Dormouse, Greater Horseshoe Bat, Lesser Horseshoe Bat |
| Aquatic mammals (3) | Common Seal, Grey Seal, Otter |
| Aquatic invertebrate | Desmoulins Whorl Snail |
| Fish species (8) | Allis Shad, Atlantic Salmon, Brook Lamprey, Bullhead, River Lamprey, Sea Lamprey, Spined Loach, Twaite Shad |
| Reptiles (2) | Sand Lizard, Smooth Snake |

The first part of the Stage 3 assessment is to identify information associated with the Natura 2000 site, so for example Table 2 lists the 81 species that were identified as needing protection (Copplestone et al., 2003). Having identified the feature species present, then the occupancy factors (time spent in different compartments of the ecosystem), and concentration factors (for each radionuclide/species combination) should be determined. It was then necessary to link the geometry of the feature species to that of the reference organism geometries provided in Copplestone et al. (2001) which can then be used as the basis of the dosimetric method. It should be noted however that these feature species may not be the most exposed to the ionising radiation and so it is important to check on the default reference organism estimated doses to ensure that there are no species being adversely impacted that would not otherwise be detected.

### 6.2. *Derivation of concentration factor (CF) values*

Probably the main source of uncertainty in the Stage 3 assessment results from the derivation of concentration factors (CFs) for the radionuclide/species combination. A review of the scientific literature revealed little data on the concentration of the radionuclides of interest in the feature species. This was not surprising given that most of the feature species are of conservation value and therefore unlikely to be included in routine sampling and monitoring programmes. Some CF data have been obtained from the literature and from species with a similar ecology but which are not of conservation value and that have been sampled.

In the absence of feature species and radionuclide specific CF values, an approach has been developed, which should provide a consistent method for selecting CF values for inclusion in the assessment. Essentially, the approach is as follows: determine if a CF is available for the species and radionuclide of interest. This is used if available. If not, a CF is selected from Copplestone et al. (2001) for a reference organism with similar ecology. If this is not available, the CF and $K_d$ (aquatic only) values for the radionuclide of interest are reviewed and expert judgement is used to determine which value should be recommended. Expert judgement is needed because, although the use of the $K_d$ to equate to CF can be viewed as generally conservative in aquatic ecosystems, certain radionuclides, e.g. $^{131}I$, are known to accumulate in organisms and in these situations the use of the $K_d$ is unlikely to be appropriate. Finally, where no CF is available for another plant or animal group, the CF for the Environment Agency recommended analogue radionuclide (Copplestone et al., 2003) has been used. It should be noted that this approach might produce some highly conservative CFs (as for example in the case of using $^{137}Cs$ as an analogue for $^{99m}Tc$ above). It is important to note the origins of the CFs used when interpreting the outputs of dose rate assessments. It is emphasised that the overall assessment must be supported by field measurements to produce site-specific data if there is any doubt over the results although this may be difficult if you have to measure the radionuclide content of a protected species.

### **6.3.** *Linking feature species geometry to that of reference organism geometries listed in Copplestone et al. (2001)*

In order to adapt the assessment methodology to enable dose rates to be predicted for a wide range of species, a method to relate the geometry of the new species to the nearest reference organism geometry was devised using a mathematical area/volume ratio relationship. This assumes that both the feature species and the reference organism geometries are ellipsoids. Full details of the method are given elsewhere (Copplestone et al., 2003) but Figure 4 demonstrates the relationship graphically. Using this method, the seventy-seven feature species identified were assigned to an appropriate reference organism geometry from Copplestone et al. (2001) which can then be used to approximate the dose rate. As an example, a bird egg was described as a geometric shape of size 40 mm × 25 mm × 25 mm with an area/volume ratio of 212.7 in Copplestone et al., 2001. A bird egg can then be used to determine the doses to feature organisms of a similar size, for example, a Dormouse (shape of 40 mm × 15 mm × 15 mm and an area/volume ratio of 165. Copplestone et al. (2003) showed that the error in the dose assessment by doing this would be small, calculated as 5% and 11.1% for internal and external exposures respectively. It should be noted however that plants are not conducive to representation by means of an ellipsoid and hence were excluded from this process. Therefore any plant species were represented by the shrub or herb geometries during Natura 2000 terrestrial site assessments.

Ideally, absorbed doses for all new geometries should be calculated using the Monte Carlo absorbed fraction functional method (Jones et al., 2003) and conducted for all reference organism geometries. However, these calculations are onerous and, as the above approximations give rise to generally small errors in the dosimetric calculations, the area/volume approach seems an acceptable way of addressing the wide range of geometries of the identified feature species.

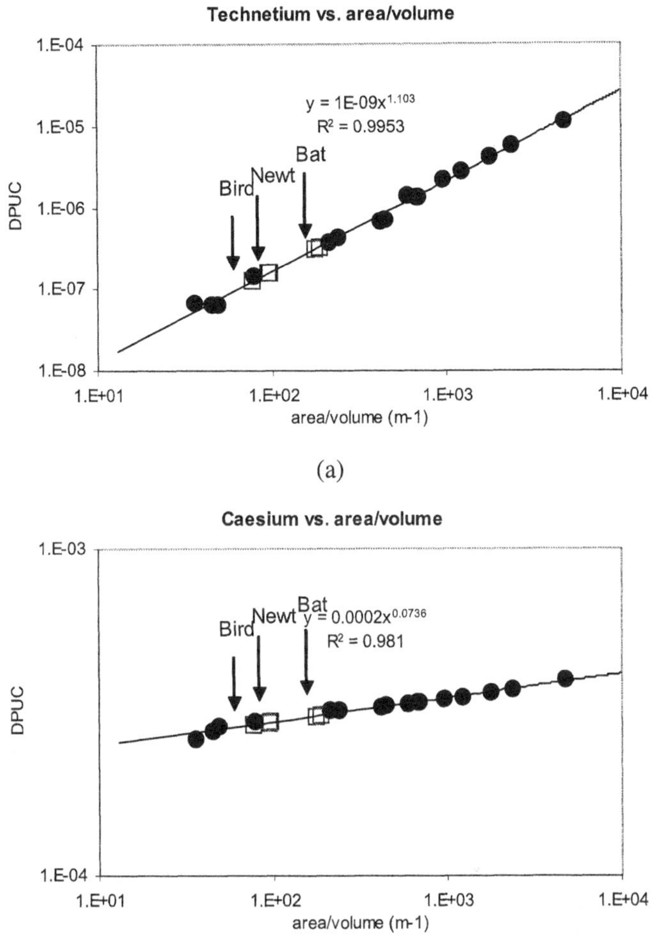


(b)

Fig. 4. Plots of $^{99}$Tc and $^{137}$Cs as examples of the observed relationship between organism area/volume ratio and the DPUC values.

## 7. The Tees estuary assessment—a case study

The Tees is a heavily industrialised river in the north-east of England that opens out onto tidal mud and sand flats at its mouth. Around the mouth of the river is a National Nature Reserve which has a range of habitats including intertidal mud and sand flats, sand dune systems, saltmarsh and a grazing marsh. These are close to several large conurbations and are internationally important for wader and wildfowl populations of birds. It is these bird populations that are of interest in the assessment.

The river receives a number of authorised discharges of radioactive substances. 22 permits to discharge directly to the river, 5 permits to discharge to sewers and 25 permits to release

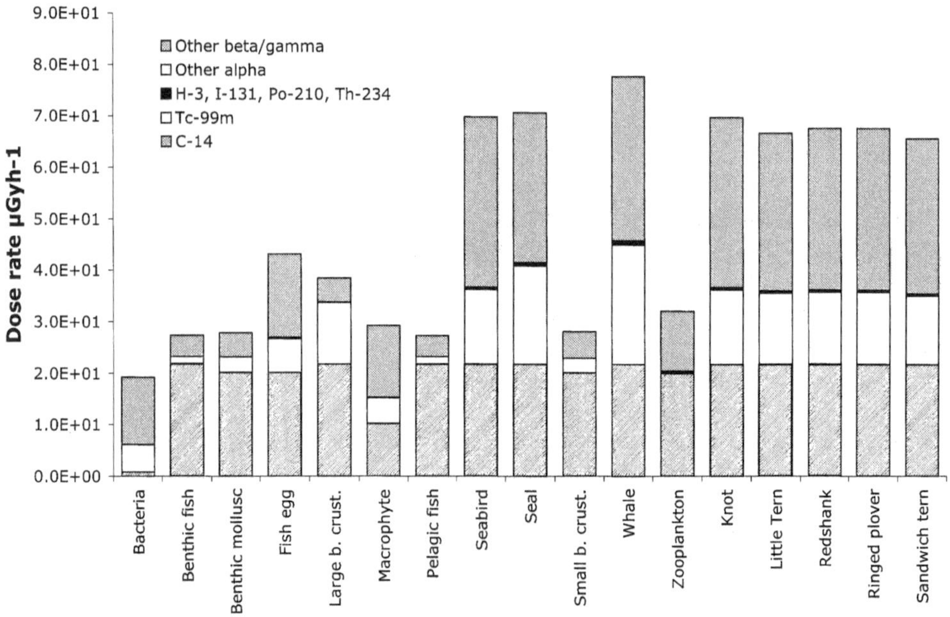


Fig. 5. Predicted dose rates in the estuarine ecosystem to the reference organisms (Copplestone et al., 2001) and feature species in the Tees case study.

to atmosphere have been identified within the catchment whose discharges are likely to reach the Natura 2000 sites at the mouth of the river. Table 2 lists the 5 feature species, all birds that were identified in the Natura 2000 sites in the Estuary. The following radionuclides have been permitted for release: (terrestrial only) $^{35}$S, $^{41}$Ar, $^{85}$Kr, (aquatic only) $^{99m}$Tc, $^{210}$Po, (both) $^{3}$H, $^{14}$C, $^{131}$I and $^{234}$Th. Figure 5 shows the predicted dose rates to the reference organisms and the five feature species.

From the assessment it is possible to identify that the air impacts on the site are around 4.8 $\mu$Gy h$^{-1}$ (worst case) whilst consideration of the impacts from the estuary shows a much higher range of estimated doses of between 20 and 80 $\mu$Gy h$^{-1}$ for reference organisms and the feature species present in the estuary.

The final step in the assessment process is to compare the predicted doses to known biological effects data. There have been a number of major compilations of radiation effects data on non-human species (e.g. UNSCEAR, 1996a, 1996b) and these have been revisited recently within the FASSET project (FASSET D4, 2003). The results of these studies have been used to suggest guideline values below which it is thought unlikely that any biological (deterministic) effects will be observed. For example the IAEA and others have suggested that a dose rate of 40 and 400 $\mu$Gy h$^{-1}$ for terrestrial animals and terrestrial plants respectively would be suitable. The recent work conducted by FASSET, which has constructed a database of radiation effects (from both acute and chronic exposure) for different non-human species grouped by umbrella endpoints (reproduction, mortality, morbidity and mutation) has concluded that, broadly, the threshold for statistically significant effects in most studies is about 100 $\mu$Gy h$^{-1}$ and that the responses increase progressively with increasing dose rate and usually become very clear at dose rates $>1000\ \mu Gy\,h^{-1}$ over a large fraction of the life-span. They do highlight the fact that the available data are fragmentary and that there may be some specific exceptions to this for tritium radiation on certain species but that further data are required.

These numbers do however allow us to start to predict the likely consequences of radiation exposure to non-human species although it should be noted that there are very few, if any, locations in the world where ionising radiation is the only environmental stressor impacting on non-human species and one possible requirement for the future is a better understanding of how ionising radiation and other environmental stressors interact to impact on non-human species.

It can be seen that in the case of the Tees estuary the permitted discharges of radioactive substances are not likely to affect the habitat or the feature species living within. This is particularly true given that the assessments have been made using the authorised limits of discharges into the Tees estuary and, given that many of these are industrial tracer tests (which allow for greater operational flexibility), it is likely that this is quite pessimistic. However, it is also likely that this site will be selected for further evaluation by measuring radionuclide activity concentrations in different environmental media and biota in order to validate the overall assessment approach.

## 8. Conclusions—significance for regulators and operators

It is clear that a system for protection of the environment from ionising radiation will be further developed, with substantive recommendations from ICRP being published in 2006 or 2007.

Moreover, in the UK and US, regulators are already requiring site operators to consider the possible impact of proposed radioactive waste discharges or disposals on the environment when submitting proposals for authorisations, and are themselves assessing the impacts of radioactive discharges (in conjunction with those of other pollutants) on specific sites. An example of the assessment approach used by the Environment Agency of England and Wales has been presented here. As mentioned previously, the pressure for this assessment comes from the needs of conservation (e.g. in the EU, the Habitats and Wild Bird Directives) and not because of a perceived or actual radiological hazard in the environment.

However, it would be wise of both regulators and operators to assume that neither ICRP, UNSCEAR, IAEA nor the ERICA project will produce a complete and comprehensive set of data and methodologies to conduct such an assessment in specific circumstances. Considerable tensions still exist on the matter of 'knowledge gaps' between those who want a 'simple, robust and proportional' system and those who believe that substantial radioecological research is required before an acceptable system can be developed. In the opinion of the authors, the ICRP recommendations to be published in 2005/06 are likely to provide only a relatively simple basis for an assessment, which would need to be significantly enhanced and extended to carry out a fully defensible assessment for a particular case. At the same time, the mere existence of the recommendations will further emphasise the requirement for such assessments to be carried out.

Significant knowledge gaps exist for radionuclide uptake by biota and for radiation effects on biota, particularly in the case of chronic exposure. However, anticipating ICRP's approach to use 'derived consideration levels' based on natural radiation exposure, gaps in knowledge on effects (which would require extensive and expensive research to fill adequately) may be adequately covered by the alternative of an adequate knowledge of the range of natural radiation exposures for particular taxa.

Future work will address some of the knowledge gaps particularly with reference to the biological effects of radiation on different non-human species and will also revolve around the testing and refinement of dosimetric methods through case studies etc. The IAEA has recently (November 2004) established a working group under their Environmental Modelling for Radiation Safety (EMRAS) programme to encourage non-human species dosimetric model testing and validation.

## Acknowledgements

This chapter has been adapted from a review conducted originally for United Kingdom Nirex Ltd in early 2004 (Jones, 2004) and also presented to the UK National Dose Assessment Working Group (NDAWG) in 2004 (Jones and Copplestone, 2004). The example assessment has been taken from the Habitats assessments conducted by Environment Agency staff in England and Wales.

## References

Allott, R., Dunn, M. (2001). NCAS Technical Report NCAS/TR/2001/019 (Environment Agency, 2001).

Brown, J.E., Jones, S.R., Saxen, R., Thørring, H., Vives, J. (2004). Radiation doses to aquatic organisms from natural radionuclides. *Journal of Radiological Protection* **24**, A63–A77.

COGEMA (2003). Assessment of marine biota doses arising from the radioactive sea discharges of the COGEMA La Hague facility (3 volumes), www.cogema.com.

Cooper, M.B. (2003). National Radioactive Waste Repository—radiological protection of non-human biota. Report ERS-009 prepared for ARPANSA by EnviroRad Services Pty, www.arpansa.gov.au.

Copplestone, D., Bielby, S., Jones, S.R., Patton, D., Daniel, P., Gize, I. (2001). Impact Assessment of Ionising Radiation on Wildlife. Environment Agency, Bristol, UK, ISBN 1 85705590 X. R&D Publication 128, June 2001.

Copplestone, D., Wood, M.D., Bielby S., Jones, S.R., Vives, J. Beresford, N.A. (2003). Habitats regulations for Stage 3 assessments: radioactive substances authorisations. R&D Technical Report P3-101/SP1a, Environment Agency, Bristol, UK, p. 100.

FASSET (2003). In: Woodhead, D., Zinger, I. (Eds.), *Radiation Effects on Plants and Animals*, p. 196. Deliverable D4 of the FASSET project (EC contract FIGE-CT-2000-00102, 2004).

Hingston, J.L., Knowles, J.F., Walker, P.J., Wood, M.D., Copplestone, D. (2004). Effects of ionising radiation on soil fauna. Environment Agency R&D Technical Report P3-101/SP7, 64 pp.

International Atomic Energy Agency (IAEA) (1988). *Assessing the Impact of Deep Sea Disposal of Low Level Radioactive Waste on Living Marine Resources*. IAEA, Vienna. Technical Report Series No. 288.

International Atomic Energy Agency (IAEA) (1992). *Effects of Ionising Radiation on Plants and Animals at Levels Implied by Current Radiation Protection Standards*. IAEA, Vienna. Technical Report Series No 332.

International Atomic Energy Agency (IAEA) (1999). *Protection of The Environment from The Effects of Ionising Radiation—A Report for Discussion*. IAEA, Vienna. IAEA-TECDOC-1091.

International Atomic Energy Agency (IAEA) (2002). *Ethical Considerations in Protecting The Environment from The Effects of Ionising Radiation*. IAEA, Vienna. IAEA-TECDOC-1270.

International Atomic Energy Agency (IAEA) (2003). *International Conference on The Protection of The Environment from The Effects of Ionising Radiation, Stockholm, Sweden, 6–10 October 2003*. IAEA, Vienna. Contributed papers, IAEA-CN-109.

International Atomic Energy Agency (IAEA) (2005). Measures to strengthen international cooperation in Nuclear, Radiation and Transport safety and waste management: Draft Plan of Activities on the Radiation Protection of the Environment. Report by the Director General to IAEA Board of Governors. GOV/2005/57.

International Commission on Radiological Protection (1991). 1990 Recommendations of the International Commission on Radiological Protection, ICRP Publication 60. *Ann. ICRP* **21** (1–3), 1–201.

International Commission on Radiological Protection (2003). A framework for assessing the impact of ionising radiation on non-human species. *Annals of the ICRP* **33** (3). ICRP Publication 91.

International Commission on Radiological Protection (ICRP) (2004). 2005 Recommendations of the International Commission on Radiological Protection, http://www.icrp.org/docs.

International Union of Radioecology (IUR) (2000). Dose and effects on non-human systems—summary of the work of the action groups of the IUR (working document). IUR, Østerås, Norway.

International Union of Radioecology (IUR) (2002). Protection of the environment: current status and future work. IUR Report 3, IUR, Østerås, Norway.

International Union of Radioecology (IUR) (2003). Protection of the Environment in the 21st Century: Radiation Protection of the Biosphere including Humankind—Statement of the International Union of Radioecology. IUR, Østerås, Norway.

Johnson, A., Humprey, C., Martin, P. (2004). Protection of the environment from the effects of ionizing radiation associated with uranium mining. In: *Protection of the Environment from the Effects of Ionizing Radiation*. Proceedings of an International Conference Stockholm, 6–10 October 2003. IAEA International Atomic Energy Agency STI/PUB/1229, pp. 193–207.

Jones S.R. (2004). Update on Results and Trends in Assessments of Impacts of Non-human Biota. Westlakes Scientific Consulting Report to Nirex, Document No 040013/01.

Jones, S.R., Patton, D., Copplestone, D., Norris, S., O'Sullivan, P. (2003). Generic Performance Assessment for a deep repository for low and intermediate level waste in the UK—a case study in assessing radiological impacts on the natural environment. *Journal of Environmental Radioactivity* **66** (1–2), 89–119.

Jones, S.R., Copplestone, D. (2004). Protection of the Environment from Ionising Radiation—Current Status and Ongoing developments. NDAWG paper 6-02, Natioanl Dose Assessment Working Group, Chilton, UK (www.ndawg.org).

Kelly, M., Thorne, M. (2003). Radionuclides Handbook. R&D Technical Report P3-101/SP1b, Environment Agency, Bristol, UK.

Mihok, S. (2004). Personal communication.

National Council on Radiation Protection and Measurements (NCRP) (1991). Effects of Ionizing Radiation on Aquatic Organisms. NCRP Report no. 109, NCRP, Bethesda, MD 20814, USA.

Nuclear Energy Agency (NEA) (2002). Radiological Protection of the Environment: The Path Forward to a New Policy? Workshop Proceedings, Taormina, Sicily, Italy, 12–14 February 2002. OECD Nuclear Energy Agency, Issy-les-Moulineaux, France (ISBN 92-64-09969-7).

Nuclear Energy Agency (NEA) (2003). Radiological Protection of the Environment—Summary Report of the Issues. OECD Nuclear Energy Agency, Issy-les-Moulineaux, France (ISBN 92-64-18497-X).

Pentreath, R.J. (1999). A system for radiological protection of the environment: some initial thoughts and ideas. *Journal of Radiological Protection* **19**, 117–128.

Sazykina, T.G. Kryshev, I.I. (2002). Marina II—assessment of the impact of radioactive substances on marine biota of North European waters (report of working subgroup D). Radiation Protection 132, Marina II. Annex F. EC Directorate—General Environment.

St-Pierre, S., Chambers, D.B., Lowe, L.M., Bontoux, J.G. (1999). Screening level dose assessment of aquatic biota downstream of the Marcoule nuclear complex in southern France. *Health Physics* **77** (3), 313–321.

Strand, P., Larsson, C.-M. (2001). In: Brechignac, F., Howard, B.J. (Eds.), *Radioactive Pollutants Impact on The Environment*. EDP Sciences, Les Ulis.

Strand, P., Oughton, D.H. (2002). Radiation Protection in the 21st Century: Ethical, Philosophical and Environment Issues. Proceedings from the Consensus Conference on Protection of the Environment, NRPA/AUN, Østerås, Norway.

Swami, L., Garisto, N.C., Chambers, D.B., Gorber, D.M. (1999). The Pickering Environmental Review—development of a priority ranking in a multi-stakeholder process. Proceedings of the Second International Symposium on Protection of the Environment from Ionising Radiation, Ottawa.

Twining, J.R., Ferris, J.M., Zinger, I., Copplestone, D. (2003). A radiological case study of quantitative, probabilistic ecological risk assessment using recently developed assessment tools. IAEA-CN-109/114.
United Nations (1992). Rio Declaration on Environment and Development.
UNSCEAR (1996a). Effects of radiation on the environment. Annex in: Sources and Effects of Ionizing Radiation. United Nations Scientific Committee on the Effects of Atomic Radiation 1996. Report to the General Assembly, with Annexes, pp. 1–86. United Nations, New York.
UNSCEAR (1996b). Sources and Effects of Ionising Radiation. Report to The General Assembly, United Nations, New York.
US Department of Energy (1993). Radiation protection of the public and the environment. Order 5400.5.
US Department of Energy (1996). Radiation protection of the public and the environment. Notice of proposed rule-making 10 CFR Part 834, extension of consultation period.
US Department of Energy (2002). DoE Standard: A Graded Approach for Evaluating Radiation Doses to Aquatic and Terrestrial Biota. DOE-STD-1153-2002.
US Department of Energy (2004). RESRAD-BIOTA: A tool for implementing a graded approach to biota dose evaluation. User's Guide, Version 1. Interagency Steering Committee on Radiation Standards, ISCORS Technical Report 2004-02, January 2004.
Wood, M.D., Knowles, J.D., Whittaker, J.H., Copplestone, D., Malcolm, H.M. Bielby, S. (2003). Developing experimental protocols for chronic irradiation studies on wildlife. R&D Technical Report P3-101/SP2, Environment Agency, Bristol, UK.

# Concluding remarks

George Shaw

*Division of Agricultural & Environmental Sciences, University of Nottingham, University Park, Nottingham, NG7 2RD, UK*

Nuclear power has now been with us for more than 60 years and the environmental consequences of man's exploitation of nuclear energy during this period have been felt in many ways. Since 1945, the development and use of nuclear technology, including weapons, civil reactors and reprocessing facilities, have been contentious and have unquestionably impacted on the quality of the environment, as described in Chapters 2 to 10 of this book. Since shortly after the advent of nuclear fuel cycle technology, the field of radioecology has endeavoured to quantify the environmental impact of the nuclear industry. This has often been in response to major incidents, such as the Chernobyl accident, which have demanded that the competent authorities in different parts of the world make their best efforts to understand, predict and mitigate any potentially deleterious effects on the human populations to which they are responsible.

Despite the omnipresence of radionuclides from the nuclear fuel cycle, nuclear weapons and the other sources listed in Table 2 of Chapter 1, the effects of nuclear energy on the environment have, in radiological terms, been shown to be relatively minor. The 20th anniversary of the Chernobyl accident (2006) has provided an opportunity for the most recent compilation and analysis of the extremely large body of data collected in many countries contaminated as a result of the accident. In particular, the environmental and health effects of Chernobyl have been reviewed by the United Nations Chernobyl Forum (IAEA, 2006). In carrying out this analysis the question of ultimate importance has been the effect of the accident on human health. While the number and types of cancers directly attributable to Chernobyl are still relatively low (Cardis et al., 2006) it is reasonable to conclude, based on the epidemiology of the Hiroshima and Nagasaki survivors, that it is still too early to draw any final conclusions on the ultimate human impacts of Chernobyl (Williams and Baverstock, 2006). However, as far as the environment itself is concerned, a consensus is emerging that the sometimes apocalyptic fears expressed following the Chernobyl accident have turned out to be largely unfounded. Certainly, radionuclides in the environment will persist with more or less predictable residence times, and with associated radiation exposures in proportion to the type and activities of radionuclides present (Smith and Beresford, 2005). In the case of the 30 km exclusion zone surrounding the Chernobyl reactor, the presence of very long lived radionuclides such

RADIOACTIVITY IN THE ENVIRONMENT
VOLUME 10 ISSN 1569-4860/DOI: 10.1016/S1569-4860(06)10012-1

as $^{239}$Pu will ensure that this area will be contaminated on a time scale of 10,000 to 100,000 years. However, even at the levels of radioactive contamination experienced in the first 20 years following the accident, the Chernobyl exclusion zone has unexpectedly become an important reserve in which nature thrives (Baker and Chesser, 2000). This is not to say that we fully understand the effects of environmental radiation sources on non-human biota and, as with human epidemiology, the possibility of latent or long-term effects on the fitness of animal and plant populations dictates that ongoing monitoring should be continued.

This book has reviewed and summarised present day understanding of the behaviour of radionuclides in a variety of terrestrial ecosystems. As a result of several decades of radioecological research we can have considerable confidence in our ability to predict the impact of radioactive releases to the environment, although there is always more to learn and certainly no excuse for complacency. Several of the authors contributing to this volume have, quite correctly, identified that resources for radioecological research are in decline. It is an unfortunate fact that too many key areas of environmental pollution science have been reactive in nature and have been reliant on funding being made available after the 'genie' has been allowed to escape from its bottle. Historically, this approach applied to environmental radioactivity, to CFCs, to pesticides and, most significantly of all, to carbon dioxide releases from fossil fuel combustion. It is to be hoped that lessons have been learned from these salutary examples and, in recent years, there has been widespread adoption of the precautionary principle in environmental regulation. As an example, the OSPAR commission aims that, by 2020, 'discharges, emissions and losses of radioactive substances are reduced to levels where the additional concentrations in the marine environment above historic levels, resulting from such discharges, emissions and losses, are close to zero' (www.ospar.org). Similarly, the problem of radioactive waste management is being tackled in a highly precautionary manner by many countries with nuclear fuel cycle legacies. This is clearly desirable from the standpoint of radiological and environmental safety but has inevitably led to argument and controversy between scientists and decision makers when attempting to decide on the most appropriate options for waste management (Baverstock and Ball, 2005). The delays which result from such disagreements are unfortunate since the successful and publicly acceptable management of radioactive wastes from the last few decades will play a key part in any decisions to develop new nuclear power plants in the future.

The key question facing nuclear energy is not so much its environmental impact as its future contribution to mankind's energy requirements. At the present time, we are faced with the increasingly urgent threat of a carbon-modified global climate and it is all the more important that radioecologists contribute to the debate about any future role of nuclear power in reducing $CO_2$ emissions. On the basis of radioecological evidence gained over the past 60 years we can conclude that civil nuclear energy has contributed positively to mankind's development while impacting negatively to a measurable, yet comparatively insignificant, degree on global environmental quality.

## References

Baker, R.J., Chesser, R.K. (2000). The Chernobyl nuclear disaster and subsequent creation of a wildlife preserve. *Environmental Toxicology and Chemistry* **19**, 1231–1232.

Baverstock, K., Ball, D.J. (2005). The UK committee on radioactive waste management. *Journal of Radiological Protection* **25**, 313–320.

Cardis, E., Howe, G., Ron, E., et al. (2006). Cancer consequences of the Chernobyl accident: 20 years on. *Journal of Radiological Protection* **26**, 127–140.

IAEA, (2006). Environmental Consequences of the Chernobyl Accident and Their Remediation: Twenty Years of Experience. Report of the Chernobyl Forum Expert Group 'Environment'. Radiological Assessment Reports Series, International Atomic Energy Agency, Vienna.

Smith, J.T., Beresford, N.A. (2005). *Chernobyl: Catastrophe and Consequences.* Springer Praxis Publishers, Chichester, UK.

Williams, D., Baverstock, K. (2006). Too soon for a final diagnosis. *Nature* **440**, 994.

# Author Index

# Keyword Index

FSC
www.fsc.org
MIX
Papier | Fördert
gute Waldnutzung
FSC® C083411

Kolibri 360 GmbH
Ferdinand-Jühlke-Straße 7
99095 Erfurt, Deutschland
produktsicherheit@kolibri360.de